URBAN GEOGRAPHY

URBAN GEOGRAPHY

Thomas Lee
Harry O'Neill
Bethany Williams

Kruger Brentt
Publishers
2023

Kruger Brentt Publishers UK. LTD.
Company Number 9728962

Regd. Office: 68 St Margarets Road, Edgware, Middlesex HA8 9UU

ISBN: 978-1-78715-108-6

For information on all our publications visit our website at http://krugerbrentt.com/

PREFACE

Urban geography is a branch of human geography concerned with various aspects of cities. An urban geographer's main role is to emphasize location and space and study the spatial processes that create patterns observed in urban areas. To do this, they study the site, evolution and growth, and classification of villages, towns, and cities as well as their location and importance in relation to different regions and cities. Economic, political and social aspects within cities are also important in urban geography.In order to fully understand each of these aspects of a city, urban geography represents a combination of many other fields within geography. Physical geography, for example, is important in understanding why a city is located in a specific area as site and environmental conditions play a large role in whether or not a city develops. Cultural geography can aid in understanding various conditions related to an area's people, while economic geography aids in understanding the types of economic activities and jobs available in an area. Fields outside of geography such as resource management, anthropology, and urban sociology are also important.

The study of urban geography can help us have a better understanding of the economics of what goes on within cities and recognize the small things that are involved in local, national, and international scales. It helps us isolate patterns so that we have a basic understanding of location and spacing within a community. There are many reasons that people study urban geography. It helps us understand everything that is occurring around us inside city limits and outside as well. It lets us have a better understanding of why things are occurring economically and why certain cities are thriving and others are shrinking.

The present book contains 14-chapters namely Urban geography : an introduction, Urbanization, The formation of urban areas, Urbanization in India, Urban planning and design, Environmental design and planning, Ecological urbanism, Urban morphology, Urbanization and migration, Urban society, Urban e-governance, Eco-cities, Urban movement and transport & Urban system and economic development. It is written in a lucid style with the most recent advances, and with theoretical discussions for easy understanding of the subject. In this book most of the relevant topics are covered adequately. This book explores the great diversity of urban worlds. It is an essential read for students, town planner and scholars of urban geography.

We are grateful to all those persons as well as various books, manuals, periodicals, magazines, journals etc. that helped in the preparation of this book. In spite of the best efforts, it is possible that some errors may have occurred into the compilation and editing of the book. Further queries, constructive suggestions and criticisms for the improvement of the book are always welcomed and shall be thankfully acknowledged.

Thomas Lee

Harry O'Neill

Bethany Williams

CONTENTS

CHAPTER

URBAN GEOGRAPHY: AN INTRODUCTION 1

1.1 INTRODUCTION

Urban geography is the study of urban places with reference to their geographical environment. Broadly speaking, the subject matter includes origin of towns, their growth and development, their functions in and around their surroundings.

The subject of urban geography has gradually taken a special place among the various branches of geography in the period after the Second World War in various foreign and Indian universities and colleges. With the increase of population globally, towns and cities have become magnets of economic, social and political processes.

The changes brought about by these processes have become instructive as well as interesting too in case of the single phenomenon, i.e., city in a spatial context. Under these circumstances, the studies of towns and cities have formed an essential part of the branch of Human Geography.

Urban geography is a branch of human geography concerned with various aspects of cities. An urban geographer's main role is to emphasize location and space and study the spatial processes that create patterns observed in urban areas. To do this, they study the site, evolution and growth, and classification of villages, towns, and cities as well as their location and importance in relation to different regions and cities. Economic, political and social aspects within cities are also important in urban geography.

In order to fully understand each of these aspects of a city, urban geography represents a combination of many other fields within geography. Physical geography, for example, is important in understanding why a city is located in a specific area as site and environmental conditions play a large role in whether or not a city develops. Cultural geography can aid in understanding various conditions related to an area's people, while economic geography aids in understanding the types of economic activities and jobs available in an area. Fields outside of geography such as resource management, anthropology, and urban sociology are also important.

1.2 MEANING OF AN URBAN PLACE

It is one of the most essential and immediate problems to decide 'what is urban?' How does it differ from its counterpart, i.e., rural? In everyday life we are aware that difference between rural and urban depends upon their nature of work – the former being engaged in agricultural operations and the latter in non-agricultural activities.

But it is a difficult task to transform the above stated meaning between the two different natures of settlements into precise and scholarly terms. This is because of the fact that 'an urban place' has been defined differently by different scholars and agencies. Even the United Nations Demographic Year Book (UN, 1990) has given a wide range of examples covering the various countries defining demographically.

UNO defines a permanent settlement with a minimum population of 20,000 as an urban place. But several countries have their own minimum such as Botswana (5,000), Ethiopia (2,000), Argentina (2,000), Israel (2,000), Czechoslovakia (5,000), Iceland (200), Norway (200), Portugal (10,000), Japan (50,000), Australia (1,000), India (5,000), etc.

But, the UN Demographic Year Book concludes: "There is no point in the continuum from large agglomerations to small clusters or scattered dwellings where urbanity disappears and rurality begins the division between urban and rural populations is necessarily arbitrary." A review of the problems of rural and urban centres as revealed by the Census Reports of various countries identifies a few bases for reckoning a place as urban.

This course is an introduction to urban geography. Geography is the study of the physical and human/social environments of the Earth, while urban inquiry focuses on the people and processes of cities and towns -- which now account, for the first time in human history, for a majority of the world's population. Urban geography, then, is concerned with the relations among people, and between people and their environments, in cities and towns across the world.

Such a simple definition is indeed helpful, but it's only a first step. "All boundary-drawing exercises are fraught with difficulties," Derek Gregory (2009, p. 287) reminds us, "and intellectual landscapes are no exception." Consider both terms in the phrase "urban geography." The "urban" is often a contested category: for some, the term includes a wide range of comparatively dense settlements, from big cities all the way to small towns and even low-density suburbs out on the edges of cities; for others, the word is only useful if it refers to the most dense parts of the biggest cities with millions of people. There is also some disagreement, moreover, on what is distinctively and fundamentally urban about the important social, economic, and political changes of our era: yes, we see a lot of fascinating things happening in cities, but does the city change any of the processes underway in society? Despite these disagreements -- and perhaps because of them -- the "urban" is at the heart of the most urgent questions being asked in many fields today. One of the most prominent urban geographers sums up the history this way:

"The study of urban places is central to many social sciences, including geography, because of their importance not only in the distribution of population within countries but also in the organization of economic production, distribution and exchange, in the structuring of social reproduction and cultural life, and in the exercise of political power. Sub-fields of the different social science disciplines were established in the decades after the Second World War to study these separate components, such as urban anthropology, economics, geography, politics, and sociology; later attempts were made to integrate these under the umbrella title of urban studies."

The second part of the phrase -- geography -- brings us into interesting territory. Many people don't really understand what geography is, other than the elementary-school memorization of countries, capitals, and commodities. Don't we already know where everything is? But even once we get past those widespread popular misconceptions, it can still be rather difficult to categorize geographers. "Like other aspects of human geography," Paul Knox and Linda McCarthy write, "urban geography is concerned with 'local variability within a general context.' This means that it is concerned with an understanding of both the distinctiveness of individual places (in this case, towns and cities) and the regularities within and between cities in terms of the spatial relationships between people and their environment." Environment means both the physical-natural and human-social worlds, reflecting the shared heritage of human and physical geography. In turn, the human-physical duality in the field hints at the sense of ambiguity that you probably detect when you talk with geographers: they're pretty hard to pigeonhole, not only because some of them study physical phenomena (climate, rivers, mountains, forests) while others focus on human processes (social, economic, political). But they also seem to go about their work in very different ways, using different tools, languages, and styles of communication. As a result, geographers might very well be best described as undisciplined members of an elusive discipline. If you're a bit unsettled by all this ambiguity, you're in good company. The eminent historian of geography David Livingstone describes efforts to summarize the field this way:

"Because the term 'geography' means, and has meant, different things to different people in different times and places, there is no agreed-upon consensus on what constitutes the project of writing the history of this enterprise. Moreover, while the story of geography as an independent scholarly discipline is inescapably bound up with the history of the professionalization of academic knowledge since around the middle of the nineteenth century, it is clear that the history of geography as a discourse not only operates without such constraints but also reaches beyond the historical and institutional confines of the modern-day discipline. Of course geography as discourse and discipline are interrelated in intimate ways -- one might even say that the purpose of a discipline is precisely to 'discipline' discourse."

In this course, our main goal is "to make sense of the ways that towns and cities have changed and are changing, with particular reference to the differences both between urban places and within them." We tackle four distinct, and yet inter-related, sets of questions.

First, how do cities and other kinds of human settlements vary across time and space? How has the evolution of cities reflected prevailing historical conditions? What are the crucial differences between cities within particular regions or countries, and between different countries? How do certain cities reflect the distinctive circumstances of their geographical context?

Second, what regularities unite seemingly different cities? What are the similarities in patterns, processes, and relationships within different cities, and in networks between cities?

Third, how do social relations shape the form of the city? How are particular activities, land uses, and social groups distributed within and among cities?

Fourth, how does the form of the city shape social relations? How do spatial constraints and locational considerations, for example, influence the way that people decide where to live and where to work? How do the geographies of cities created in previous generations influence the decisions of today's corporations, investors, and governments as they gradually create new urban geographies?

These are:

1. A place designated by administrative status;
2. A minimum population;
3. A minimum population density;
4. A concept of contiguity to include or exclude under suburban area or loosely scattered settlement;
5. A proportion engaged in non-agricultural occupations; and
6. A functional character.

In case of our country (India), the census of 1981 has identified the following places as urban:

1. Centres having Municipality, City Board, Cantonment Board/Notified Town Area;
2. A minimum population of 5,000;
3. 75 percent males engaged in non-agricultural activities;
4. A minimum population density of 400 persons per square km or 1,000 persons per square mile; and

5. Centres defined by urban amenities prescribed by the Director, Provincial Census.

Two important facts must be borne in mind before accepting the meaning of urban and rural. One is the fact that it is rather impossible now to identify a dividing line between the rural and urban – the two being merged to create a sort of diffusion and present a landscape which is neither purely agricultural nor engaged wholly in tertiary activities.

Industrialization has brought into being a large number of settlements which are not certainly villages but are nucleated settlements of agricultural population. Another problem is about the concept of what is urban, which is not static and is subject to change with time as well as with space.

The proportion of population engaged in agricultural activities is the most effective measure. But capitalization of agriculture in modern times and rural depopulation by commuting urban workers have made the criterion of proportion irrelevant.

Thus, to conclude the discussion one reaches to the point that with the changing nature of both rurality and urbanity, there has developed the functional overlap between the two. Therefore, the distinction between what is urban and what is rural has lost its meaning in reality.

1.3 DEFINITION OF A CITY

An essential component within urban geography is defining what a city or urban area actually is. Although a difficult task, urban geographers generally define the city as a concentration of people with a similar way of life-based on job type, cultural preferences, political views, and lifestyle. Specialized land uses, a variety of different institutions, and use of resources also help in distinguishing one city from another.

In addition, urban geographers also work to differentiate areas of different sizes. Because it is hard to find sharp distinctions between areas of different sizes, urban geographers often use the rural-urban continuum to guide their understanding and help classify areas. It takes into account hamlets and villages which are generally considered rural and consist of small, dispersed populations, as well as cities and metropolitan areas considered urban with concentrated, dense populations.

Attributes of a Town:

What are the attributes or characteristics qualifying a town?

Several qualities of a city or town may be summarized as:

a. Town is a kind of settlement having a social organization of much greater scope than a simple rural establishment.

b. It does not represent a mere greater number of people agglomerated in a vast area. But it represents a stage of civilization quite different from a locality expressing rural way of life.

c. Cities and towns have their historical origin Blache has pointed out that cities characteristically possess mythical halo surrounding their genesis (ritual, eponymous hero, etc.).

d. Towns and cities are creatures of commerce, and politics accompanying the earliest developments such as: Babylon, Athens, London, Paris, Delhi, etc.

Emrys Jones has also expressed various attributes of towns and cities which also resemble whatever has already been discussed above:

...a town is a physical agglomeration of streets and houses, a centre of commerce and administration, a kind of society, even a cultural frame of mind of urbanite or a way of life.

1.4 SCOPE OF URBAN STUDIES AND DEFINITIONS

Urban geography studies urban centre in the context of geographical factors. The factors operate spatially to explain processes – economic, socio-cultural and also political. But the subject of urban geography has its limited scope in the sense that it deals with these processes in relation to only one phenomenon, i.e., town or city. Some of the general principles on which a town is based form the subject-matter.

Commonly, it includes in the very beginning, consideration about the origin of an urban place. The genesis about a town is invariably related to its history. Who is behind its origin? What is that which makes a town to take its root where it is, and why it is there? Town site or the ground on which it is sited has some specific and geographic attributes. These need explanation to bring forth personality of a town.

Another point which has been emphasized by D. Stamp to cover the scope of urban geography is the study of the actual town itself, i.e., town as an entity. He further has added that influence of the town on its surrounding area too forms a significant aspect of the study. This means that 'townscape' and also hinterland including 'umland' are vital issues for studying urban geography.

One of the pioneer scholars in urban studies in India, R.L. Singh has stressed on three broad categories under the scope, viz.

(a) The physical structure of the city,

(b) The stage of its historical development, and

(c) The process influencing the structure.

Dickinson defines urban geography as the study of a city commanding the surrounding region. He describes the city as a king among the surrounding towns. His trait for cities of all ages has been institutional supremacy for their surrounding territory.

Their existence depends upon the resources of the surrounding areas, and also, by virtue of their interaction through their physical, social and economic infrastructure. Their interdependence with their surrounding regions is the spatial reality.

Raymond E. Murphy points out the dual role of urban geographer, i.e.,

(i) To analyze cities as entities in terms of locations, characters, growth, and relations to the surrounding countryside, as well as,

(ii) To discuss patterns of the city's interior – land use, social and cultural patterns, patterns of circulation, and above all, natural patterns of environment – all as they exist in interrelation and interaction in the urban area.

Harold Carter opined that since the geographer is concerned with the analysis of the variable character of the earth's surface, and thus, "the populations and the buildings agglomerated together to make up towns constitute the special interest of the urban geographer". Since considerable population of the world live in towns, and the problems of the urban environment are paramount, the study of urban geography is important and its relevance to applied geography needs no further stress.

Towns and cities have their wide impact on human life and activities. Overall growth rate of city population has been faster during the last two-three decades. It is only after Second World War that the study of urban geography got due recognition in the universities in India and abroad.

Prior to that period, it was taught as a theme within human geography where its scope was restricted to the description of site-situation of towns including their description as a part of settlements.

Since the publication of the major work of Doxiadis, urban geography has made much headway in and outside India. Brian J.L. Berry also encouraged the urban studies by introducing urban systems as consequences of economic development. In the present circumstances the scope of urban studies has reached far-flung areas and is not restricted to its site-situation structural approach.

The ICSSR Report of the Fourth Survey of Research in Geography, covering the period 1976-82 in India has pointed out various themes of urban phenomena bringing into light the scope of the subject. These include trends and patterns of urbanization; rural-urban migration; urban systems and hierarchical orders; morphology; economic base; land use; functional housing classification; slums and squatter settlements; rural-urban fringe, surrounding areas of influence, umland and interaction between a city and surrounding settlements; urban environment; pollution; poverty; crime and quality of life; urban services and amenities; urban politics and administration; tourism; urban planning and problems including urban metropolises.

N. Baransky, the founder of Soviet economic geography, has pointed out that study of cities has a wide scope in the sense that it has now become the subject matter of historians, geographers, statisticians, economists and sociologists. Similarly, planners and plan designers are interacted in cities, each in their own way, as well as architects, financial specialists and representatives of a number of special fields.

He further advocates that the studies of cities may vary in terms of their territorial scope and can be studied in a global context, in the context of a country, or in the context of an individual region. One may carry out comparative studies of cities belonging to a particular category.

Finally, one may engage in a geographical study of a particular city constituting the subject of a monograph. Baransky stresses that from an economic-geographic point of view a city together with its network of roads constitutes the skeleton on which everything else hangs that defines the relevant territory, and endows it with a specific configuration. About planning, Baransky has opined that cities may be viewed as applied urban micro-geography.

1.4.1 Bases and Concepts

Cooley, in his treatise on transportation, made it clear – "why is a town there, where it is?" He has pointed out the geographical qualities which govern the site of a town is 'binary'. On the one hand, it is bountiful with respect to its resources and facilities of production, while on the other, it is equipped with transportational facilities.

1.4.2 Site-Situation Concept

Taylor's various classes and types of cities are the product of their natural site like hills; cuestas; mountain-corridors; passes; plateaux; eroded domes; ports, including fiords, rias, river-estuaries, and roadsteads; rivers, falls, meanders, terraces, deltas, fans, valleys, islands; lakes, etc. All of these are 'controlled' primarily by the topography of sites.

Dickinson's point of view about a city is that of natural beginning. But with the passage of time, town's natural setting is changed by its utilization of the available resources and its adaptability with the locality and the surrounding region. Its growth and expansion sometimes diffuses natural site to the extent to make it beyond recognition. In this context there was little scope for the development of a true urban geography.

The aim was restricted and hardly it was possible to explain a complex economic function and social system. Crowe, writing on methodology, pointed out that treatment of towns as "indicative of the inability of geographers to penetrate beyond the superficial".

He further stressed that the application of the 'site and situation' formula was meaningless "where site had nothing but historical interest as situation was viewed in terms of routes and not currents of movement". Such state of affairs rejected the stereotyped concept of 'site and situation'.

1.4.3 Concept of Ecology

Plant ecology came into being to influence geographical phenomena during the two World Wars. Robert Park opined that because of increase in population and city region there came a change in the ecological processes of a city. Urban ecology influenced the relationship with the surrounding areas of a city and had its sway over the people and their environment.

Park, in 1925 published a book The City in which he introduced the typical process of the expansion of the city. The process of expansion was best illustrated by Burgess for number of American cities in general and of Chicago in particular. He made it clear that during the course of time city – land use tended to display a zonal organization concentrically arrayed about the city centre.

Encircling the downtown area there is normally an area of transition which is being invaded by business and light manufacture, a third area is inhabited by the workers in industries who have escaped from the area of deterioration but who desire to live within easy access of their work. Beyond this zone is the area of high-class residences, and further out beyond the city limits, is the commuters' zone – suburban areas or satellite cities.

The concept of Burgess, of course, initiated the specialized spatial patterns where residential zones of varied ecology made their headway characterized by successive areas with different land uses. Nevertheless it is a model in the proper sense, but it was criticized on the ground that by the late 20th century it was anachronistic and limited to large Western industrial cities.

It is true that urban ecological and social patterns are determined by geographical, economic, industrial and transportational factors but these do not have their evenly spread impact everywhere and in all times on socio-economic life of cities.

1.4.4 Behaviouralism and Urban Centres

Berry's contention about city and its growth is related to its consumer's behaviour who are the befitting choice makers for using land. It depends upon three variables, viz.,

1. Value of residential unit – buying cost or rent?
2. Quality of residence, and
3. Relationship with place of work and neighbourhood?

Family income is an important component for the choice of site in the city, and this depends upon the ability of behavior to interact with space by its users. But as a general tendency, it is evident that people of almost the same income group agglomerate make their choice for an even space.

In the context of India, social relationship and behavior are the products of caste-ridden values and culture. These have given rise to 'mohallas' of the same communities. The city, of course, has its geographical base, but nonetheless important is behaviour of its citizens and their socio-cultural background. It is because of this behaviouralism of the residents that cities have been able to infuse in them their own smell.

1.4.5 Concept of Radicalism

An important aspect of the urban world is the concept of 'total change' brought into play by development of metropolises. This has changed totally the consumers' ideas. These ranged from the relative deprivation of immobile residents of small-country towns who are isolated from standard urban services which are concentrated at only big cities.

The people of inner city are better placed than the residents of the peripheral sectors. These sorts of people have to depend on the private institutions. The situation becomes worse when these institutions exploit the common masses, and finally, control the city.

These radical reactions also affect urban planning and capitalist city overpowers the rights of the masses. Radical urban concept "stresses the constraints that society as a whole, and particularly certain groups within it, impose on the behavior of individuals".

Radicals believe in the replacement root and branch of the operation of 'laissez faire' market forces and the inequities generated by them. They offer alternative system of 'socialist city' – the city of planning controls and central directions ensure egalitarianism as well as plenty for all.

But recent events have revealed that the radicals' proposals are no more than an Utopian dream. These bring about economic conditions even worse than those of market forces. The rigid totalitarian controls are far from reality. They are responsible for breeding massive bureaucracies.

Concluding the discussion, urban geography is more of an amalgam of different approaches. It is nearer perhaps to a multidisciplinary analysis of its object rather than a systematic geographical study. The town cannot be viewed as an object of study as a geologist would appraise a piece of rock.

The urban environment has a vast domain of world's people spending their lives with a central theme of humanism. The prime objective is the fact that people of varied nature within a city aspire for the good life with a degree to which it may be met spatially.

1.5 HISTORY OF URBAN GEOGRAPHY

The earliest studies of urban geography in the United States focused on site and situation. This developed out of the man-land tradition of geography which focused on the impact of nature on humans and vice versa. In the 1920s, Carl Sauer became influential in urban geography as he motivated geographers to study a city's population and economic aspects with regard to its physical location. In addition, central place theory and regional studies focused on the hinterland (the rural outlying are supporting a city with agricultural products and raw materials) and trade areas were also important to early urban geography.

Throughout the 1950s and 1970s, geography itself became focused on spatial analysis, quantitative measurements and the use of the scientific method. At the same time, urban geographers began quantitative information like census data to compare different urban areas. Using this data allowed them to do comparative studies of different cities and develop computer-based analysis out of those studies. By the 1970s, urban studies were the leading form of geographic research.

Shortly thereafter, behavioral studies began to grow within geography and in urban geography. Proponents of behavioral studies believed that location and spatial characteristics could not be held solely responsible for changes in a city. Instead, changes in a city arise from decisions made by individuals and organizations within the city.

By the 1980s, urban geographers became largely concerned with structural aspects of the city related to underlying social, political and economic structures. For example, urban geographers at this time studied how capital investment could foster urban change in various cities.

Throughout the late 1980s until today, urban geographers have begun to differentiate themselves from one another, therefore allowing the field to be filled with a number of different viewpoints and focuses. For example, a city's site and situation is still regarded as important to its growth, as is its history and relationship with its physical environment and natural resources. People's interactions with each other and political and economic factors are still studied as agents of urban change as well.

1.5.1 Themes of Urban Geography

Although urban geography has several different focuses and viewpoints, there are two major themes that dominate its study today. The first of these is the study of problems relating to the spatial distribution of cities and the patterns of movement and links that connect them across space. This approach focuses on the city system. The second theme in urban geography today is the study of patterns of distribution and interaction of people and businesses within cities. This theme mainly looks at a city's inner structure and therefore focuses on the city as a system.

In order to follow these themes and study cities, urban geographers often break down their research into different levels of analysis. In focusing on the city system, urban geographers must look at the city on the neighborhood and citywide level, as well as how it relates to other cities on a regional, national and global level. To study the city as a system and its inner structure as in the second approach, urban geographers are mainly concerned with the neighborhood and city level.

1.6 AREAS OF STUDY

There are essentially two approaches to urban geography. The study of problems relating to the spatial distribution of cities themselves and the complex patterns of movement, flows and linkages that bind them in space. Studies in this category are concerned with the city system. Secondly there is the study of patterns of distribution and interaction within cities, essentially the study of their inner structure. Studies in this category are concerned with the city as a system. A succinct way to define urban geography that recognizes the link between these two approaches within the subject is then, that "urban geography is the study of cities as systems within a system of cities."

1.6.1 Site and Situation

Site describes the location of a city with respect to its soil, water supply and relief, or more still the actual point on which a settlement is built while situation describes the surrounds of the city in terms of other settlements, rivers, mountains and communication. Locations for cities are usually chosen for good reasons. Benefits of certain locations can include:

- **A wet area:** Water is a constant necessity for urban areas and is difficult to transport. For this reason many cities are located near or adjacent to rivers.
- **A dry area:** In wet areas a dry area offers protection from flooding and marshland.
- **Easy access to building materials:** Stone, wood or clay are necessary for the construction of cities and are difficult to transport long distances.
- **A strategic defensive position:** Historically many cities have been constructed on high ground in order to make attack more difficult and to give a good view of surrounding land (for example, Quebec City). River meanders are also used as partial moats. Some cities were also built in swampy areas for the same reason (for example, Paris).
- **Fuel supply:** Most cities were initially constructed near wood for burning and cooking. Today many cities are constructed near coal, oil and gas mines to make use of those resources (for example: Newcastle, Glasgow, Pittsburgh, Essen).
- **A food supply:** Cities need some nearby land to be suitable for animal grazing or crop growing
- **A travel intersection point and bridging points:** It is often useful for a city to be located at the intersection of rivers, roads or train lines in order to facilitate travel and trade. Bridging points are shallow areas that allow easy construction of bridges, (for example: London, Cologne).
- Historically many cities grew at so-called "break-of-bulk" points along navigable rivers, where a local obstacle such as rapids required trade goods to be transshipped from larger boats to smaller boats, for example: Chicago, Montreal.
- **Shelter and aspect:** It is desirable to construct cities located on the side of a slope that is protected from incoming winds, and in a direction that receives maximum sun exposure.

1.7 CITIES AS CENTERS OF MANUFACTURING AND SERVICES

Cities differ in their economic makeup, their social and demographic characteristics and the roles they play within the city system. These differences can be traced back to regional variations in the local resources on which growth was based during the early development of the urban pattern and in part the subsequent shifts in the competitive advantage of regions brought about by changing locational forces affecting regional specialization within the framework of the market economy. Recognition of different city types necessitates their classification, and it is to this important aspect of urban geography that we now turn. Emphasis is on functional town classification and the basic underlying dimensions of the city system.

The purpose of classifying cities is twofold. On the one hand, it is undertaken in order to search reality for hypotheses. In this context, the recognition of different types of cities on the basis of, for example, their functional specialization may enable the identification of spatial regularities in the distribution and structure of urban functions and the formulation of hypotheses about the resulting patterns. On the other hand, classification is undertaken to structure reality in order to test specific hypotheses that have already been formulated. For example, to test the hypotheses that cities with a diversified economy grow at a faster rate then those with a more specialized economic base, cities must first be classified so that diversified and specialized cities can be differentiated.

The simplest way to classify cities is to identify the distinctive role they play in the city system. There are three distinct roles. 1. Central places functioning primarily as service centers for local hinterlands. 2. Transportation cities performing break-of-bulk and allied functions for larger regions. 3. Specialized-function cities are dominated by one activity such as mining, manufacturing or recreation and serving national and international markets. The composition of a cities labor force has traditionally been regarded as the best indicator of functional specialization, and different city types have been most frequently identified from the analysis of employment profiles. Specialization in a given activity is said to exist when employment in it exceeds some critical level.

The relationship between the city system and the development of manufacturing has become very apparent. The rapid growth and spread of cities within the heartland-hinterland framework after 1870 was conditioned to a large extent by industrial developments and that the decentralization of population within the urban system in recent years is related in large part to the movement of employment in manufacturing away from the traditional industrial centers. Manufacturing is found in nearly all cities, but its importance is measured by the proportion of total earnings received by the inhabitants of an urban area. When 25 percent or more of the total earnings in an urban region are derived from manufacturing, that urban areas is arbitrarily designated as a manufacturing center.

The location of manufacturing is affected by myriad economic and non-economic factors, such as the nature of the material inputs, the factors of production, the market and transportation costs. Other important influences include agglomeration and external economies, public policy and personal preferences. Although it is difficult to evaluate precisely the effect of the market on the location of manufacturing activities, two considerations are involved: the nature of and demand for the product and transportation costs.

1.7.1 Urban Area

An urban area is characterized by higher population density and vast human features in comparison to areas surrounding it. Urban areas may be cities, towns or conurbations, but the term is not commonly extended to rural settlements such as villages and hamlets.

Urban areas are created and further developed by the process of urbanization. Measuring the extent of an urban area helps in analyzing population density and urban sprawl, and in determining urban and rural populations.

Unlike an urban area, a metropolitan area includes not only the urban area, but also satellite cities plus intervening rural land that is socio-economically connected to the urban core city, typically by employment ties through commuting, with the urban core city being the primary labor market. In fact, urbanized areas agglomerate and grow as the core population/economic activity center within a larger metropolitan area or envelope.

Metropolitan areas tend to be defined using counties or county sized political units as building blocks. Counties tend to be stable political boundaries; economists prefer to work with economic and social statistics based on metropolitan areas. Urbanized areas are a more relevant statistic for determining per capita land usage and densities.

1.7.2 Definitions

They vary somewhat amongst different nations. European countries define urbanized areas on the basis of urban-type land use, not allowing any gaps of typically more than 200 meters, and use satellite photos instead of census blocks to determine the boundaries of the urban area. In less developed countries, in addition to land use and density requirements, a requirement that a large majority of the population, typically 75%, is not engaged in agriculture and/or fishing is sometimes used.

Australia

In Australia, urban areas are referred to as "urban centres" and are defined as population clusters of 1000 or more people, with a density of 200 or more persons per square kilometre.

Canada

In Canada, an urban area is an area that has more than 400 people per square kilometre and has more than 1,000 people. If two or more urban areas are within two kilometres of each other, they are merged into a single urban area. The boundaries of an urban area are not influenced by municipal or even provincial boundaries.

China

In China, an urban area is an urban district, city and town with a population density higher than 1,500 persons per square kilometre. As for urban districts with a population density lower than 1,500 persons per square kilometre, only the population that lives in streets, town sites, and adjacent villages is counted as urban population.

France

In France, an urban area is a zone (aire urbaine) encompassing an area of built-up growth (called an "urban unit"—close in definition to the North American urban area) and its commuter belt (couronne périurbaine). Although the official INSEE translation of aire urbaine is "urban area", most North Americans would find the same as being similar in definition to their metropolitan area.

Japan

In Japan urbanized areas are defined as contiguous areas of densely inhabited districts (DIDs) using census enumeration districts as units with a density requirement of 4,000 inhabitants per square kilometre (10,000 /sq mi).

New Zealand

Statistics New Zealand defines New Zealand urban areas for statistical purposes as a settlement with a population of a thousand people or more.

Norway

Statistics Norway defines urban areas ("tettsteder") similarly to the other Nordic countries. Unlike in Denmark and Sweden, the distance between each building has to be of less than 50 meters, although exceptions are made due to parks, industrial areas, rivers, and similar. Groups of houses less than 400 metres from the main body of an urban area are included in the urban area.

Poland

In Poland, official "urban" population figures simply refer to those localities which have the status of towns (miasta). The "rural" population is that of all areas outside the boundaries of these towns. This distinction may give a misleading impression in some cases, since some localities with only village status may have acquired larger and denser populations than many smaller towns.

Sweden

Urban areas in Sweden (tätorter) are statistically defined localities, totally independent of the administrative subdivision of the country. There are 1,940 such localities in Sweden, with a population ranging from 200 to 1,252,000 inhabitants.

England and Wales

The United Kingdom's Office of National Statistics produced census results from urban areas since 1951, since 1981 based upon the extent of irreversible urban development indicated on Ordnance Survey maps. The definition is an extent of at least 20 hectares and at least 1,500 census residents. Separate areas are linked if less than 200 metres apart. Included are transportation features.

United States

In the United States there are two categories of urban area. The term urbanized area denotes an urban area of 50,000 or more people. Urban areas under 50,000 people are called urban clusters. Urbanized areas were first delineated in the United States in the 1950 census, while urban clusters were added in the 2000 census. There are 1,371 urban areas and urban clusters with more than 10,000 people.

The U.S. Census Bureau defines an urban area as: "Core census block groups or blocks that have a population density of at least 1,000 people per square mile (386 per square kilometer) and surrounding census blocks that have an overall density of at least 500 people per square mile (193 per square kilometer)."

The concept of Urbanized Areas as defined by the U.S. Census Bureau are often used as a more accurate gauge of the size of a city, since in different cities and states the lines between city borders and the urbanized area of that city are often not the same. For example, the city of Greenville, South Carolina has a city population under 60,000 but an urbanized area over 300,000, while Greensboro, North Carolina has a city population over 200,000 but an urbanized area population of around 270,000 — meaning that Greenville is actually "larger" for some intents and purposes, but not for others, such as taxation, local elections, etc.

The largest urban area in the United States is that of New York City, with its city proper population exceeding 8 million and its metropolitan area population almost 19 million. The next four largest urban areas in the U.S., are those of Los Angeles, Chicago, Miami and Philadelphia. About 70 percent of the population of the United States lives within the boundaries of urbanized area (210 out of 300 million). Combined, these areas occupy about 2 percent of the United States. The majority of urbanized area residents are suburbanites; core central city residents make up about 30 percent of the urbanized area population (about 60 out of 210 million).

1.8 URBAN ECOLOGY

Urban ecology is the subfield of ecology which deals with the interaction of plants, animals and humans with each other and with their environment in urban or urbanizing settings. Analysis of urban settings in the context of ecosystem ecology (looking at the cycling of matter and the flow of energy through the ecosystem) may ultimately help us to design healthier, better managed communities.

Urban ecology also involves the study of the effects of urban development patterns on ecological conditions. Emphasis may be placed on planning communities via design and building materials in order to promote a healthy and biodiverse urban ecosystem.

By 2030 it is estimated that 60% of the global population will live in a metropolitan setting.

Interactions between non-living factors, such as sunlight and water, and biological factors, such as plants and microbes, take place in all environments including cities. Concentrating humans and the resources they consume in metropolitan areas alters such things as soil drainage, water flow, and light availability. For example, sidewalks and rooftops can change an area's hydrology by increasing storm water runoff and can contribute to higher urban temperatures by storing heat energy and acting as an artificial heat sink.

Urban ecology does not necessarily make value judgments about whether urban environments are 'good' or 'bad'. Some species of animals such rats, pigeons, and cockroaches, for example, are able to survive and thrive in a non-natural urban setting. Attempting to understand the factors that make some species successful in urban environments while others perish is a common topic of research.

1.8.1 Urban Field

The urban field is a form of urban habitat of relatively high density involving a good transportation system and a broad array of economic, social and recreational opportunities. Even though the sections of the urban field may still be in agricultural use, the area is nonetheless urbanized because anywhere within it a person is able to connect his/her home to telephones, radio and television facilities, electricity, gas, water supply systems and a network of freeway and primary roads.

1.8.2 Criteria

An urban field is a field in the centre of an urban area, with a boundary wall and road frontage. Planning permission is not essential for a field to be classed as an 'urban field', although Outline Planning Permission is desirable. It is centered on and dominated to a certain extent by a metropolitan area of at least 200,000-300,000 people. It's outer limits can be defined by two things.

1. The maximum time or distance that most people are prepared to commute.
2. The time or distance that most people are prepared to spend traveling to or from weekly or weekend recreational activities. The daily commute perspective defines the "hard-core" of the urban field and results in regions of about a 40-50 mile radius from the central metropolitan area. The weekend recreational area perspective results in a much wider field of about 100 miles with far less determinate boundaries.

1.8.3 Extent

The actual extent of the urban field depends upon the freeway facilities available and the density of metropolitan development. In the South, Midwest and on the Prairies the fields may be quite extensive. In the East and the area bordering the lower Great Lakes where metropolitan densities are rather high, the urban field are much more restricted and may run into each other to form "galaxies". In these higher density areas the extent of the field, as defined as criterion 1, may be only 20–30 miles and the recreational field about 75 miles.

1.8.4 Implications

The planning implications of the urban field have become quite obvious. These large areas are inhabited and used by urbanites but they involve vast quantities of land and often very sensities environmental conditions. The management questions involve the preservation of valuable agricultural land and the protection of this land from abuse. A related issue that is becoming more important involves the pricing of farms and farmland out of the realm of wealthy individuals seeking weekend retreats or good speculative investments. The management of recreational resources and the provision of land for campsites and second homes has also become a question of concern, along with questions of conservation. Thus, the definition of an urban area according to the extent of its urban field has raised an enormous number of questions related to the management of agricultural, natural and recreational resources around a major metropolitan center.

1.9 URBAN ANTHROPOLOGY

Urban Anthropology, a subset of Anthropology, is a relatively new and developing field. Traditional anthropology was divided into four main fields: Sociocultural Anthropology, Archaeology, Linguistics, and Physical Anthropology. As research increased in these areas during the twentieth century, however, more anthropologists focused their scope of research. Smaller, more specific subject areas arose that did not fit into the traditional main fields, and were therefore defined as individual subfields. A relatively new subfield that emerged in the 1960s and 70s, urban anthropology is often concerned with issues of urbanization, poverty, and neoliberalism. Ulf Hannerz quotes a 1960s remark that anthropologists were "a notoriously agoraphobic lot, anti-urban by definition". Various social processes in the Western World as well as in the "Third World" (the latter being the habitual focus of attention of anthropologists) brought the attention of "specialists in 'other cultures'" closer to their homes.

Urban anthropology is heavily influenced by sociology. The traditional difference between sociology and anthropology was that the former was traditionally conceived as the study of civilized populations, whilst anthropology was approached as the study of primitive populations. There were, in addition, methodological differences between these two disciplines — sociologists would normally study a large population sample while anthropologists relied on fewer informants with deeper relations.

However, as interest increased in urban societies, methodology between these two fields and subject matters began to blend, leading some to question the differences between Urban Sociology and Urban Anthropology. The lines between the two fields have blurred with the interchange of ideas and methodology, to the advantage and advancement of both disciplines. However, a key difference is that Urban Sociology frequently studies smaller societies, whilst Urban Anthropology tends to focus on larger, more complex societies.

While Urban Anthropology is a newly acknowledged field, anthropologists have been conducting work in the area well before. For instance, numerous early scholars have attempted to define exactly what the city is and pinpoint the ways in which urbanism sets apart modern city lifestyles from what used to be regarded as the "primitive society". It is increasingly acknowledged in Urban Anthropology that, although there are significant differences in the characteristics and forms of organisation of urban and non-urban communities, there are also important similarities, insofar as the city can also be conceived in anthropological studies as a form of community. Urban Anthropology is an expansive and continuously evolving area of research. With a different playing field, anthropologists have had to modify their methods and even readdress traditional ethics in order to adjust to different obstacles and expectations.

1.9.1 History of The Discipline

The field of Anthropology first began developing in the 19th century as a result of the expansion of the West and discovering new frontiers and, as a result, new civilizations. A new sort of anthropologists budded, those who employed the method of fieldwork and immersed themselves in these new cultures that were mainly viewed as primitive. It was a comparative approach; mainly, the comparison was between the primitive non-western culture to the more civilized western culture. While Anthropology soundly became associated with the study of primitive cultures, several factors began leading the field more towards the study of urban cultures. A strong influence in this direction was with the discovery of practically the entire world, there were few undiscovered people to study. The first step in this process began with World War I, after which developing nations began to emerge. Anthropologists began studying these "peasant societies" that were actually different from the "folk societies" they traditionally would research. Folk societies were namely villages that were self-contained. Robert Redfield was a prominent anthropologist who studied both folk and peasant societies. While researching peasant societies of developing nations, such as India, he discovered that these societies were actually not the same as folk societies. They were not self-contained; for example, they were economically linked to forces outside of their own community. They were actually part of a bigger society—the city.

This discovery led into more anthropologists studying societies that began more identified as cities, such as the indigenous populations of Mexico. This crossover was instrumental in the development of Urban Anthropology as an independent field in the 1960s. Prior to this, other fields expressed interest in the study of the city, such as archaeology which studied the origins of the city and anthropologists who studied preindustrial cities. Their efforts were largely unrelated, though their findings still hold relevant for urban anthropologists. Archaeology focused on the excavation of urban centers, from which findings showed the shift from small settlements to larger urban sites. This work helped define what characteristics constituted an area as "urban," specifically physical factors such as street plans and building structures.

Several prominent anthropologists and sociologists have played key roles, along with specific studies, in laying a strong foundation for what Urban Anthropology would become, a study of the city as a community. Louis Wirth, a sociologist, published an essay titled "Urbanism as a Way of Life." This essay proved to be essential in distinguishing urbanism as an unique society that could be studied from three perspectives: "a physical structure, as a system of social organization, and as a set of attitudes and ideas". Lloyd Warner led the Community Study Approach and was one of the first anthropologists to transition from studying primitive cultures, the aborigines in his case, to studying urban cities using similar methods. The Community Study Approach was an important influence leading to studying the city as a community. Warner's case was about New England and titled Yankee City, and studies such as his proved the city to be made up of urban communities with a new method of network studies. William Whyte was another anthropologist who expanded Warner's methods for small urban centers to larger neighborhoods. Another influence of the study of the city was the study conducted by the Chicago School of Urban Ecology. As early as the 1920s, the school defined the city, in terms of urban ecology, as "made up of adjacent ecological niches accompanied by human groups in rings surrounding the core".

1.9.2 Methods, Techniques, and Ethics

Anthropologists typically have one significant difference from their affiliated field of science: their method of gathering information. Scientists prefer Research design, where defined independent and dependent variables are used. Anthropologists, however, prefer the ethnographic method, which is broader and does not oversimplify a case. With Urban Anthropology, the subject is exactingly broad as it is, there needs to be a degree and channel of control. For this reason, Urban Anthropologists find it easier to incorporate research design in their methods and usually define the city as either the independent variable or the dependent variable. So, the study would be conducted on either the city as the factor on some measure, such as immigration, or the city as something that is responding to some measure.

A common technique used by anthropologists is "myth debunking". In this process, anthropologists present a specific question and conduct a study to either verify or negate its validity. Research design is actually an important part of this process, allowing anthropologists to present a specific question and answer it. Being able to hone into such a broad subject specifically while remaining holistic is largely the reason why this technique is popular among anthropologists. Another technique is based on how anthropologists conduct their studies; they either use single case studies or controlled comparisons. By using case studies, they present and analyze a single urban society. The more sophisticated method is using controlled comparisons, where different societies are compared with controlled variables, so that the associations are more valid and not merely correlations. In order to conduct either type of study, the anthropologist must define a basic unit, which is the ethnographic target population. The target population can be central to the research question, but not necessarily; for example, when studying migrant immigrations, the people are being studied, not the neighborhoods. Common ways to define target populations that are central to the research design are by spatial boundaries, common cultures, or common work.

Ethics largely remain the same for all anthropologists. Still, working in an urban setting and a more complex society raises new issues. The societies that anthropologists are now studying are more similar to their own, and the familiarity raises issues concerning objectivity. The best idea is for an anthropologist to identify his or her own values explicitly and adapt to the society based on what he or she is studying. With primitive societies, it would have been acceptable for an anthropologist to enter the society and explain at the beginning their intentions of studying the society. In urban cultures, however, they are not in what are considered alien cultures. Therefore, an anthropologist finds that a more detailed explanation of their intentions is needed and often find that their intent must be explained multiple times throughout the study.

1.9.3 Main Areas of Study

There are two main ways to go about researching Urban Anthropology: by examining the types of cities or examining the social issues within the cities. These two methods are overlapping and dependent of each other. By defining different types of cities, one would use social factors as well as economic and political factors to categorize the cities. By directly looking at the different social issues, one would also be studying how they affect the dynamic of the city.

There are four central approaches to the anthropological study of cities. The first is the Urban Ecology model in which the community and family network are central. The second is based on power and knowledge, specifically of how the city is planned. The third approach is studying local and supralocal and the link between the two degrees of units in the city. The last approach focuses on cities where political economy is central to the city's infrastructure. Low uses several prominent studies from urban anthropologists to compile a list of the different types of cities that do not fall into only one category, and what factors individualize them. These types of cities include those focused on religious, economic, and social processes. An example of the religious city is what Low calls the "sacred city" in which religion is central to the daily life processes of the city. An example of an economic centered city image is the "Deindustrialized city". In America, this type of city is usually found in areas where coal mining was the main industry in the city, and once coalmines were shut down, the city became a ghost city rampant with unemployment and displaced workers. Globalization has been studied as a force that severely affects these areas, and anthropological studies have greatly increased the knowledge of the implications.

Other types of cities include, but are certainly not limited to the contested city, in which urban resistance is a key image; the gendered city, dominant in urbanizing areas such as Africa where women find themselves newly employed in low-wage labor; postmodern city, that is centered around capitalism; and fortress city, where different populations within the city are separated, usually based on socioeconomic factors. The main reasons for the current studies focusing on types on cities are to understand the patterns in which cities are now developing in, to study theoretical cities that may come about in the future based on these current trends, and to increase the implications of anthropological studies. An important note is that anthropological studies have serious implications on the understanding of urban society: with the rapid rate of globalization, many peasant societies are quickly attempting to modernize their cities and populations, but at an expense of the interests of the people within the cities. Studies can illustrate these negative effects and project how the overall city will fare poorly in the future.

The other method of studying Urban Anthropology is by studying various factors, such as social, economic, and political processes, within the general city. Focuses on these factors include studies on rural-urban migration, kinship in the city, problems that arise from urbanism, and social stratification. These studies are largely comparative between how these relations function in an urban setting versus how they function in a rural setting. When studying kinship, anthropologists have been focusing on the importance of extended family for urban natives versus migrants. Studies have shown, generally, that the more "native" one becomes with the urban city, the less importance is placed on maintaining familial relations. Another important and commonly studied aspect of the urban society is poverty, which is believed to be a problem that arises out of urbanism. Urban Anthropologists study several aspects individually, and attempt to tie different aspects together, such as the relationship between poverty and social stratification.

1.9.4 Nature and Scope of Urban Geography

Urban Geography is the study and analysis of urban systems with reference to their geographical environment. Broadly speaking, the subject matter includes the origin of towns, their growth and development, their functions in and around their surroundings.

In the case of our country (India), the census of 1981 has identified the following places as urban:

1. Centres having Municipality, City Board, Cantonment Board/Notified Town Area;

2. A minimum population of 5,000;
3. 75 percent of males engaged in non-agricultural activities;
4. A minimum population density of 400 persons per square km or 1,000 persons per square mile; and
5. Centres defined by urban amenities prescribed by the Director, Provincial Census.

Two important facts must be borne in mind before accepting the meaning of urban and rural.

One is the fact that it is rather impossible now to identify a dividing line between the rural and urban – the two being merged to create a sort of diffusion and present a landscape which is neither purely agricultural nor engaged wholly in tertiary activities.

Industrialization has brought into being a large number of settlements which are not certainly villages but are nucleated settlements of the agricultural population. Another problem is about the concept of what is urban, which is not static and is subject to change with time as well as with space.

The proportion of the population engaged in primary economic activities is the most effective measure. But capitalization of agriculture in modern times and rural depopulation by commuting urban workers have made the criterion of proportion irrelevant.

Thus, to conclude the discussion one reaches to the point that with the changing nature of both rurality and urbanity, there has developed the functional overlap between the two. Therefore, the distinction between what is urban and what is rural has lost its meaning in reality.

Attributes of a Town:

What are the attributes or characteristics qualifying a town?

Several qualities of a city or town may be summarized as:

(a) Town is a kind of settlement having a social organization of much greater scope than a simple rural establishment.

(b) It does not represent a mere greater number of people agglomerated in a vast area. But it represents a stage of civilization quite different from a locality expressing a rural way of life.

(c) Cities and towns have their historical origin Blache pointed out that cities characteristically possess mythical halo surrounding their genesis (ritual, eponymous hero, etc.).

(d) Towns and cities are creatures of commerce, and politics accompanying the earliest developments such as Babylon, Athens, London, Paris, Delhi, etc.

CHAPTER

URBANIZATION 2

2.1 INTRODUCTION

Urbanization (also spelled "urbanisation") is the physical growth of urban areas as a result of global change. Urbanization is also defined by the United Nations as movement of people from rural to urban areas with population growth equating to urban migration. The United Nations projected that half of the world's population would live in urban areas at the end of 2008.

Urbanization is closely linked to modernization, industrialization, and the sociological process of rationalization.

By the early 1900s both Great Britain and the United States had become predominantly urbanized nations; since that time, urbanization has been occurring around the globe at a rapid rate. Today, as many as 50 percent of the world's population lives in urban areas, compared to only a few percent just 200 years ago.

Sociologists studying urbanization trends note three distinct historical stages in the development of cities: preindustrial, industrial, and metropolitan-megalopolitan stages.

2.1 A BRIEF HISTORY OF URBANIZATION

One of the most significant changes over the centuries has been urbanization, or the shift from rural areas to large cities. Urbanization has had important consequences for many aspects of social, political, and economic life.

The earliest cities developed in ancient times after the rise of horticultural and pastoral societies made it possible for people to stay in one place instead of having to move around to find food. Because ancient cities had no sanitation facilities, people typically left their garbage and human waste in the city streets or just outside the city wall (which most cities had for protection from possible enemies). This poor sanitation led to rampant disease and high death rates. Some cities eventually developed better sanitation procedures, including, in Rome, a sewer system. Still, the world remained largely rural until the industrialization of the nineteenth century. We return to industrialization shortly.

During the American colonial period, cities along the eastern seaboard were the centers of commerce and politics. Boston, New York, and Philadelphia were the three largest cities in population size. Yet they were tiny in comparison to their size today. In 1790, the year after George Washington became the first president of the new nation, New York's population was only 33,131; Philadelphia's was 28,522; and Boston's was 18,230. Today, of course, cities of this size are called small towns. New York's population is vastly higher, at about 8.2 million; Philadelphia's is 1.5 million; and Boston's is 618, 000.

US cities became more numerous and much larger during the nineteenth century because of two trends. The first was immigration, as waves of immigrants from Ireland and then Italy and other nations began coming to the United States during the 1820s. The second was industrialization, as people moved to live near factories and other sites of industrial production. These two trends were momentous: People crowded together as never before, and they crowded into living conditions that were often squalid. Lack of sanitation continued to cause rampant disease, and death rates from cholera, typhoid, and other illnesses were high.

Crime also became a significant problem, as did riots and other mob violence beginning in the 1830s. This type of mass violence was so common that the 1830s have been called the "turbulent era". Most of this mass violence was committed by native-born whites against African Americans, Catholics, and immigrants. Native whites resented their presence and were deeply prejudiced against them. During the three decades beginning in 1830, almost three-fourths of US cities with populations above 20,000 had at least one riot. This wave of mass violence in the nation's cities led Abraham Lincoln to lament, "Accounts of outrages committed by mobs form the everyday news of the times Whatever their causes be, it is common to the whole country".

American cities grew even more rapidly after the Civil War as both industrialization and immigration continued. By the early years of the twentieth century, US cities on the East Coast were almost unimaginably crowded, and their living conditions continued to be wretched for many of their residents. Their city governments, police forces, and business worlds were also notoriously corrupt. In 1904, Lincoln Steffens, a renowned "muckraking" journalist, published his classic work, The Shame of the Cities (Steffens, 1904), which was a collection of six articles he had written for McClure's Magazine. In this book, Steffens used biting prose to attack the municipal corruption of the times in Chicago, Philadelphia, St. Louis, and other cities. In the original articles that compose the book, he named names: He listed by name people who gave and received bribes and those who were corrupt in other ways. A decade earlier, another muckraker, Jacob Riis, had published How the Other Half Lives: Studies among the Tenements of New York

(Riis, 1890), a book of searing photographs of poverty in the largest US city. The books by Steffens and Riis remain as vivid reminders of what cities were like a century ago, and perhaps are still like today in some respects.

As Americans moved west after the Civil War and during the twentieth century, western cities appeared almost overnight and expanded the pace of urbanization. Continued industrialization, immigration, and general population growth further increased the number and size of US cities. Internal migration had a similar impact, as waves of African Americans moved from the South to Chicago and other northern cities.

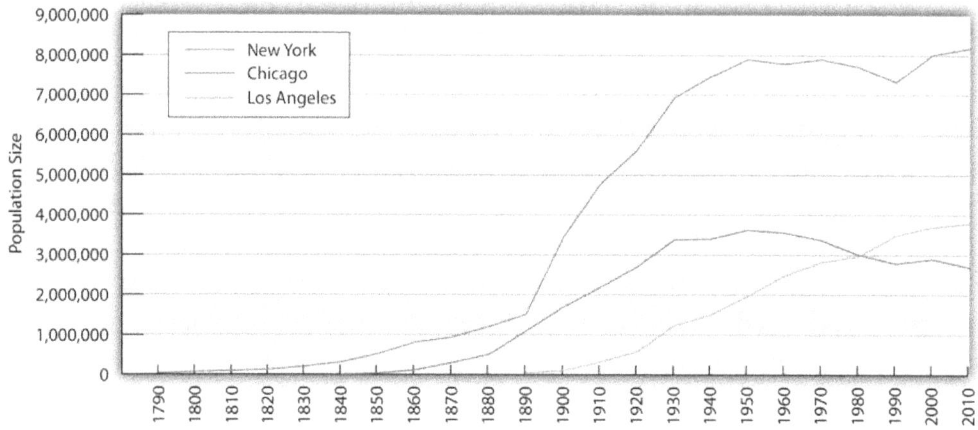

Fig. 1. Populations of Chicago, New York, and Los Angeles, 1790–2010

Note: New York annexed Brooklyn in 1898; therefore, New York's population beginning in 1900 includes Brooklyn's population.

Sources: Gibson, C. (1998). Population of the 100 largest cities and other urban places in the United States: 1790–1990. Washington, DC: US Census Bureau; US Census Bureau. (2012). Statistical abstract of the United States: 2012. Washington, DC: US Government Printing Office. Retrieved from http://www.census.gov/compendia/statab.

Chicago and Los Angeles first appear in the graph when they began to rank in the largest one hundred cities.

Note that the populations of New York and Chicago show some decline after 1950. This decline reflects two other trends affecting cities in the past half-century: (1) the movement of people from cities to suburbs; and (2) the movement of Americans from northern cities to southern and southwestern cities. Reflecting this second trend, and also reflecting increases in immigration from Mexico and Asia, southern and southwestern cities have grown rapidly during the past few decades. For example, during the 1970–2010 period, the populations of Albuquerque, New Mexico, and Phoenix, Arizona, more than doubled, while the populations of Cleveland, Ohio, and Detroit, Michigan, both fell by about half.

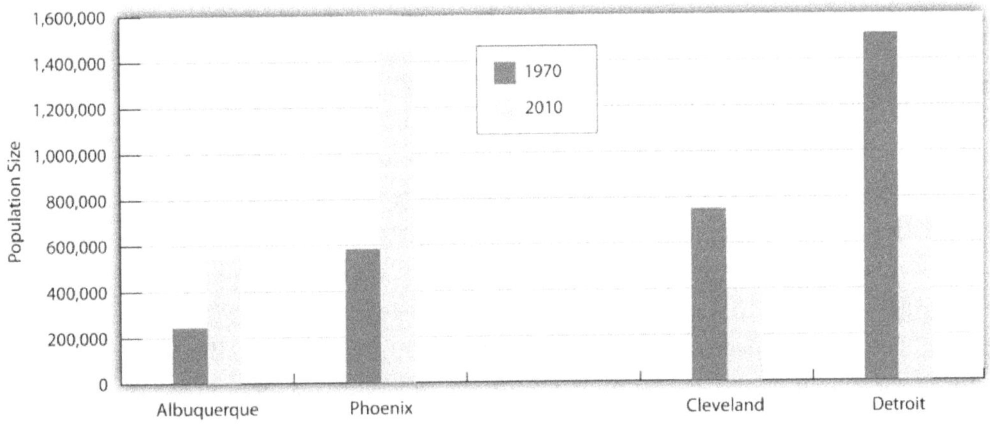

Fig. 2. Population Change from 1970 to 2010 for Selected Cities

Source: US Census Bureau. (2012). Statistical abstract of the United States: 2012. Washington, DC: US Government Printing Office. Retrieved from http://www.census.gov/compendia/statab.

This trend in urbanization aside, the fact remains that the United States has become much more urbanized since its formation. Today, more than three-fourths of the US population lives in an urban area (defined generally as an incorporated territory with a population of at least 2,500), and less than one-fourth lives in a rural area. "Urbanization in the United States (Percentage Living in Urban Areas)" shows, the degree of urbanization rose steadily through the nineteenth and twentieth centuries before slowing down by the end of the last century.

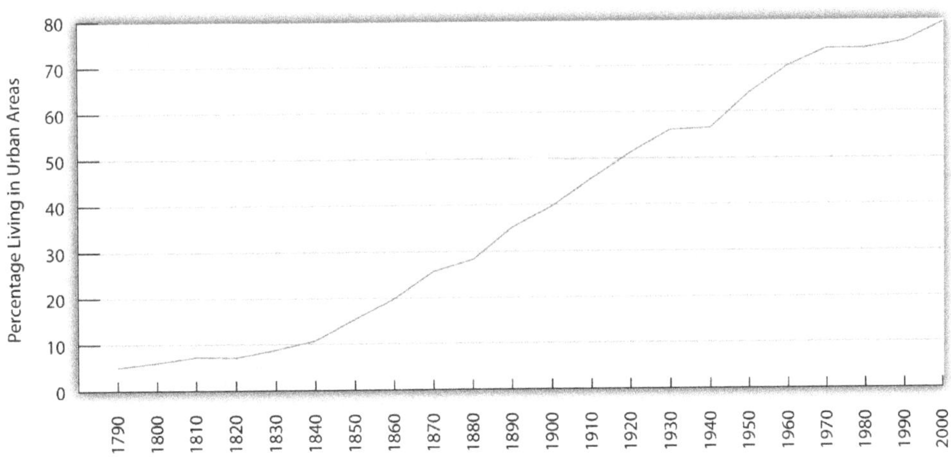

Fig. 3. Urbanization in the United States (Percentage Living in Urban Areas)

2.2.1 The Early Days of Urbanization

Cities have been around for thousands of years. Have you thought about how cities were formed? After the Neolithic Revolution, people shifted from a nomadic lifestyle of hunting and gathering to a lifestyle of farming and settlement. Oftentimes, people gathered around water sources, such as rivers and lakes, and built non-nomadic or permanent settlements. Gradually, these settlements expanded, and small cities formed. Densely populated settlements emerged, along with the specialization and division of labor among people living in the city. Trade, bartering, or other forms of economic exchange also took place. Architecture, centralized administrations, and political structures also all became part of a city. A city is complex and encompasses all these different meanings. Two hundred years ago, the only city that had more than one million people was Beijing, China. But today, there are more than 500 cities with over a million people. This was due to the Industrial Revolution, which contributed to the development of faster means of transportation and communication methods. This ability for people, resources, and information to move around rapidly allows cities to exist.

As more cities formed, an increasing proportion of the population moved into the cities. This is called urbanization. According to statistics, in 1800, only 3% of the world's population lived in cities. A tipping point was reached in 2008, and there were officially more people living in cities than rural areas. Urbanization has been occurring especially rapidly in less developed parts of the world. As cities and their infrastructure become more mature in developing countries, more people move into the cities from rural areas. Why would people want to move to a city? First of all, people have become less attached to land as a means of production. This means that people aren't necessarily farmers anymore, and their jobs are not associated with farming. Therefore, moving into a city would be better for them, since cities offer a wide variety of jobs. Second of all, cities, because of the division of labor, offer all types of products and services for their residents. Thirdly, with the development of schools, people are becoming more educated or at least have better access to education. Therefore, human capital has become more important—things such as knowledge and creativity, instead of labor capital, which is based on a person being able to make money in exchange for their labor.

2.2.2 What are Cities Like in the US?

In the beginning, the US was largely rural. This is hard to imagine since 80% of Americans live in cities today! As we can of the top 5 growing cities are in the West, while most of the shrinking cities are in the Midwest. This potentially has to do with the shift in industries in the Midwest. New Orleans is a special case, not having recovered from Hurricane Katrina.

City growth 2000-2011

Top 5 Growing Cities	Top 5 Shrinking Cities
Raleigh, NC	New Orleans, LA
Las Vegas, NV	Youngstown, OH
Provo, UT	Cleveland, OH
Cape Coral—Ft. Myers, FL	Detroit, MI
Greeley, CO	Flint, MI

As cities expand, what happens in areas in between cities? Edge cities have appeared. When you are driving down the highway, you see houses, stores, and hotels located near major highway intersections. Those clusters are edge cities. Often they do not have their own mayors and belong to a bigger city or town. What about within cities, how cities change internally? One prominent phenomenon is gentrification. This is an interesting urban development pattern in the US. The middleclass move into a more under-maintained area of a city due to the increase in the number of residents and surge in rent or housing prices in expensive city centers. In doing so, the poor who primarily lived in the under-maintained areas are displaced as the middleclass drive up rent prices and buy up property. You can see this phenomenon in San Francisco and surrounding areas such as Oakland. As housing prices in San Francisco increases, middleclass residents move outwards to Oakland across the bay. As Oakland prices also increase, people look again to move outwards, further expanding into other small cities nearby. Harlem in New York City is another example. The "unsafe" image of Harlem 10 years ago is a very different one from the Harlem today full of amenities—shops, restaurants, and other services.

Another interesting city growth phenomenon in the US is suburbanization. In countries such as China, the term 'suburbs' is not a familiar one. This may have to do with its relatively late urbanization in the late 1980s. In the US, when inner cities become too crowded, residents started migrating to the suburbs, which are communities that are adjacent to but outside of the city. This is also called urban sprawl or suburban sprawl, which describes how people move from central urban areas to low-density areas outside of the city. Suburban communities are mostly car-dependent. Some large cities are so 'sprawled' out that they are almost connected to other cities that used to seem further away.

2.2.3 Global Urbanization

If the United States has urbanized during the last two centuries, so has much of the rest of the world. Only 3 percent of the world's population lived in urban areas in 1800. By a century later in 1900, 14 percent of the world's population lived in urban areas, and twelve cities had populations over 1 million. Just a half-century later in 1950, the world's urban population had doubled to 30 percent, and the number of cities over 1 million grew six times to eighty-three cities.

Today, more than half the world's population lives in urban areas, and the number of cities over 1 million stands at more than four hundred. By 2030, almost two-thirds of the world's population is projected to live in urban areas. The number of megacities—cities with populations over 10 million—rose from three in 1975 to sixteen in 2000, and is expected to reach twenty-seven by 2025.

Despite all this growth, the degree of urbanization still varies around the world. In general, wealthy nations are more urban than poor nations, thanks in large part to the latter's rural economies. Still, urbanization in poor nations is proceeding rapidly. Most megacities are now in, and will continue to be in, nations that are relatively poor or desperately poor. The number of urban residents in these nations will increase greatly in the years ahead as people there move to urban areas and as their populations continue to grow through natural fertility. Fertility is a special problem in this regard for two reasons. First, women in poor nations have high fertility rates. Second, poor nations have very high proportions of young people, and these high rates mean that many births occur because of the large number of women in their childbearing years.

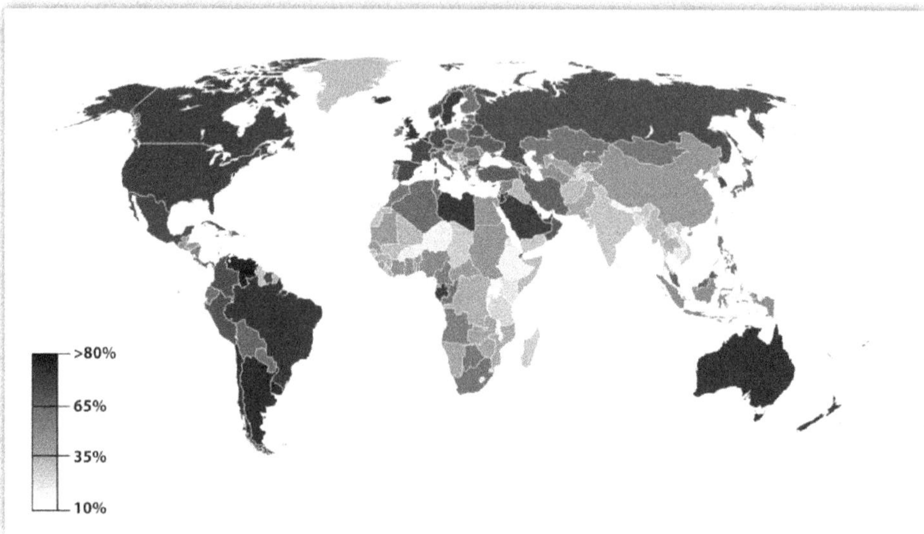

Fig. 4. Percentage of World Population Living in Urban Areas

Source: Adapted from http://en.wikipedia.org/wiki/File:Urban_population_in_2005_world_map.PNG.

Rapid urbanization poses both opportunities and challenges for poor nations. The opportunities are many. Jobs are more plentiful in cities than in rural areas and incomes are higher, and services such as health care and schooling are easier to deliver because people are living more closely together. In another advantage, women in poor nations generally fare better in cities than in rural areas in terms of education and employment possibilities.

In large cities in poor nations, as this scene illustrates, many people live in deep poverty and lack clean water and sanitation.

But there are also many challenges. In the large cities of poor nations, homeless children live in the streets as beggars, and many people lack necessities and conveniences that urban dwellers in industrial nations take for granted. As the United Nations Population Fund (2007) warns, "One billion people live in urban slums, which are typically overcrowded, polluted and dangerous, and lack basic services such as clean water and sanitation." The rapid urbanization of poor nations will compound the many problems these nations already have, just as the rapid urbanization in the industrial world more than a century ago led to the disease and other problems discussed earlier. As cities grow rapidly in poor nations, moreover, these nations' poverty makes them ill equipped to meet the challenges of urbanization. Helping these nations meet the needs of their cities remains a major challenge for the world community in the years ahead. In this regard, the United Nations Population Fund (2007) urges particular attention to housing: "Addressing the housing needs of the poor will be critical. A roof and an address in a habitable area are the first step to a better life. Improving access to basic social and health services, including reproductive health care, for poor people in urban slums is also critical to breaking the cycle of poverty."

Life in the megacity of Mumbai (formerly called Bombay) in India illustrates many of the problems facing large cities in poor nations. Mumbai's population exceeds 12.4 million, with another 8 million living in the greater metropolitan area; this total of more than 20 million ranks Mumbai's metropolitan population as the fourth highest in the world. An author who grew up in Mumbai calls his city an "urban catastrophe." He continued, "Bombay is the future of urban civilization on the planet. God help us". A recent news story illustrated his bleak assessment with this description of life in Mumbai: "The majority of Mumbai's population now lives in slums, up from one-sixth in 1971—a statistic that reflects a lack of decent affordable housing, even for those gainfully employed. Congested, overcrowded, and polluted, Mumbai has become a difficult place to live. The life expectancy of a Mumbaikar is now seven years shorter than an average Indian's, a remarkable statistic in a country still populated by poor villagers with little or no access to health care".

2.2.4 Preindustrial Cities

For the vast majority of human history, as far as anyone knows, people roamed about in search of sustenance. While they gathered edible plants, fished, and hunted, our ancestors could never find enough food in one area to sustain themselves for an extended period of time. Consequently, they had to keep moving until they could find another place in which to settle temporarily.

Eventual technological improvements—such as simple tools and information on how to farm and raise animals—allowed people to settle in one place. They built villages, with perhaps only a few hundred people living in each, and, for the following 5,000 years, produced just enough food for themselves—with nothing more in reserve.

About 5,000 years ago, however, humans developed such innovations as irrigation, metallurgy, and animal-drawn plows. These developments allowed farmers to produce an excess of food beyond their immediate needs. The resulting surplus of food led some people to make their living in other ways: for instance, by making pottery, weaving, and engaging in other nonagricultural activities that they could sell or exchange with others for the surplus food. As a result, people moved off the farms, commerce developed, and cities began to form.

Preindustrial cities—which first arose on fertile lands along rivers in the Middle East, Egypt, and China—were quite small compared to today's cities. Most preindustrial cities housed fewer than 10,000 inhabitants. Others, like Rome, may have contained as many as several hundred thousand people.

Preindustrial cities differed significantly from today's cities. The residential and commercial districts were not as sharply separated as they are today. Most traders and artisans worked at home, although people with the same trades tended to live in the same areas of town. People in cities also segregated themselves from one another according to class, ethnicity, and religion—with little or no chance for social mobility or interaction with other groups.

2.2.5 Industrial Cities

Between 1700 and 1900, increasing numbers of people moved into cities, resulting in an urban revolution. For example, in 1700 less than 2 percent of British people lived in cities, but by 1900 the majority of them did so. The United States and other European countries soon attained similar levels of urbanization, driven by the Industrial Revolution.

Industrialization produced the mechanization of agriculture, which, in turn, limited the amount of work available on farms. This lack of employment forced farm laborers to move to cities to find work. This migration of workers from rural to urban areas then gave rise to the industrial city.

The industrial city was larger, more densely populated, and more diverse than its preindustrial counterpart. It contained many people of varying backgrounds, interests, and skills who lived and worked together in a defined amount of space. The industrial city also served as a commercial center, supporting many businesses and factories. The latter attracted large numbers of immigrants from other countries hoping to better themselves by securing stable work and finding a "fresh start."

2.2.6 Metropolis and Megalopolis Cities

As larger and larger industrial cities spread outward in the early 1900s, they formed metropolises (large cities that include surrounding suburbs, which are lands outside the city limits, usually with separate governance). While some suburbs become distinct cities in and of themselves, they retain strong geographic, economic, and cultural ties to their "parent" city. Many metropolitan areas house a million or more residents.

The upper and middle classes ultimately brought about the so-called flight to the suburbs. As economic woes increasingly plagued cities in the latter half of the 1900s, many families decided to move out of their inner-city neighborhoods and into the suburbs. The ability to afford an automobile also influenced this migration. Beginning in the 1970s, most suburbs were largely “bedroom communities,” which means that suburban residents commuted into the city to work and shop, and then returned to the suburb at night.

Commuting presented a downside, but most people felt that escaping “urban ghettoization,” or the tendency for the quality of life in inner cities to decline, was well worth any hassles, given the fact that suburbs tended to offer nicer and larger homes, better schools, less crime, and less pollution than cities provided.

Today, suburbs continue to grow and develop. Many have become economic centers in their own right. Offices, hospitals, and factories coexist with shopping malls, sports complexes, and housing subdivisions. In this way, many suburbs have essentially become small (and, in some cases, not so small) cities. Demographically, suburbs tend to attract “whiter” and more affluent residents than do cities. Yet not all suburbs and suburbanites are alike. Even within a suburb, families of varying ethnic and religious backgrounds exist.

Because of all this growth, many suburbs have developed “urban” problems, such as air and water pollution, traffic congestion, and gangs. To escape these problems, some people have chosen to move to rural areas. Others have chosen to return to and revive their cities by renovating and remodeling buildings and neighborhoods. Such an interest in urban renewal (also called gentrification) has turned some slums into decent areas in which to live, work, and raise a family.

The vast urban complex known as a megalopolis was created as suburbs continued to grow and merge with other suburbs and metropolitan areas. That is, some suburbs and cities have grown so large that they end up merging with other suburbs and cities, forming a virtually continuous region.

One example of a megalopolis is the hundreds of miles of almost uninterrupted urbanization from Boston to Washington, D.C. The typical megalopolis consists of literally millions of people.

2.2.7 Movement in Urbanization

As more and more people leave villages and farms to live in cities, urban growth results. The rapid growth of cities likeChicago in the late 19th century and Mumbai a century later can be attributed largely to rural-urban migration. This kind of growth is especially commonplace in developing countries. This growth can also be attributed to new job opportunities.

The rapid urbanization of the world’s population over the twentieth century is described in the 2005 Revision of the UN World Urbanization Prospects report. The global proportion of urban population rose dramatically from 13% (220 million) in 1900, to 29% (732 million) in 1950, to 49% (3.2 billion) in 2005. The same report projected that the figure is likely to rise to 60% (4.9 billion) by 2030.

2.2.8 Percentage of World Population: Urban vs Rural

According to the UN State of the World Population 2007 report, sometime in the middle of 2007, the majority of people worldwide will be living in towns or cities, for the first time in history; this is referred to as the arrival of the "Urban Millennium" or the 'tipping point'. In regard to future trends, it is estimated 93% of urban growth will occur in developing nations, with 80% of urban growth occurring in Asia and Africa.

Urbanization rates vary between countries. The United States and United Kingdom have a far higher urbanization level than China, India, Swaziland or Niger, but a far slower annual urbanization rate, since much less of the population is living in a rural area.

- Urbanization in the United States never reached the Rocky Mountains in locations such as Jackson Hole, Wyoming; Telluride, Colorado; Taos, New Mexico; Douglas County, Colorado andAspen, Colorado. The state of Vermont has also been affected, as has the coast of Florida, theBirmingham-Jefferson County, AL area, the Pacific Northwest and the barrier islands of North Carolina.
- In the United Kingdom, two major examples of new urbanization can be seen in Swindon, Wiltshireand Milton Keynes, Buckinghamshire. These two towns show some of the quickest growth rates in Europe.

2.3 CAUSES OF URBANIZATION

Urbanization occurs naturally from individual and corporate efforts to reduce time and expense in commuting and transportation while improving opportunities for jobs, education, housing, and transportation. Living in cities permits individuals and families to take advantage of the opportunities of proximity, diversity, and marketplace competition.

People move into cities to seek economic opportunities. A major contributing factor is known as "rural flight". In rural areas, often on small family farms, it is difficult to improve one's standard of living beyond basic sustenance. Farm living is dependent on unpredictable environmental conditions, and in times of drought, flood or pestilence, survival becomes extremely problematic. In modern times, industrialization of agriculture has negatively affected the economy of small and middle-sized farms and strongly reduced the size of the rural labor market.

Cities, in contrast, are known to be places where money, services and wealth are centralized. Cities are where fortunes are made and where social mobility is possible. Businesses, which generate jobs and capital, are usually located in urban areas. Whether the source is trade or tourism, it is also through the cities that foreign money flows into a country. It is easy to see why someone living on a farm might wish to take their chance moving to the city and trying to make enough money to send back home to their struggling family.

There are better basic services as well as other specialist services that aren't found in rural areas. There are more job opportunities and a greater variety of jobs. Health is another major factor. People, especially the elderly are often forced to move to cities where there are doctors and hospitals that can cater for their health needs. Other factors include a greater variety of entertainment (restaurants, movie theaters, theme parks, etc.) and a better quality of education, namely universities. Due to their high populations, urban areas can also have much more diverse social communities allowing others to find people like them when they might not be able to in rural areas.

These conditions are heightened during times of change from a pre-industrial society to an industrial one. It is at this time that many new commercial enterprises are made possible, thus creating new jobs in cities. It is also a result of industrialization that farms become more mechanized, putting many labourers out of work. This is currently occurring fastest in India.

2.3.1 Economic Effects of Urbanization

As cities develop, effects can include a dramatic increase in costs, often pricing the local working class out of the market, including such functionaries as employees of the local municipalities. For example, Eric Hobsbawm's bookThe age of the revolution: 1789-1848 (published 1962 and 2005) chapter 11, stated "Urban development in our period [1789-1848] was a gigantic process of class segregation, which pushed the new labouring poor into great morasses of misery outside the centres of government and business and the newly specialised residential areas of the bourgeoisie. The almost universal European division into a 'good' west end and a 'poor' east end of large cities developed in this period." This is likely due the prevailing south-west wind which carries coal smoke and other airborne pollutants downwind, making the western edges of towns preferable to the eastern ones. Similar problems now affect the developing world, rising inequality resulting from rapid urbanisation trends. The drive for rapid urban growth and often efficiency can lead to less equitable urban development, think tanks such as the Overseas Development Institute have even proposed policies that encourage labour intensive growth as a means of absorbing the influx of low skilled and unskilled labour.

Urbanization is often viewed as a negative trend, but can in fact, be perceived simply as a natural occurrence from individual and corporate efforts to reduce expense in commuting and transportation while improving opportunities for jobs, education, housing, and transportation. Living in cities permits individuals and families to take advantage of the opportunities of proximity, diversity, and marketplace competition.

2.3.2 Environmental Effects of Urbanization

The urban heat island has become a growing concern and is increasing over the years. The urban heat island is formed when industrial and urban areas are developed and heat becomes more abundant. In rural areas, a large part of the incoming solar energy is used to evaporate water from vegetation and soil. In cities, where less vegetation and exposed soil exists, the majority of the sun's energy is absorbed by urban structures and asphalt. Hence, during warm daylight hours, less evaporative cooling in cities allows surface temperatures to rise higher than in rural areas. Additional city heat is given off by vehicles and factories, as well as by industrial and domestic heating and cooling units.This effect causes the city to become 2 to 10 °F (1 to 6 °C) warmer than surrounding landscapes.Impacts also include reducing soil moisture and intensification of carbon dioxide emissions.

In his book Whole Earth Discipline, Stewart Brand argues that the effects of urbanization are on the overall positive for the environment. Firstly, the birth rate of new urban dwellers falls immediately to replacement rate, and keeps falling. This can prevent overpopulation in the future. Secondly, it puts a stop to destructive subsistence farming techniques, like slash and burn agriculture. Finally, it minimizes land use by humans, leaving more for nature.

2.3.3 Changing Forms in Urbanization

Different forms of urbanization can be classified depending on the style of architecture and planning methods as well as historic growth of areas.

In cities of the developed world urbanization traditionally exhibited a concentration of human activities and settlements around the downtown area, the so-called in-migration. In-migration refers to migration from former colonies and similar places. The fact that many immigrants settle in impoverished city centres led to the notion of the "peripheralization of the core", which simply describes that people who used to be at the periphery of the former empires now live right in the centre.

Recent developments, such as inner-city redevelopment schemes, mean that new arrivals in cities no longer necessarily settle in the centre. In some developed regions, the reverse effect, originally called counter urbanisation has occurred, with cities losing population to rural areas, and is particularly common for richer families. This has been possible because of improved communications, and has been caused by factors such as the fear of crime and poor urban environments. It has contributed to the phenomenon of shrinking cities experienced by some parts of the industrialized world.

When the residential area shifts outward, this is called suburbanization. A number of researchers and writers suggest that suburbanization has gone so far to form new points of concentration outside the downtown both in developed and developing countries such as India. This networked, poly-centric form of concentration is considered by some an emerging pattern of urbanization. It is called variously exurbia, edge city, network city, or postmodern city. Los Angeles is the best-known example of this type of urbanization.

Rural migrants are attracted by the possibilities that cities can offer, but often settle in shanty towns and experience extreme poverty. In the 1980s, this was attempted to be tackled with the urban bias theory which was promoted by Michael Lipton who wrote: "...the most important class conflict in the poor countries of the world today is not between labour and capital. Nor is it between foreign and national interests. It is between rural classes and urban classes. The rural sector contains most of the poverty and most of the low-cost sources of potential advance; but the urban sector contains most of the articulateness, organization and power. So the urban classes have been able to win most of the rounds of the struggle with the countryside...". Most of the urban poor in developing countries able to find work can spend their lives in insecure, poorly paid jobs. According to research by the Overseas Development Institute pro-poor urbanisation will require labour intensive growth, supported by labour protection, flexible land use regulation and investments in basic services.'

Urbanization can be planned urbanization or organic. Planned urbanization, i.e.: planned community or the garden city movement, is based on an advance plan, which can be prepared for military, aesthetic, economic or urban design reasons. Examples can be seen in many ancient cities; although with exploration came the collision of nations, which meant that many invaded cities took on the desired planned characteristics of their occupiers. Many ancient organic cities experienced redevelopment for military and economic purposes, new roads carved through the cities, and new parcels of land were cordoned off serving various planned purposes giving cities distinctive geometric designs. UN agencies prefer to see urban infrastructure installed before urbanization occurs. Landscape planners are responsible for landscape infrastructure (public parks, sustainable urban drainage systems, greenways etc.) which can be planned before urbanization takes place, or afterward to revitalize an area and create greater livability within a region. Concepts of control of the urban expansion are considered in the American Institute of Planners.

2.3.4 Counter Urbanization

Counter urbanization is a demographic and social process whereby people move from urban areas to rural areas. It first took place as a reaction to inner-city deprivation and overcrowding. Initial study of counter urbanisation was carried out by human geographer Brian Berry. More recent research has documented the social and political drivers of counter urbanisation, and the impacts of the same, in developing countries, which are currently undergoing processes of mass-urbanisation, such as China. It is one of the causes that can lead to shrinking cities.

2.4 SUBURBANIZATION

As people gain wealth around the world, they all usually tend to do the same thing: spread out. A common dream shared among people of all cultures is to have a piece of land to call their own. The suburbs are the place that many urban dwellers turn to because it offers the space needed to satisfy these dreams.

Suburbanization (or suburbanisation) a term used to describe the growth of areas on the fringes of major cities. It is one of the many causes of the increase in urban sprawl. Many residents of metropolitan regions work within the central urban area, choosing instead to live in satellite communities called suburbs and commute to work via automobile or mass transit. Others have taken advantage of technological advances to work from their homes, and chose to do so in an environment they consider more pleasant than the city. These processes often occur in more economically developed countries, especially in the United States, which is believed to be the first country in which the majority of the population lives in the suburbs, rather than in the cities or in rural areas. Proponents of containing urban sprawl argue that sprawl leads to urban decay and a concentration of lower income residents in the inner city.

2.4.1 What are Suburbs?

Suburbs are the communities surrounding cities that are usually made up of single-family homes, but are increasingly including multifamily homes and places like malls and office buildings. Emerging in the 1850s as a result of a fast rising urban population and improving transportation technology, suburbs have remained a popular alternative to the city even today. As of 2000, about half the population of the United States lived in suburbs.

Suburbs are generally spread out over greater distances than other types of living environments. For instance, people may live in the suburb in order to avoid the density and untidiness of the city. Since people have to get around these vast stretches of land automobiles are common sights in suburbs. Transportation (including, to a limited extent, trains and buses) plays an important role in the life of a suburban resident who generally commutes to work.

People also like to decide for themselves how to live and what rules to live by. Suburbs offer them this independence. Local governance is common here in the form of community councils, forums, and elected officials. A good example of this is a Home Owners Association, a group common to many suburban neighborhoods that determines specific rules for the type, appearance, and size of homes in a community.

People living in the same suburb usually share similar backgrounds with regard to race, socioeconomic status, and age. Often, the houses that make up the area are similar in appearance, size, and blueprint, a layout design referred to as tract housing, or cookie-cutter housing.

2.4.2 History of Suburbs

Though they appeared on the outskirts of many world cities in the early 1800s, it was only after the general implementation of electric railways in the late 1800s that suburbs began to grow extensively, especially in the United States. Such a relatively cheap and quick method of transportation made it practical to travel from home to work (in the inner city) on a daily basis.

Early examples of suburbs include areas created for lower class citizens outside of Rome, Italy during the 1920s, streetcar suburbs in Montreal, Canada created during the late 1800s, and the picturesque Llewellyn Park, New Jersey, created in 1853.

Henry Ford was also a big reason why suburbs caught on the way they did. His innovative ideas for making cars cut manufacturing costs, reducing the retail price for customers. Now that an average family could afford a car, more people could go to and from home and work everyday. Additionally, the development of the Interstate Highway System further encouraged suburban growth.

The government was another player that encouraged movement out of the city. Federal legislation made it cheaper for someone to construct a new home outside of the city than to improve upon a preexisting structure in the city. Loans and subsidies were also provided to those willing to move to new planned suburbs (usually wealthier white families).

In 1934 the United States Congress created the Federal Housing Administration (FHA), an organization intended to provide programs to insure mortgages. Poverty struck everyone's life during the Great Depression (beginning in 1929) and organizations like the FHA helped to ease the burden and stimulate growth.

Rapid growth of suburbia characterized the post-World War II era for three chief reasons:

- The economic boom following World War II
- The need for housing returning veterans and baby boomers relatively cheaply
- Whites fleeing the desegregation of urban cities brought on by the civil rights movement(the "White Flight")

Some of the first and most famous suburbs in the post-war era were the Levittown developments in the Megalopolis.

2.4.3 Causes and Effects of Suburbanization

Suburbanization can be linked to a number of different push and pull factors. Push factors include the congestion and population density of the cities, pollution caused by industry and high levels of traffic and a general perception of a lower quality of life in inner city areas. Pull factors include more open spaces and a perception of being closer to "nature", lower suburban house prices and property taxes in comparison to the city, and the increasing number of job opportunities in the suburban areas. Improvements in transportation infrastructure encourage suburbanization, as people become increasingly able to live in a suburb and commute in to the nearby town or city to work. Developments in railways, bus routes and roads are the main improvements that make suburbanization more practical. The increase in the number and size of highways is a particularly significant part of this effect.

Government policies can have a significant effect on the process. In the United States, for instance, policies of the Federal government in the post-World War II era, such as the building of an efficient network of roads, highways and superhighways, and the underwriting of mortgages for suburban one-family homes, had an enormous influence on the pace of suburbanization in that country. In effect, the government was encouraging the transfer of the middle-class population out of the inner cities and into the suburbs, sometimes with devastating effects on the viability of the city centers. However, some argue that the effect of Interstate Highway Systems on suburbanization is overstated. Researchers of this vein believe city center populations would have declined even in the absence of highway systems, contending that suburbanization is a long-standing and almost universal process. They primarily argue that as incomes rise, most people want the range and choice offered by automobiles. In addition, there is no significant evidence directly linking the development of highway systems to declining urban populations.

Insurance companies also fueled the push out of cities, as in many cases, it redlined inner-city neighborhoods, denying mortgage loans there, and instead offering low rates in the suburban areas. More recently, some urban areas have adopted "green belt" policies which limit growth in the fringe of a city, in order to encourage more growth in the urban core. It began to be realized that a certain amount of population density in the center city is conducive to creating a good, working urban environment. Race also played a role in American suburbanization. During World War I, the massive migration of African Americans from the South resulted in an even greater residential shift toward suburban areas.

The cities became seen as dangerous, crime-infested areas, while the suburbs were seen as safe places to live and raise a family, leading to a social trend known in some parts of the world as white flight. This phenomenon runs counter to much of the rest of the world, where slums mostly exist outside the city, rather than within them. With the increasing population of the older, more established suburban areas, many of the problems which were once seen as purely urban ones have manifested themselves there as well.

Some social scientists suggest that the historical processes of suburbanization and decentralization are instances of white privilege that have contributed to contemporary patterns of environmental racism. Recent developments in communication technology, such as the spread of broadband services, the growth of e-mail and the advent of practical home video conferencing, has enabled more people to work from home rather than commuting. Although this can occur either in the city or in the suburbs, the effect is generally decentralizing, which works against the largest advantage of the center city, which is easier access to information and supplies due to centralization. Similarly, the rise of efficient package express delivery systems, such as (in the United States) FedEx and UPS, which take advantage of computerization and the availability of an efficient air transportation system, also eliminates some of the advantages that were once to be had from having a business located in the city.

Industrial, warehousing, and factory land uses have also moved to suburban areas. Cheap telecommunications removes the need for company headquarters to be within quick courier distance of the warehouses and ports. Urban areas suffer from traffic congestion, which creates costs in extra driver costs for the company which can be reduced if they were in a suburban area near a highway. As with residential, lower property taxes and low land prices encourage selling industrial land for profitable brownfield redevelopment. Suburban areas also offer more land to use as a buffer between industrial and residential and retail space to avoid NIMBY sentiments and gentrification pressure from the local community when residential and retail is adjacent to industrial space in an urban area. Suburban municipalities can offer tax breaks, specialized zoning, and regulatory incentives to attract industrial land users to their area, such as City of Industry, California.

The overall effect of these developments is that businesses as well, and not just individuals, now see an advantage to locating in the suburbs, where the cost of buying land, renting space, and running their operations, is cheaper than in the city. This continuing dispersal from a single city center has led to other recent phenomena in American suburbs, the advent of edge cities and exurbs, arising out of clusters of office buildings built in suburban commercial centers around shopping malls and higher density developments. With more and more jobs for suburbanites being located in these areas rather than in the main city core that the suburbs grew out of, traffic patterns, which for decades centered on people commuting into the center city to work in the morning and then returning home in the evening, have become more complex, with the volume of intra-suburban traffic increasing tremendously. By 2000, half of the US population lived in suburban areas.

2.4.4 Effects on Psychological Health

Historically it was believed that living in highly urban areas resulted in social isolation, social disorganization, and psychological problems, and that living in suburbs would be more conducive to overall happiness, due to lower population density, lower crime, and a more stable population. A study based on data from 1974, however, found this not to be the case, finding that people living in suburbs had neither greater satisfaction with their neighborhood nor greater satisfaction with the quality of their lives as compared to people living in urban areas.

2.5 ECONOMIC IMPACTS OF SUBURBANIZATION

The economic impacts of suburbanization have become very evident since the trend began in the 1950s. Changes in infrastructure, industry, real estate development costs, fiscal policies, and diversity of cities have been easily apparent, as "making it to the suburbs", mainly in order to own a home and escape the chaos of urban centers, have become the goals of many American citizens. These impacts have many benefits as well as side effects and are becoming increasingly important in the planning and revitalization of modern cities.

2.5.1 Suburbanization's Impact on Industry in the City

The days of industry dominating the urban cores of cities are gone. Companies are increasingly looking to build industrial parks in less populated areas, largely to match the desires of employees to work in more pleasant areas. Also contributing to the flight of industry from the city are government economic policies that provide incentives for companies to build new structures, as well as the lack of incentives to build on "brownfield" land (previously used industrial land). As companies continue to build on pieces of land in less populated areas, it will become increasingly difficult to build in high-density areas. The main factor that causes this is that if any of the side effects of industry (noise, excessive lights, heavy traffic) affect residential areas, there is resistance from the homeowners. Another potential impact of industry leaving the city is that generally, when industry is separated from an urban area by some open space, as well as infrastructure, the open space between the city and the company becomes more intensely used. As this land becomes used more and more, the value of properties very often increases, causing many landowners in that open space to sell their land.

2.5.2 Consequences on Infrastructure

As suburbanization continues and America continues to sprawl, the cost of the required water lines, sewer lines, and roads could cost more than $21,000 per residential and non residential development unit, costing the American government $1.12 Trillion over the next twenty five years. Along with the costs of infrastructure, existing infrastructure suffers, as most of the government's money that is dedicated to improving infrastructure goes to paying for the new necessities in areas further out from the urban core. As a result, the government will often forgo maintenance on previously built infrastructure.

2.5.3 Impact on Real Estate Development Costs

For residential properties, suburbanization allows for home prices to decrease, so people can drive until they can find an area in which they can afford to buy a home. However, these homes may lack certain things such as parks and access to public transit. Also, the prices of homes in downtown center usually decreases as well to compete with the inexpensive homes in the suburbs. One of the main benefits of living in the suburbs is that one gets a much larger piece of land than one would in the city. Therefore, as the size of lots increases, the supply of housing is more limited.

2.5.4 Fiscal Impact of Suburbanization

The fiscal deficit grows as a result of suburbanization, mainly because in less densely populated areas, property taxes tend to be lower. Also, because of the typical spread pattern of suburban housing, the lack of variety of housing types, and the greater distance between homes, real estate development and public service costs increase, which in turn increase the federal deficit.

2.5.5 Suburbanization's Effect on Diversity within the City

In the 1950s when the trend of suburbanization took hold, many of the people who left the city for the suburbs were white. As a result, there was a rise in black home ownership in central cities. As white households left for the suburbs, housing prices in transition neighborhoods fell, which often lowered the cost of home ownership for black households. This trend was stronger in older and denser cities, especially in the northeast and Midwest, because new construction was generally more difficult.

2.5.6 Current Trends in Suburbs

In the United States more jobs are now located in the suburbs than in central cities as a result of the movement of commercial and industrial parks from the inside to the outside of the city. Express highways are constantly being constructed to and from major hubs or edge cities, and it is on these roads where new suburbs are being developed.

In other parts of the world suburbs do not resemble the affluence of their American counterparts. Due to extreme poverty, crime, and lack of infrastructure suburbs in developing parts of the world are characterized by higher density and lower standards of living.

One issue arising from suburban growth is the disorganized, reckless manner in which neighborhoods are built, called sprawl. Because of the desire for larger plots of land and the rural feel of the countryside, new developments are infringing upon more and more of the natural, uninhabited land. The unprecedented growth of population in the past century will continue to fuel the expansion of suburbs in the coming years.

2.6 URBAN SPRAWL

Urban sprawl, also known as suburban sprawl, is a multifaceted concept, which includes the spreading outwards of a city and its suburbs to its outskirts to low-density and auto-dependent development on rural land, high segregation of uses (e.g. stores and residential), and various design features that encourage car dependency.

Discussions and debates about sprawl are often obfuscated by the ambiguity associated with the phrase. For example, some commentators measure sprawl only with the average number of residential units per acre in a given area. But others associate it with decentralization (spread of population without a well-defined centre), discontinuity (leapfrog development, as defined below), segregation of uses, and so forth.

The term urban sprawl generally has negative connotations due to the health, environmental and cultural issues associated with the phrase. Residents of sprawling neighbourhoods tend to emit more pollution per person and suffer more traffic fatalities. Sprawl is controversial, with supporters claiming that consumers prefer lower density neighbourhoods and that sprawl does not necessarily increase traffic.

Sprawl is characterized by several land use patterns that usually occur in unison:

2.6.1 Single Use Zoning

This refers to a situation where commercial, residential, institutional and industrial areas are separated from one another. Consequently, large tracts of land are devoted to a single use and are segregated from one another by open space, infrastructure, or other barriers. As a result, the places where people live, work, shop, and recreate are far from one another, usually to the extent that walking, transit use and bicycling are impractical, so all these activities generally require an automobile.

2.6.2 Low-density Zoning

Sprawl consumes much more land per-capita than traditional urban developments because zoning laws generally require that new developments are of low density. The exact definition of "low density" is arguable, but a common example is that of single family homes on large lots, with four or fewer units per net acre. Buildings usually have fewer stories and are spaced farther apart, separated by lawns, landscaping, roads or parking lots. Lot sizes are larger, and because more automobiles are used much more land is designated for parking. The impact of low density development in many communities is that developed or "urbanized" land is increasing at a faster rate than the rate of population.

Overall density is often lowered by "leap-frog development". This term refers to the relationship, or lack thereof, between subdivisions. Such developments are typically separated by large green belts, i.e. tracts of undeveloped land, resulting in an average density far lower even than the low density described in the previous paragraph. This is a 20th and 21st century phenomenon generated by the current custom of requiring a developer to provide subdivision infrastructure as a condition of development. Usually, the developer is required to set aside a certain percentage of the developed land for public use, including roads, parks and schools. In the past, when a local government built all the streets in a given location, the town could expand without interruption and with a coherent circulation system, because it had condemnation power. Private developers generally do not have such power (although they can sometimes find local governments willing to help), and often choose to develop on the tracts that happen to be for sale at the time they want to build, rather than pay extra or wait for a more appropriate location.

2.6.3 Car-Dependent Communities

Areas of urban sprawl are also characterized as highly dependent on automobiles for transportation, a condition known as automobile dependency. Most activities, such as shopping and commuting to work, require the use of a car as a result of both the area's isolation from the city and the isolation the area's residential zones have from its industrial and commercial zones. Walking and other methods of transit are not practical; therefore, many of these areas have few or no sidewalks. In many suburban communities, stores and activities that are in close proximity "as the crow flies" require automobiles, because the different areas are separated by fences, walls, and drainage ditches. Some critics argue that excessive parking requirements exacerbate car dependency.

2.6.4 Job Sprawl and Spatial Mismatch

Job Sprawl is another land use symptom of urban sprawl and car-dependent communities. It is defined as low-density, geographically spread-out patterns of employment, where the majority of jobs in a given metropolitan area are located outside of the main city's Central Business District (CBD), and increasingly in the suburban periphery. It is often the result of urban disinvestment, the geographic freedom of employment location allowed by predominantly car-dependent commuting patterns of many American suburbs, and many companies' desire to locate in low-density areas that are often more affordable and offer potential for expansion. Spatial mismatch is related to job sprawl and economic Environmental Justice. Spatial Mismatch is defined as the situation where poor urban, predominantly minority citizens are left without easy access to entry-level jobs, as a result of increasing job sprawl and limited transportation options to facilitate a reverse commute to the suburbs.

Job sprawl has been documented and measured in various ways. It has been shown to be a growing trend in America's metropolitan areas. The Brookings Institution has published multiple articles on the topic. In 2005, author Michael Stoll defined job sprawl simply as jobs located more than 5-mile (8.0 km) radius from the CBD, and measured the concept based on year 2000 U.S. Census data.Other ways of measuring the concept with more detailed rings around the CBD include a 2001 article by Edward Glaeser and Elizabeth Kneebone's 2009 article, which show that sprawling urban peripheries are gaining employment while areas closer to the CBD are losing jobs. These two authors used three geographic rings limited to a 35-mile (56 km) radius around the CBD: 3 miles (4.8 km) or less, 3 to 10 miles (16 km), and 10 to 35 miles (56 km). Kneebone's study showed the following nationwide breakdown for the largest metropolitan areas in 2006: 21.3% of jobs located in the inner ring, 33.6% of jobs in the 3-10 mile ring, and 45.1% in the 10-35 mile ring. This compares to the year 1998 - 23.3%, 34.2%, and 42.5% in those respective rings. The study shows CBD employment share shrinking, and job growth focused in the suburban and exurban outer metropolitan rings.

In terms of measurement, spatial mismatch can be thought of as the percentage of people who would have to move in order to be distributed in the same way as jobs. Stoll's research shows that a substantially higher percentage of African Americans (53.5%) experience spatial mismatch than European Americans (35.6%). On average, more than half of African American citizens would need to move to accomplish similar distribution to jobs. Latinos (45.8%) experience spatial mismatch as well, though to a lesser extent than African Americans.

2.6.5 Benefits of Sprawl

There are many benefits to suburbanization, specifically for the homeowner. The household has the opportunity to have a larger lot and live in a less populated area.

For companies, escaping the city allows them to use more land, satisfy their employee's requests for more quiet work environments, and allows them to develop their business on cheaper land.

2.7 DEVELOPMENTS CHARACTERISTIC OF SPRAWL

2.7.1 Housing Subdivisions

Housing subdivisions are large tracts of land consisting entirely of newly-built residences. Prominent New Urbanist architectural firm Duany Plater-Zyberk & Company claim that housing subdivisions "are sometimes called villages, towns, and neighbourhoods by their developers, which is misleading since those terms denote places that are not exclusively residential." They are also referred to as developments.

Subdivisions often incorporate curved roads and cul-de-sacs. Such subdivisions may offer only a few places to enter and exit the development, causing traffic to use high volume collector streets. All trips, no matter how short, must enter the collector road in a suburban system.

2.7.2 Strip Malls

Shopping centres are locations consisting of retail space. In the U.S. and Canada, these vary from strip malls, which refer to collections of buildings sharing a common parking lot, usually built on a high-capacity roadway with commercial functions (i.e., a "strip"). Similar developments in the UK are called Retail Parks. Strip malls/retail parks contain a wide variety of retail and non-retail functions that also cater to daily use (e.g. video rental, takeout food, laundry services, hairdresser). Strip malls consisting mostly of big box stores or category killers are sometimes called "power centers" (U.S.). These developments tend to be low-density; the buildings are single-storey and there is ample space for parking and access for delivery vehicles. This character is reflected in the spacious landscaping of the parking lots and walkways and clear signage of the retail establishments. Some strip malls are undergoing a transformation into Lifestyle centres; entailing investments in common areas and facilities (plazas, cafes) and shifting tenancy from daily goods to recreational shopping. European countries such as France, Belgium and Germany have implemented size restrictions for superstores found in strip malls in an effort to limit sprawl.

2.7.3 Shopping Malls

Another prominent form of retail development in areas characterized by "sprawl" is the shopping mall. Unlike the strip mall, this is usually composed of a single building surrounded by a parking lot that contains multiple shops, usually "anchored" by one or more department stores. The function and size is also distinct from the strip mall. The focus is almost exclusively on recreational shopping rather than daily goods. Shopping malls also tend to serve a wider (regional) public and require higher-order infrastructure such as highway access and can have floorspaces in excess of a million square feet (ca. 100,000 m^2). Shopping malls are often detrimental to downtown shopping centres of nearby cities since the shopping malls act as a surrogate for the city centre. Some downtowns have responded to this challenge by building shopping centres of their own.

In the 1970s, the Ontario government created the Ontario Downtown Renewal Programme, which helped finance the building of several downtown malls across Ontario (such as the aforementioned Eaton Centre). The program was created to reverse the tide of small business leaving downtowns for larger sites surrounding the city.

2.7.4 Fast Food Chains

Fast food chains are common in suburban areas. They are often built early in areas with low property values where the population is expected to boom and where large traffic is predicted, and set a precedent for future development. Eric Schlosser, in his book Fast Food Nation, argues that fast food chains accelerate suburban sprawl and help set its tone with their expansive parking lots, flashy signs, and plastic architecture (65). Duany Plater Zyberk & Company believe that this reinforces a destructive pattern of growth in an endless quest to move away from the sprawl that only results in creating more of it.

2.7.5 Smart Growth and the Compact City

The term 'smart growth' has been particularly used in North America. The terms 'compact city' or 'urban intensification' are often used to describe similar concepts, in Europe and particularly the UK where it has influenced Government policy and planning practice in recent years.

The first urban growth boundary in the U.S. was in Fayette County, Kentucky in 1958. Fifteen years later, the state of Oregon enacted a law in 1973 limiting the area urban areas could occupy, through urban growth boundaries. As a result, Portland, the state's largest urban area, has become a leader in smart growth policies that seek to make urban areas more compact (they are called urban consolidation policies). After the creation of this boundary, the population density of the urbanized area increased somewhat (from 1,135 in 1970 to 1,290 per km^2 in 2000) USA Urbanized Areas 1950-1990USA Urbanized Areas 2000. While the growth boundary has not been tight enough to vastly increase density, the consensus is that the growth boundaries have protected great amounts of wild areas and farmland around the metro area.

Many parts of the San Francisco Bay Area have also adopted urban growth boundaries; 25 of its cities and 5 of its counties have urban growth boundaries. Many of these were adopted with the support and advocacy of Greenbelt Alliance, a non-profit land conservation and urban planning organization.

In other areas, the design principles of District Regionalism and New Urbanism have been employed to combat urban sprawl.

2.8 CRITICISM OF URBAN SPRAWL

Arguments opposing urban sprawl run the gamut from the more concrete effects such as health and environmental issues to more abstract consequences involving neighborhood vitality.

2.8.1 Health and Environmental Impact

Urban sprawl is associated with a number of negative environmental and public health outcomes, with the primary result being increased dependence on automobiles.

Increased Pollution and Reliance on Fossil Fuel

In the years following World War II, when vehicle ownership was becoming widespread, public health officials recommended the health benefits of suburbs due to soot and industrial fumes in the city center. However, air in modern suburbs is not necessarily cleaner than air in urban neighborhoods. In fact, the most polluted air is on crowded highways, where people in suburbs tend to spend more time. On average, suburban residents generate more per capita pollution and carbon emissions than their urban counterparts because of their increased driving.

Increase in Traffic and Traffic-Related Fatalities

A heavy reliance on automobiles increases traffic throughout the city as well as automobile crashes, pedestrian injuries, and air pollution. Motor vehicle crashes are the leading cause of death for Americans between the ages of five and twenty-four and is the leading accident-related cause for all age groups. Residents of more sprawling areas are at greater risk of dying in a car crash.

Delays in Emergency Medical Services Response Times

Research covered in the Journal of Economic Issues and State and Local Government Review shows a link between sprawl and emergency medical services response delays.

Increased Obesity

The American Journal of Public Health and the American Journal of Health Promotion, have both stated that there is a significant connection between sprawl, obesity, and hypertension. Many urbanists argue that this is due to less walking in sprawl-type developments. Living in a car centered culture forces inhabitants to drive everywhere, thus walking far less than their urban (and generally healthier) counterparts.

Decrease in Social Capital

Urban sprawl may be partly responsible for the decline in social capital in the United States. Compact neighborhoods can foster casual social interactions among neighbors, while sprawl creates barriers. Sprawl tends to replace public spaces with private spaces such as fenced-in backyards.

2.8.2 Decrease in Land and Water Quantity and Quality

Due to the larger area consumed by sprawling suburbs compared to urban neighborhoods, more farmland and wildlife habitats are displaced per resident. As forest cover is cleared and covered with impervious surfaces (concrete and asphalt) in the suburbs, rainfall is less effectively absorbed into the ground water aquifers. This threatens both the quality and quantity of water supplies. Sprawl increases water pollution as rain water picks up gasoline, motor oil, heavy metals, and other pollutants in runoff from parking lots and roads. Sprawl fragments the land, which increases the risk of invasive species spreading into the remaining forest.

2.8.3 Increased Infrastructure Costs

Living in larger, more spread out spaces generally makes public services more expensive. Since carusage becomes endemic and public transport often becomes significantly more expensive, city planners are forced to build large highway and parking infrastructure, which in turn decreases taxable land and revenue, and decreases the desirability of the area adjacent to such structures. Providing services such as water, sewers, and electricity is also more expensive per household in less dense areas.

2.8.4 Increased Personal Transportation Costs

Residents of low-density areas spend a higher proportion of their income on transportation than residents of high density areas. The RAC estimates that the average cost of operating a car in the UK is £5,000 a year, most of which stems from financing costs and depreciation. In comparison, a yearly underground ticket for a suburban commuter in London (where the average wage is higher than the national average) costs £1,000-1,500, which, because of subsidies, do not cover financing for the rail or depreciation of the infrastructure. In the Euro-15, rail transit requires $69 billion euro in subsidies while road transportation nets $107 billion euro in additional taxes.

2.9 URBAN SPRAWL AND AUTOMOBILE DEPENDENCY

Whether urban sprawl does increase problems of automobile dependency and whether conversely, policies of smart growth can reduce them have been fiercely contested issues over several decades. An influential study in 1989 by Peter Newman and Jeff Kenworthy compared 32 cities across North America, Australia, Europe and Asia. The study has been criticised for its methodology but the main finding that denser cities, particularly in Asia, have lower car use than sprawling cities, particularly in North America, has been largely accepted although the relationship is clearer at the extremes across continents than it is within countries where conditions are more similar.

Within cities, studies from across many countries (mainly in the developed countries) have shown that denser urban areas with greater mixture of land use and better public transport tend to have lower car use than less dense suburban and ex-urban residential areas. This usually holds truc even after controlling for socio-economic factors such as differences in household composition and income.This does not necessarily imply that suburban sprawl causes high car use, however. One confounding factor, which has been the subject of many studies, is residential self-selection: people who prefer to drive tend to move towards low density suburbs, whereas people who prefer to walk, cycle or use transit tend to move towards higher density urban areas, better served by public transport. Some studies have found that, when self-selection is controlled for, the built environment has no significant effect on travel behaviour. More recent studies using more sophisticated methodologies have generally refuted these findings: density, land use and public transport accessibility can influence travel behaviour, although social and economic factors, particularly household income, usually exert a stronger influence.

2.9.1 The Paradox of Intensification

Reviewing the evidence on urban intensification, smart growth and their effects on travel behaviour Melia et al. (2011) found support for the arguments of both supporters and opponents of smart growth measures to counteract urban sprawl. Planning policies that increase population densities in urban areas do tend to reduce car use, but the effect is a weak one, so doubling the population density of a particular area will not halve the frequency or distance of car use.

These findings led them to propose the paradox of intensification, which states:

Ceteris paribus, urban intensification which increases population density will reduce per capita car use, with benefits to the global environment, but will also increase concentrations of motor traffic, worsening the local environment in those locations where it occurs.

2.9.2 Debate Over Traffic and Commute Times

Those not opposed to low density development argue that traffic intensities tend to be less, traffic speeds faster and, as a result, ambient air pollution is lower. Kansas City, Missouri is often cited as an example of ideal low-density development, with congestion below the mean and home prices below comparable Midwestern cities. Wendell Cox and Randal O'Toole are the leading figures supporting lower density development.

Longitudinal (time-lapse) studies of commute times in major metropolitan areas in the United States have shown that commute times decreased for the period 1969 to 1995 even though the geographic size of the city increased.

2.9.3 Risk of Increased Housing Prices

There is also some concern that Portland-style anti-sprawl policies will increase housing prices. Some research suggests Oregon has had the largest housing affordability loss in the nation, but other research shows that Portland's price increases are comparable to other Western cities.

In Australia, it is claimed by some that housing affordability has hit "crisis levels" due to "urban consolidation" policies implemented by state governments. In Sydney, the ratio of the price of a house relative to income is 9:1. The issue has at times been debated between the major political parties.

2.9.4 Freedom

There are some sociologists such as Durkheim who suggest there is a link between population density and the number of rules that must be imposed. The theory goes that as people are moved closer together geographically their actions are more likely to noticeably impact others around them. This potential impact requires the creation of additional social or legal rules to prevent conflict. A simple example would be as houses become closer together the acceptable maximum volume of music decreases, as it becomes intrusive to other residents.

2.9.5 Crowding and Increased Aggression

Numerous studies link increased population density with increased aggression. Some people believe that increased population density encourages crime and anti-social behavior. It is argued that human beings, while social animals, need significant amounts of social space or they become agitated and aggressive. However, the relationship between higher densities and increased social pathology has been largely discredited

2.10 SUBURBAN COLONIZATION

The suburban colonization process is observed in larger cities that have suffered population and political power loss to suburbs. Other colonialism is often studied for the effects upon those already inhabiting the colonized area, but students of suburban colonization tend to take greater interest in the effects upon the metropole.

2.10.1 Services and Taxes Favour Suburbs

As hinterlands acquire more population and more power then, according to the one man one vote principle, they get more votes in representative bodies, notably metropolitan regions or greater urban areas such as the Greater Toronto Area or Greater London. Suburban votes then come to outweigh inner city votes, just as, a century earlier, urbanization or urban colonialization diminished the power of rural voters.

Decisions of these bodies accordingly begin to favour people who live in suburbs, providing more car-oriented and commuter services and more favourable property tax rates for single family homes astenants in downtown apartment buildings pay higher rates. In urban areas that are growing rapidly, services may be developed that favour urban sprawl, such as large trunk sewers, express highways orshopping malls, as other services such as youth recreation disappear from downtown areas. This increases population drain to the suburbs as quality of life drops, but the increased population may then drive more people further out to the hinterlands which increases the political rewards (especiallypolitical donations from real estate developers building greenfield developments) for sprawl.

2.10.2 Urban Bankruptcy Requires Outside Aid

In very extreme cases, where cities are unable to recover costs of serving a vast suburban hinterland and are politically controlled by a larger jurisdiction, such as Manhattan within New York state, cities may go bankrupt as New York City in fact did in the 1970s. This had been predicted by urbanists including Jane Jacobs who had fought Robert Moses and his plan for the Cross-Manhattan Expressway system which was eventually defeated. The City only recovered with federal aid and urban autonomy rights including the right to levy its own income tax which it still has.

2.10.3 Suburban Flight Polarizes Communities

Cities with impoverished downtown services can suffer riots or major unrest, as Los Angeles andDetroit did in the 1960s to 1980s. Such incidents speed the flight of middle class residents to the suburbs and sometimes to gated community developments where they are insulated from urban problems, and consume a very different range of services than downtown residents, which again are favoured strongly by political representatives.

2.10.4 Forced Mergers Further Reduce Downtown Power

In some cases, notably Toronto and Montreal in the 1990s, a larger political unit will force smaller urban units to merge against the will of residents, and this further increases the hold of the outer suburban regions as they hold a majority of seats in the new aggregated city council. Where a strong mayor system applies, the larger number of suburban residents will likely also control that post, and the need to campaign over a larger urban area will tend to exclude grassroots candidates or anti-poverty activist candidates not funded nor supported by wealthier suburban voters or real estate developers. Those who speak for the city may live on its outer edges. Mayors may be former mayors of former suburban cities such as Mel Lastman, former mayor of North York who became Mayor of Toronto once those cities (and three others) were merged in 1998.

The political consequences of both mergers were severe. In Quebec, the Parti Québécois government was defeated by Jean Charest who permitted Montreal to hold a referendum in which it was permitted to de-amalgamate politically and regain the separate pre-merger urban identities. In Toronto no such relief occurred but a Province of Toronto movement emerged under Jane Jacobs (who had moved to Toronto in the 1960s and again fought expressways penetrating the downtown there, notably theSpadina Expressway and Front Street Extension), 2000 Lastman opponent Tooker Gomberg and Mayor in 2003 David Miller.

2.10.5 Theoretical Analyses

Joel Garreau in Edge City described the growth of cities on the edge of major urban areas, which became population and power centres in themselves.

Dale Johnston in Lost in the Suburbs described a cultural and political gap that occurred in New Jersey and Ontario in the early 1990s when suburban voters began to outnumber urban or rural voters, and began to perceive that they were paying taxes to provide urban areas with services that were not duplicated in their community. Meanwhile, suburban communities would export problems to the cities, typically in the form of drug addicts, homelessness, smog, prostitution and other crimes serving suburban residents, and the need to accommodate a large number of commuters and their sewage and parking requirements. As downtown residents and suburban voters became estranged, each perceived themselves subsidizing the other, and accordingly a common solution, called in both New Jersey and Ontario the Common Sense Revolution, transferred funds from urban needs to suburban sprawl, triggering a decline in urban quality of life in both places, as population further spread out and downtowns became more hostile to suburban visitors.

2.11 PSEUDO-URBANIZATION

Pseudo-urbanization is the condition in which a large city has formed in an area without a functional infrastructure to support it. As the population of an urbanized area grows, the city's infrastructure must grow with it, or else shortages will develop, typically in housing, education, transportation, clean water and waste removal services, or other services such as law enforcement. Overpopulation in urban areas is often characterized by shanty towns, where such services are inadequate or wholly absent. A city in which significant growth in the absence of adequate infrastructure has taken place will be deemed "pseudo-urbanized".

Urbanization in the third world tends to consist primarily of pseudo-urbanization. This happens largely because of so-called "rural push": factors which push people from the countryside into the cities, without the city being prepared to accept them. Rural-urban migrants in the third world usually move into the cities due to poverty-related reasons, leading to a demographic explosion and a progressive concentration of poor migrants in the cities. This is a finite process, as one city can only hold so many people due to limited infrastructure and available resources.

CHAPTER 3

THE FORMATION OF URBAN AREAS

3.1 INTRODUCTION

Early cities developed in a number of regions, from Mesopotamia to Asia to the Americas. The very first cities were founded in Mesopotamia after the Neolithic Revolution, around 7500 BCE. Mesopotamian cities included Eridu, Uruk, and Ur. Early cities also arose in the Indus Valley and ancient China. Among the early Old World cities, one of the largest was Mohenjo-daro, located in the Indus Valley (present-day Pakistan); it existed from about 2600 BCE, and had a population of 50,000 or more. In the ancient Americas, the earliest cities were built in the Andes and Mesoamerica, and flourished between the 30th century BCE and the 18th century BCE.

Ancient cities were notable for their geographical diversity, as well as their diversity in form and function. Theories that attempt to explain ancient urbanism by a single factor, such as economic benefit, fail to capture the range of variation documented by archaeologists. Excavations at early urban sites show that some cities were sparsely populated political capitals, others were trade centers, and still other cities had a primarily religious focus. Some cities had large dense populations, whereas others carried out urban activities in the realms of politics or religion without having large associated populations. Some ancient cities grew to be powerful capital cities and centers of commerce and industry, situated at the centers of growing ancient empires. Examples include Alexandria and Antioch of the Hellenistic civilization, Carthage, and ancient Rome and its eastern successor, Constantinople (later Istanbul).

3.2 THE FORMATION OF CITIES

Why did cities form in the first place? There is insufficient evidence to assert what conditions gave rise to the first cities, but some theorists have speculated on what they consider pre-conditions and basic mechanisms that could explain the rise of cities. Agriculture is believed to be a pre-requisite for cities, which help preserve surplus production and create economies of scale. The conventional view holds that cities first formed after the Neolithic Revolution, with the spread of agriculture. The advent of farming encouraged hunter-gatherers to abandon nomadic lifestyles and settle near others who lived by agricultural production. Agriculture yielded more food, which made denser human populations possible, thereby supporting city development. Farming led to dense, settled populations, and food surpluses that required storage and could facilitate trade. These conditions seem to be important prerequisites for city life. Many theorists hypothesize that agriculture preceded the development of cities and led to their growth.

A good environment and strong social organization are two necessities for the formation of a successful city. A good environment includes clean water and a favorable climate for growing crops and agriculture. A strong sense of social organization helps a newly formed city work together in times of need, and it allows people to develop various functions to assist in the future development of the city (for example, farmer or merchant). Without these two common features, as well as advanced agricultural technology, a newly formed city is not likely to succeed.

Cities may have held other advantages, too. For example, cities reduced transport costs for goods, people, and ideas by bringing them all together in one spot. By reducing these transaction costs, cities contributed to worker productivity. Finally, cities likely performed the essential function of providing protection for people and the valuable things they were beginning to accumulate. Some theorists hypothesize that people may have come together to form cities as a form of protection against marauding barbarian armies.

A 'settlement' is an environment of household shelters: it is to a community what a shelter is to a household or family. Urban settlement options support settlements in (small and large) towns and cities.

A human settlement derives from the structured landscape of a territory. It takes into consideration spatial allocation of functions while maintaining equilibrium between the needs of the population, the availability and allocation of resources, economic dynamics, the amelioration of living conditions, the provision of services and enhancing transportation networks, as well as recreational spaces. A settlement must address the needs of the community at large and be designed with the active involvement of persons of concern, partners, and all sectors.

Settlement refers to the physical spaces and environments in which households are sheltered, and how one shelter relates to others. The term is generally used in the context of displaced populations to describe the temporary or sometimes permanent living arrangements of displaced families. In this context settlements can range from planned camps to dispersed accommodation in host villages/neighbourhoods, collective centres, spontaneous camps, rental accommodation, etc.

An urban settlement is where displaced populations settle within an urban agglomeration such as a town or city. A master plan usually divides towns or cities into zones regulated by norms based on specific sectors such as housing, hygiene, habitat, and environment. Zones are inclusive of residential areas, services and infrastructures, and spaces for administrative, commercial and industrial activities.

Land availability is limited in cities and towns; persons of concern often settle in informal areas or marginalized neighbourhoods which have inherent issues of access, availability of services, lack of sanitation, and limited space for shelters. Displaced population may blend into the urban poor which makes targeted assistance challenging, but ever more important to ensure resources reach the intended recipients.

This section looks at common urban settlement options and the development of urban settlement strategies. These require analysis that should be undertaken during the preparedness planning phase, in which settlement options are identified in collaboration with stakeholders. Responses should take account of the capacity of displaced communities, and resources offered by the city and its communities. Prior to determining suitable options, ensure that the following information is available:

- Spatial analysis that describe the availability, uses, and suitability of land
- Evaluation of the absorption capacity of hosting areas and the extent of natural resources
- Sources of water and their capacity
- Market assessments including - infrastructure, logistical resources, storage capacity, the availability of construction materials, and the feasibility of setting up supply chains into affected areas

3.2.1 Collective Centres

A variety of pre-existing buildings or structures may be used as collective centres - community centres, town halls, hotels, gymnasiums, warehouses, unfinished buildings, disused factories. These facilities are seldom fit for habitation and must be rehabilitated and/or upgraded to meet the shelter needs of residents. Collective centres should be used only as short-term accommodation to gain time to provide more suitable shelter.

Considerations

- Families and communities may be able to be kept together maintaining existing support mechanisms

- If the building is connected to the water and sanitation system it may only require upgrades to achieve adequate standards. If buildings are in good condition, it is very cost effective and can be easily winterized
- Collective centres can get overcrowded; psychosocial problems can arise if displaced individuals remain in collective centres for too long without privacy and independence
- Increased security, fire and communicable disease risk due to the concentration of people

3.2.2 Host Neighbourhoods

Displaced people may live with and amongst local households, on land or in properties that local people own. Hosts may be relatives, distant family members, friends or acquaintances, or people previously unknown to those who have been displaced.

Considerations

- Host arrangements provide immediate shelter for persons of concern.
- Access to spontaneous community support mechanisms can encourage self-reliance, independence and a sense of belonging. Host population may have limited resources; and already living below the poverty line. Absorption capacity maybe limited and competition for resources can be fierce in urban areas.
- Long-term accommodation with host families in dense urban areas can be over-crowded, with detrimental effects to health and welfare of both host and displaced families.
- Housing may already be substandard; host families may be in need of improved shelter.

Short-Term Tenancy (Land, A House, An Apartment, A Room)

Persons of concern may rent from the local population via formal or informal agreements. As with host neighbourhoods rental accommodation may be available immediately, and persons of concern will have greater independence and self-reliance. However, in a competitive market, refugees, IDPs, and returnees may be vulnerable to discrimination and exploitation by unscrupulous land lords; rental accommodation that is available and affordable is often substandard; the relationship between landlord and tenant may be exploitative.

3.2.3 Dispersed Self-Settlement Without Legal Status

Persons of concern may settle in scattered locations across large areas, on land or buildings which they have no approval to occupy. Urban land, housing, services, and resources are all limited, and settling without permission in urban areas is extremely problematic for persons of concern; the threat of forced eviction, violent conflict with local populations, exploitation and abuse, and denial of rights is greater in dense urban areas, especially in informal settlements and slums towards which displaced populations are likely to gravitate.

The local population may also need support, for example in ensuring that resources and communal service infrastructure are not overburdened. Full consultation with formal and informal authorities is necessary to avoid conflict with existing inhabitants and plans. It is important that traditional or informal land tenure is recognized and supported to ensure the best protection of the displaced.

Considerations

- Access to livelihoods may be possible in urban areas
- Persons of concern have some degree of choice on where to settle, and may settle within local communities that share cultural ties with them
- Gathering data from and communicating with persons of concern will be difficult and costly
- Persons of concern are less likely to report security or protection concerns to authorities for fear of eviction or abuse
- Reaching formal agreements to occupy property may be more difficult if HLP legal framework is inadequate, or if the property is in informal settlements which may not have legal recognition from the state and therefore no formal property titles

3.2.4 Self-Settled, Unplanned Camps

Spontaneous settlements or unplanned camps occur when groups of displaced people populate areas without assistance or guidance from local government or the humanitarian community. Such settlements are located on land the displaced population does not officially have the right to occupy. This constitutes the establishment of an entirely new settlement within the boundaries of this land.

Considerations

- Spontaneous settlements in urban areas will likely occupy undesirable land in high risk areas such as along ravines or hillside slopes, or the limited public areas such as public squares or parks
- Access to adequate supply of water supply, sanitation and other infrastructure is unlikely
- An assessment will be needed in order to determine if the population can and should be relocated to another settlement such as a planned camp or if arrangements can be made in host neighbourhoods
- Some spontaneous settlements, even in urban and peri urban areas can be formalized and upgraded if the site is suitable and approval is granted by the

authorities. For more information see entries on spontaneous settlement strategy considerations and on managing and supporting spontaneous settlements

3.3 THE STRUCTURE OF CITIES

Urban structure is the arrangement of land use, explained using different models.

3.3.1 Urban Structure Models

We've talked about cities and urban growth, but what happens within a city? People who study how residents in the cities adapt to their environments are called urban ecologists. They are very much like ecology biologists who study the relationship between animal behaviors in connection with their environment. Urban ecologists focus on the social and spatial context—how industries are distributed within the city or how the city is spatially organized, and how this relates to city residents. Like ecologists, urban ecologists think that cities form as a result of natural growth. This natural growth refers to urban expansion, immigration, and succession (new people moving in and replacing those that move out), and this is very much related to geography and the natural environment of a city. Cities may become disorganized when newcomers or immigrants move in, but it will gradually become organized again after a period of time. During these transitional changing periods, the city may suffer from more crime and other social problems in areas with the most movement of incoming and outgoing people.

Below are three models of urban growth developed by urban sociologists in the US. In general, a city grows from the center, then outwards. In the concentric zone model, Zone 1 is the Central Business District (CBD). Zone 2 is a transitional area containing rooming houses and deteriorating housing which breed poverty, disease, and vice. Zone 3 is the area thrifty workers have moved in order to escape the transitional Zone 2 yet maintain convenient access to their work. Zone 4 contains more expensive apartments, hotels, single-family home, etc. Commuters live in Zone 10 which consists of suburbs or satellite cities that have popped up around transportation routes. Have you ever been to Chicago? This is similar to the early city setup of Chicago, and most other traditional US cities with clear central business districts. But as transportation became more developed, 'zones' have also become less obvious. For example, Los Angeles is an example of a city without clear 'zones' since it is so spread out.

The sector model is an updated version of the concentric zone model. Concentric zones can contain different sectors, one of working-class homes, another of expensive housing, and one of businesses, etc., all competing for the same land. This model takes into account transportation developments. Cities also expand outwards, but along railways, highways, and water. Lower-income housing can be found along railroads or bordering manufacturing and industrial sectors. Chicago today is closer to this model than the original concentric zone model. Calgary in Canada is also another example of a sector model city.

In a multiple nuclei model, a city may have several centers of 'nuclei'. The model says that even if a city did begin with one CBD, it can still develop multiple smaller business districts. This model is applicable to a lot of large cities today. For example, Mumbai and Shanghai are both examples of multiple nuclei cities. Each nucleus in these cities contains a specialized activity—such as clusters of fast-food restaurants or retail districts. Areas with similar activities cluster together to draw consumers, or because land-use is similar in adjacent areas.

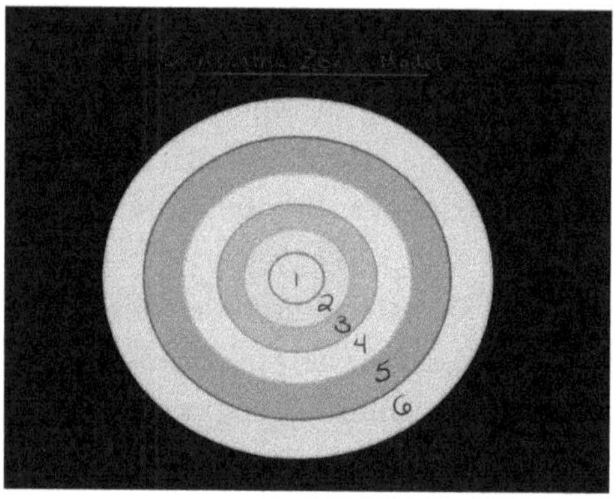

Fig. 1. Concentric zone model diagram.

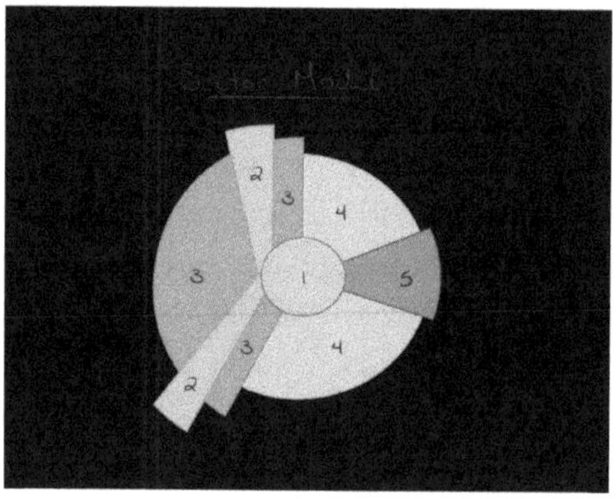

Fig. 2. Sector model diagram.

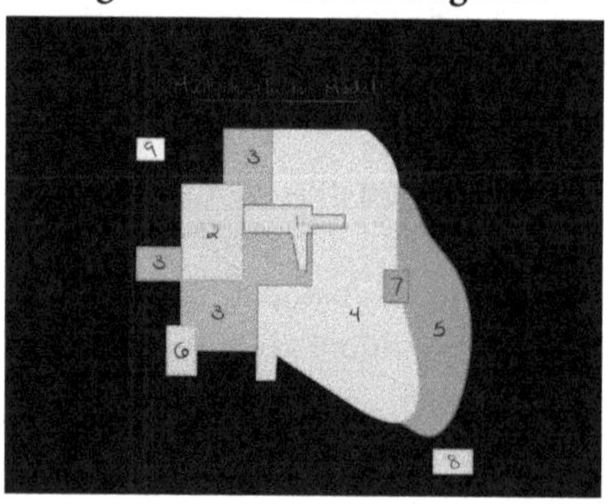

Fig. 3. Multiple nuclei model.

Legend

1. Central business district
2. Wholesale and light manufacturing
3. Low-class residential
4. Medium-class residential
5. High-class residential
6. Heavy manufacturing
7. Outlying business district
8. Residential suburb
9. Industrial suburb
10. Commuter zone

Of course no city perfectly fits these models. With geography, development of transportation, and innovative businesses, cities are becoming increasingly diverse. But overall, these models are great for understanding urban planning and development.

3.3.2 Grid

In grid models, land is divided by streets intersect at right angles, forming a grid. Grid plans are more common in North American cities than in Europe, where older cities tend to be build on streets that radiate out from a central square or structure of cultural significance. Grid plans facilitate development because developers can subdivide and auction off large parcels of land. The geometry yields regular lots that maximize use and minimize boundary disputes. However, grids can be dangerous because long, straight roads allow faster automobile traffic. In the 1960s, urban planners moved away from grids and began planning suburban developments with dead ends and cul-de-sacs.

3.4 CONCENTRIC RING MODEL

The concentric ring model was postulated in 1924 by sociologist Ernest Burgess, based on his observations of Chicago. It draws on human ecology theories, which compared the city to an ecosystem, with processes of adaptation and assimilation. Urban residents naturally sort themselves into appropriate rings, or ecological niches, depending on class and cultural assimilation. The innermost ring represents the central business district (CBD), called Zone A. It is surrounded by a zone of transition (B), which contains industry and poorer-quality housing. The third ring (C) contains housing for the working-class—the zone of independent workers' homes. The fourth ring (D) has newer and larger houses occupied by the middle-class. The outermost ring (E), or commuter's zone, is residential suburbs.

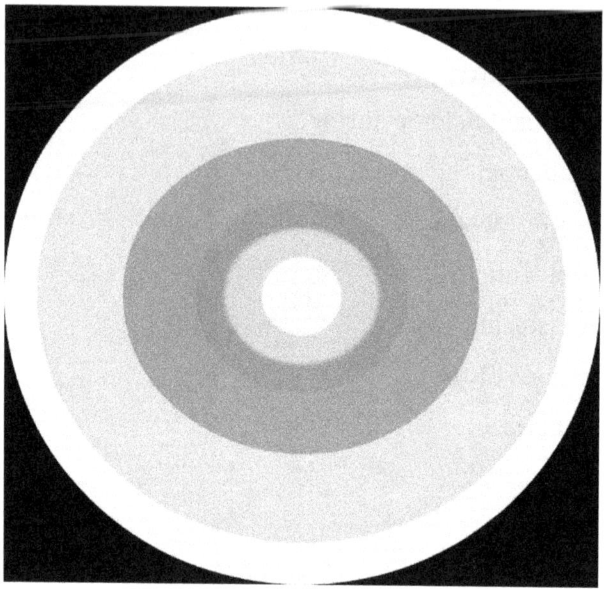

Fig. 4. Concentric Zone Model: The Concentric Ring Model described the city as a series of concentric rings, each home to a different group and social function.

This model's general applicability has been challenged. It describes an American geography in which the inner city is poor while suburbs are wealthy—elsewhere, the converse is the norm. In new, western U.S. cities such as Los Angeles, advances in transportation and communication have blurred these "zones. " Further, the model fails to account for topographical and physical features of the landscape. Even in Chicago, the concentric rings were semi-circles, interrupted by Lake Michigan.

3.4.1 Sectoral

In 1939, the economist Homer Hoyt adapted the concentric ring model by proposing that cities develop in wedge-shaped sectors instead of rings. Certain areas of a city are more attractive for various activities, whether by chance or geographic/environmental reasons. As these activities flourish and expand outward, they form wedges, becoming city sectors. Like the concentric ring model, Hoyt's sectoral model has been criticized for ignoring physical features and new transportation patterns that restrict or direct growth.

3.4.2 Multiple Nuclei

The multiple nuclei model was developed in 1945 to explain city formation after the spread of the automobile. People have greater movement due to increased car ownership, allowing for the specialization of regional centers. A city contains more than one center around which activities revolve. Some activities are attracted to particular nodes while others try to avoid them. For example, a university node may attract well-educated residents, pizzerias, and bookstores, whereas an airport may attract hotels and warehouses. Incompatible activities will avoid clustering in the same area.

3.4.3 Irregular Pattern

The irregular pattern model was developed to explain urban structure in the Third World. It attempts to model the lack of planning found in many rapidly built Third World cities. This model includes blocks with no fixed order; urban structure is not related to an urban center or CBD.

3.4.4 Alternate Uses of "Urban Structure"

Urban structure can also refer to urban spatial structure; the arrangement of public and private space in cities and the degree of connectivity and accessibility. In this context, urban structure is concerned with the arrangement of the CBD, industrial and residential areas, and open space.

A city's Central Business District (CBD), or downtown, is the commercial and often geographic heart of a city. In North America, this is referred to as "downtown" or "city center. " The downtown area is often home to the financial district, but usually also contains entertainment and retail. CBDs usually have very small resident populations, but populations are increasing as younger professional and business worqkers move into city center apartments.

An industrial park is an area zoned and planned for the purpose of industrial development. They are intended to attract business by concentrating dedicated infrastructure to reduce the per-business expenses. They also set aside industrial uses from urban areas to reduce the environmental and social impact of industrial uses and to provide a distinct zone of environmental controls specific to industrial needs.

Urban open spaces provide citizens with recreational, ecological, aesthetic value. They can range from highly maintained environments to natural landscapes. Commonly open to public access, they may be privately owned. Urban open spaces offer a reprieve from the urban environment and can add ecological value, making citizens more aware of their natural surroundings and providing nature to promote biodiversity. Open spaces offer aesthetic value for citizens who enjoy nature, cultural value by providing space for concerts or art shows, and functional value—for example, by helping to control runoff and prevent flooding.

3.5 GROWTH MACHINE THEORY

However, some people who study urbanization argue that a city is a growth machine rather than the outcomes of a natural process, unlike what urban ecologists argue. For them, urban growth is not just organized by spatial geography, and cities will not just reorganize itself after periods of disorganization. The growth machine theory points out that urban growth is driven by a coalition of interest groups who all benefit from a city's continuous growth and expansion. For them, the growth of cities is social, political, and largely planned and intended. For example, real estate interests may be involved—when urban growth happens, some groups benefit because properties increase in value. How groups lobby or manipulate the government or other groups, determine how cities grow and take shape. How people within the city are distributed is not so much related to geography, transportation, or space, but rather to the social actions of interest groups.

3.5.1 The Process of Urbanization

Urbanization is the process of a population shift from rural areas to cities, often motivated by economic factors.

Urbanization and Rural Flight

Urbanization is the process of a population shift from rural areas to cities. During the last century, global populations have urbanized rapidly:

- 13% of people lived in urban environments in the year 1900
- 29% of people lived in urban environments in the year 1950

One projection suggests that, by 2030, the proportion of people living in cities may reach 60%.

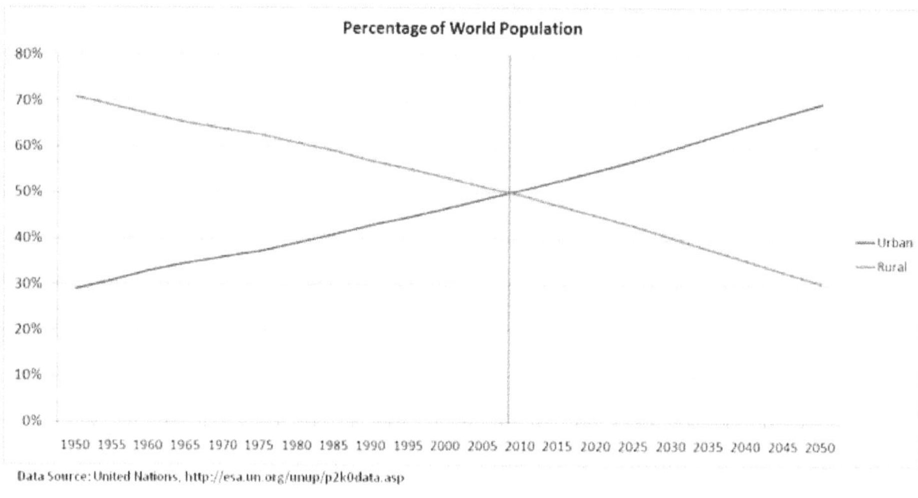

Fig. 5. Rural and Urban World Population: Over time, the world's population has become less rural and more urban.

Urbanization tends to correlate positively with industrialization. With the promise of greater employment opportunities that come from industrialization, people from rural areas will go to cities in pursuit of greater economic rewards.

Another term for urbanization is "rural flight. " In modern times, this flight often occurs in a region following the industrialization of agriculture—when fewer people are needed to bring the same amount of agricultural output to market—and related agricultural services and industries are consolidated. These factors negatively affect the economy of small- and middle-sized farms and strongly reduce the size of the rural labor market. Rural flight is exacerbated when the population decline leads to the loss of rural services (such as business enterprises and schools), which leads to greater loss of population as people leave to seek those features.

As more and more people leave villages and farms to live in cities, urban growth results. The rapid growth of cities like Chicago in the late nineteenth century and Mumbai a century later can be attributed largely to rural-urban migration. This kind of growth is especially commonplace in developing countries.

Urbanization occurs naturally from individual and corporate efforts to reduce time and expense in commuting, while improving opportunities for jobs, education, housing, entertainment, and transportation. Living in cities permits individuals and families to take advantage of the opportunities of proximity, diversity, and marketplace competition. Due to their high populations, urban areas can also have more diverse social communities than rural areas, allowing others to find people like them.

3.3.2 Economic and Environmental Effects of Urbanization

Urbanization has significant economic and environmental effects on cities and surrounding areas. As city populations grow, they increase the demand for goods and services of all kinds, pushing up prices of these goods and services, as well as the price of land. As land prices rise, the local working class may be priced out of the real estate market and pushed into less desirable neighborhoods – a process known as gentrification.

Growing cities also alter the environment. For example, urbanization can create urban “heat islands,” which are formed when industrial and urban areas replace and reduce the amount of land covered by vegetation or open soil. In rural areas, the ground helps regulate temperatures by using a large part of the incoming solar energy to evaporate water in vegetation and soil. This evaporation, in turn, has a cooling effect. However in cities, where less vegetation and exposed soil exists, the majority of the sun’s energy is absorbed by urban structures and asphalt. During the day, cities experience higher surface temperatures because urban surfaces produce less evaporative cooling. Additional city heat is given off by vehicles and factories, as well as industrial and domestic heating and cooling units. Together, these effects can raise city temperatures by 2 to 10 degrees Fahrenheit (or 1 to 6 degrees Celsius).

3.5.3 Suburbanization and Counterurbanization

Recently in developed countries, sociologists have observed suburbanization and counterurbanization, or movement away from cities. These patterns may be driven by transportation infrastructure, or social factors like racism. In developed countries, people are able to move out of cities while still maintaining many of the advantages of city life (for instance, improved communications and means of transportation). In fact, counterurbanization appears most common among the middle and upper classes who can afford to buy their own homes.

Race also plays a role in American suburbanization. During World War I, the massive migration of African Americans from the South resulted in an even greater residential shift toward suburban areas. The cities became seen as dangerous, crime-infested areas, while the suburbs were seen as safe places to live and raise a family, leading to a social trend known in some parts of the world as "white flight. " Some social scientists suggest that the historical processes of suburbanization and decentralization are instances of white privilege that have contributed to contemporary patterns of environmental racism.

In the United States, suburbanization began in earnest after World War II, when soldiers returned from war and received generous government support to finance new homes. Suburbs, which are residential areas on the outskirts of a city, were less crowded and had a lower cost of living than cities. Suburbs grew dramatically in the 1950s when the U.S. interstate highway system was built, and automobiles became affordable for middle class families. Around 1990, another trend emerged known as counterurbanization, or "exurbanization". The wealthiest individuals began living in nice housing far in rural areas (as opposed to forms).

Suburbanization may be a new urban form.Rather than densely populated centers, cities may become more spread out, composed of many interconnected smaller towns. Interestingly, the modern U.S. experience has gone from a largely rural country, to a highly urban country, to a country with significant suburban populations.

3.6 U.S. URBAN PATTERNS

The U.S. Census Bureau classifies areas as urban or rural based on population size and density.

Different international, national, and local agencies may define "urban" in various ways. For example, city governments often use political boundaries to delineate what counts as a city. Other definitions may consider total population size or population density. Different definitions may also set various thresholds, so that in some cases, a town of just 2,500 may count as an urban city, whereas in other contexts, a city may be defined as having at least 50,000 people. Other agencies may define "urban" based on land use: places count as urban if they are built up with residential neighborhoods, industrial sites, railroad yards, cemeteries, airports, golf courses, and similar areas. Using this sort of definition, in 1997, the U.S. Department of Agriculture tallied over 98,000,000 acres of "urban" land.

In spite of these competing definitions, in the United States "urban" is officially defined following guidelines set by the U.S. Census Bureau. The Census Bureau defines "urban areas" as areas with a population density of at least 1,000 people per square mile and at least 2,500 total people. Urban areas are delineated without regard to political boundaries. Because this definition does not consider political boundaries, it is often used as a more accurate gauge of the size of a city than the number of people who live within the city limits. Often, these two numbers are not the same. For example, the city of Greenville, South Carolina has a city population under 60,000 and an urbanized area population of over 300,000, while Greensboro, North Carolina has a city population over 200,000 and an urbanized area population of around 270,000. That means that Greenville is actually "larger" for some intents and purposes, but not for others, such as taxation, local elections, etc.

As of December, 2010, about 82% of the population of the United States lived within the boundaries of urbanized area. Combined, these areas occupy about 2% of the land area of the United States. The majority of urbanized area residents are suburbanites; core central city residents make up about 30% of the urbanized area population (about 60 million out of 210 million). In the United States, the largest urban area is New York City, with over 8 million people within the city limits and over 19 million in the urban area. The next five largest urban areas in the United States are Los Angeles, Chicago, Washington, D.C., Philadelphia, and Boston.

Fig. 6. American urban areas by size: This map shows major urban areas in America.

3.6.1 The Rural Rebound

During the 1970s and again in the 1990s, the rural population rebounded in what appeared to be a reversal of urbanization.

The rural rebound refers to the movement away from cities to rural and suburban areas. Urbanization tends to occur along with modernization, yet in the most developed countries many cities are now beginning to lose population. In the United States in the 1970s, demographers observed that the rural population was actually growing faster than urban populations, a phenomenon they labeled the "rural rebound. " This trend reversed in the 1980s, due in part to a recession that hit farmers particularly hard. But again in the 1990s, rural populations appeared to be gaining at the expense of cities. Indeed, in the last 50 years, about 370 cities worldwide with more than 100,000 residents have undergone population losses of more than 10%, and more than 25% of the depopulating cities are in the United States.

Rather than moving to rural areas, most participants in the so-called the rural rebound migrated into new, rapidly growing suburbs. The rural rebound, then, may be more evidence of the importance of suburbanization as a new urban form in the most developed countries.

3.6.2 Suburbanization

Suburbanization is a general term that refers to the movement of people from cities to surrounding areas. However, the suburbanization that took place after 1970 was different from the suburbanization that had occurred earlier, after World War II. In this more recent wave of suburbanization, people moved beyond the nearby suburbs to farther-away towns. Sociologists have invented several new categories to describe these new types of suburban towns; two of the most notable are ex-urbs and edge cities.

The expression exurb (for "extra-urban") refers to a ring of prosperous communities beyond a city's suburbs. Often, these communities are commuter towns or bedroom communities. Commuter towns are primarily residential; most of the residents commute to jobs in the city. They are sometimes called bedroom communities because residents spend their days away in the cities and only come home to sleep. In general, commuter towns have little commercial or industrial activity of their own, though they may contain some retail centers to serve the daily needs of residents. Although most exurbs are commuter towns, most commuter towns are not exurban.

Exurbs vary in wealth and education level. In the United States, exurban areas typically have much higher college education levels than closer-in suburbs, though this is not necessarily the case in other countries. They typically have average incomes much higher than nearby rural counties, reflecting the urban wages of their residents. Although some exurbs are quite wealthy even compared to nearer suburbs or the city itself, others have higher poverty levels than suburbs nearer the city. This may happen especially where commuter towns form because workers in a region cannot afford to live where they work and must seek residency in another town with a lower cost of living. For example, during the "dot com" bubble of the late twentieth century, housing prices in California cities skyrocketed, spawning exurban growth in adjacent counties.

3.6.3 White Flight

Sociologists have posited many explanations for counterurbanization, but one of the most debated is whether suburbanization is driven by white flight. The term white flight was coined in the mid-twentieth century to describe suburbanization and the large-scale migration of whites of various European ancestries, from racially mixed urban regions to more racially homogeneous suburban regions. During the first half of the twentieth century, discriminatory housing policies often prevented blacks from moving to suburbs; banks and federal policy made it difficult for blacks to get the mortgages they needed to buy houses, and communities used restrictive housing covenants to exclude minorities.

White flight during this period contributed to urban decay, a process whereby a city, or part of a city, falls into disrepair and decrepitude. Symptoms of urban decay include depopulation, abandoned buildings, high unemployment, crime, and a desolate, inhospitable landscape. White flight contributed to the draining of cities' tax bases when middle-class people left, exacerbating urban decay caused in part by the loss of industrial and manufacturing jobs as they moved into rural areas or overseas where labor was cheaper.

More recently, the concept has been extended to newer forms of suburbanization, including migration from urban to rural areas and to exurbs. In a similar vein, some demographers have described the rural rebound, and the newest waves of suburbanization, as a form of ethnic balkanization, in which different ethnic groups (not only whites) sort themselves into racially homogeneous communities. These phenomena, however, are not so clearly driven by the restrictive policies, laws, and practices that drove the white flight of the first half of the century.

3.7 MODELS OF URBAN GROWTH

Models of urban growth try to balance the advantages and disadvantages of cities' large sizes.

Cities are dynamic places—they grow, shrink, and change. Sociologists have developed different theories for thinking about how urban populations change.

3.7.1 Growth Machine Theory

The growth machine theory of urban growth says urban growth is driven by a coalition of interest groups who all benefit from continuous growth and expansion. First articulated by Molotch in 1976, growth machine theory took the dominant convention of studying urban land use and turned it on its head.

The field of urban sociology had been dominated by the idea that cities were basically containers for human action, in which actors competed among themselves for the most strategic parcels of land, and the real estate market reflected the state of that competition. Growth machine theory reversed the course of urban theory by pointing out that land parcels were not empty fields awaiting human action, but were associated with specific interests—commercial, sentimental, and psychological. In other words, city residents were not simply competing for parcels of land; they were also trying to fulfill their particular interests and achieve specific goals. In particular, cities are shaped by the real estate interests of people whose properties gain value when cities grow. These actors make up what Molotch termed "the local growth machine. "

Urban Sprawl

Whether explained by older theories of natural processes or by growth machine theory, the fact of urban growth is undeniable: throughout the twentieth century, cities have grown rapidly. In some cases, that growth has been poorly controlled, resulting in a phenomenon known as urban sprawl. Urban sprawl entails the growth of a city into low-density and auto-dependent rural land, high segregation of land use (e.g., retail sections placed far from residential areas, often in large shopping malls or retail complexes), and design features that encourage car dependency.

Urban sprawl's segregated land use means that the places where people live, work, shop, and relax are far from one another, which usually makes walking, public transit, or bicycling impractical. As a result, residents must use an automobile. Urban sprawl tends to include low population density: single family homes on large lots instead of apartment buildings, single story or low-rise buildings instead of high-rises, extensive lawns and surface parking lots, and so on.

Critics of urban sprawl argue that it creates an inhospitable urban environment and that it encroaches on rural land, potentially driving up land prices and displacing farmers or other rural residents. Urban sprawl is also associated with negative environmental and public health effects, many of which are related to automobile dependence: increases in personal transportation costs, air pollution and reliance on fossil fuel, increases in traffic accidents, delays in emergency medical services response times, and decreases in land and water quantity and quality.

Urban Decay

Some have suggested that urban sprawl is driven by consumer preference; people prefer to live in lower density, quieter, more private communities that they perceive as safer and more relaxed than urban neighborhoods. Such preferences echo a common strain of criticism of urban life, which tends to focus on urban decay. According to these critics, urban decay is caused by the excessive density and crowding of cities, and it drives out residents, creating the conditions for urban sprawl.

3.7.2 Broken Windows

An alternative theory suggests that density does not cause crime, and crime does not cause people to leave the city; when people leave, city neighborhoods are abandoned and neglected, resulting in crime and decay. This theory, known as the "broken windows theory," argues that small indicators of neglect, such as broken windows and unkempt lawns, promote a feeling that an area is in a state of decay. Anticipating decay, people likewise fail to maintain their own properties.

Responses to Decay

Cities have responded to urban decay and urban sprawl by launching urban renewal programs. Two specific types of urban renewal programs—New Urbanism and smart growth—attempt to make cities more pleasant and livable.

Smart growth programs draw urban growth boundaries to keep urban development dense and compact. In addition to increasing the density of cities, urban growth boundaries can protect the surrounding farmland and wild areas. Smart growth programs often incorporate transit-oriented development goals to encourage effective public transit systems and make bicyclers and pedestrians more comfortable.

New Urbanism is an urban design movement that promotes walkable neighborhoods with a range of housing options and job types. As an approach to urban planning, it encompasses principles such as traditional neighborhood design and transit-oriented development. A neighborhood designed along New Urbanist principles would have a discernible center (such as a square or a green) with a transit stop nearby. Most homes would be within a five-minute walk of the center and would provide a variety of housing options, including houses, row houses, and apartments to encourage the mixing of younger and older people, singles and families, and poor and wealthy.

3.8 UNDERLYING PRINCIPLES AND STANDARDS

- Settlement and shelter designs should reflect the needs of persons of concern, their cultural habits and their capacities, and should also attempt to build on existing resources and enhance access to infrastructure.
- Settlement and shelter interventions need to be planned and implemented to mitigate, to the extent possible, the impact on the natural environment and to prevent hazard risks such as landslides, floods and earthquakes, among others.
- Inclusive and meaningful participation of persons of concern in accordance with UNHCR's Age, Gender and Diversity approach, is essential to ensure that men, women, girls and boys have their voice heard, identify their needs, and have the opportunity to contribute to the search of adequate solutions.
- Accessibility to land constitutes a fundamental element of the realization of the right to adequate housing, and must also provide sustainable and non-discriminatory access to facilities essential for health, nutrition, security and comfort.
- Durable solutions are the ultimate goal, taking into consideration appropriate technology, capacity-building of both refugees and local communities, and use of local skills, materials, techniques and knowledge.

- Refugees and other persons of concern should be empowered to participate actively in decisions that concern them at all stages. An inclusive approach fosters ownership and acceptance of programmes and improves maintenance of shelters and settlements. It facilitates communication and can generate information and support that may be crucial to a programme's success and sustainability.
- International human rights law and refugee law recognize the right of every individual, including refugees, to move freely.
- Persons of concern should be supported to become self-reliant, enabling them to contribute to their host country and find long term solutions for themselves.
- Settlements policy and decisions should be driven primarily by the best interest of persons of concern.
- Persons of concern should have access to essential services in all types of settlements. These services include water, sanitation, roads and infrastructure, community spaces, shelter, health, nutrition, education, food, and livelihoods.
- Settlements should provide sufficient space for shelter and associated basic services. Though in host situations, for example, it can be difficult to ensure accommodations meet standards, interventions should aim to achieve minimum international or national standards.
- Housing Land and Property (HLP) considerations are fundamental in planning and implementing settlement activities. Mechanisms should be in place to protect persons of concern from forced eviction, exploitation or abuse derived from a lack of tenure security.

3.8.1 Protection Risks

- In protracted situations, deteriorating living conditions of families hosting large number of persons might lead to health and psychosocial problems, as well as risks of stigmatization, harassment, economic or sexual exploitation, and violence against the displaced families.
- In areas where refugees are not welcome, both host and displaced families might become targets of retaliation by parties to the conflict or by surrounding communities.
- Those settled spontaneously on private or public land are often under constant threat of eviction by landlords or authorities. Monitoring and responding to harassment and threats may not be achieved in a timely manner if settlements are scattered and legal tenure has not been clarified.
- Prolonged stay in camps or collective centres can result in stress and tensions and can lead to social conflict and friction with host communities, between families, clans or ethnic groups.

- Persons of concern in collective centres are often under threat of being evicted by landlords. Those occupying schools, religious, and other public buildings are under increased pressure to leave.
- The presence of ethnically, culturally, religiously or linguistically different groups may give rise to tensions.
- A proliferation of high quality shelters in an area where housing standards are low can create tension with local communities.
- High population density significantly increases health risks.

The above protection risks are applicable to rural and urban settlements. Persons of concern will encounter a number of specific risks associated with the urban environment:

- Pubic areas are scarce in cities; relations with the local community might deteriorate if the persons of concern occupy already limited number of public spaces such as parks or public squares.
- Criminal groups can be prolific in urban areas; persons of concern can be targeted in dispersed or groups settlements.
- Economies in urban environment are primarily cash based; agriculture is nonexistent; water points require payment. Access to food, water, and other necessities will come at a cost, which may force persons of concern to adopt negative coping mechanisms.
- Displaced children in urban areas face great risks. Lack of access to education can be caused by lack of resources, fear of the local community, or the need for children to complement household income. Lack of parental supervision and access to schools, and the overall poverty can lead girls and boys to try and fend for themselves and exposing them to child labour, sex work and theft. For girls especially the risks of early sex, exploitative sex and sex work is greater in cities and towns.
- Persons of concern may be accused and blamed for neighbourhoods' problems such as conflict between families, criminal acts – often despite lack of evidence, thefts, etc. Verbal abuse or accusations can become physical abuse, and persons of concern may not receive protection by the authorities.
- Inadequate housing can forced families to live in overcrowded conditions, or to separate. Children may be sent to live with other neighbours exposing them to neglect and abuse.

CHAPTER

URBANIZATION IN INDIA 4

4.1 INTRODUCTION

Urbanization in India is mainly due to liberalization of its economy after the 1990s, which gave rise to the development of the private sector. Presently, although urbanization is taking place at a fast rate in India, only one-third of its population lives in urban areas.

According to the 2011 census, there are 53 cities in India with a population of a million or more; by 2031, that number will rise to 87. Some of these metropolitan areas will become major economic powerhouses that have higher GDP than the current GDP of countries such as Israel, Portugal and the UAE.

Cities provide major opportunities for sustainable development, given that they have large numbers of people in a small area, and offer significant economies of scale which provide jobs, housing and services.

There is a need to fully realize the potential of Indian cities for ecological, economic and social sustainability. But this urbanization can only be harnessed and sustained by inclusive planning that provides affordable transportation, continuous water supply, modern sewage treatment and a good solid waste management system. The government of India has been channelling funds to renew its urban spaces through various schemes, and has been tweaking its approaches to urbanization by cataloging projects that were successful and those that failed. It is also beginning to see the value of Private–Public Partnerships.

Urban settlements are generally compact and larger in size and engaged in a variety of non-agricultural, economic and administrative functions. As mentioned earlier, cities are functionally linked to rural areas around them. Thus, exchange of goods and services is performed sometimes directly and sometimes through a series of market towns and cities. Thus, cities are connected directly as well as indirectly with the villages and also with each other.

4.2 HISTORY

Post-independence, India faced high rates of poverty, unemployment, and a stagnant economy. The first Prime Minister of India, Pandit Jawaharlal Nehru, focused on the domain of science and technology. The mixed economy system was adopted, resulting in the growth of the Public sector in India crippling down the development of Indian economy leading to what is popularly known as Hindu rate of growth. The South Asian region though predominantly rural (accounted for 69.9% rural population as of 2010), has recorded much higher annual growth of urban population. India, the leading country in South Asia has shown an unprecedented increase in the urban population in the last few decades and its urban population has increased about 14 fold from 1901 to 2011. This growth is mainly uneven but not skewed and not concentrated to a single city of the country. India shares most characteristic features of urbanization in the developing countries where the rate of urbanization is faster than the developed countries. For instance, in 1971 there were only about 150 cities whose population was more than one lakh, now this figure has reached to 500. The urban population of India has increased from 25.85 million in 1901 to 377.11 million in 2011.

4.2.1 Urban Settlement in Ancient India

Investigating transformations in the distribution and density of past settlements is crucial to the identification of "signature landscapes," which are those generated by specific social, cultural, and economic processes within specific physical environments. Comparative research has revealed an array of signature landscapes that have been associated with the emergence, transformation, and dissolution of social complexity across the globe. The identification and analysis of such landscapes contribute a large-scale dimension to models of social change, revealing interactions between societies and their dynamic and transforming environments. These investigations have the potential to transform these models, casting into high relief social processes that are dispersed across a broader landscape and may be hidden or obscured at the level of an archaeological excavation at a single site.

Patterns in settlement distribution, especially the frequency with which sites appear within a given area or environment, play a useful role in these studies by revealing settings that people favored as prevailing social conditions changed through time. However, archaeological surveys are also often constrained to specific areas by the logistics of fieldwork, limiting the scale of their interpretation and analyses. To investigate large-scale changes in settlement distribution, it is necessary to assemble and analyze large synthetic datasets built over many years by multiple teams. Successfully integrating datasets requires recognizing the limitations and errors incumbent to the production of each constituent survey project.

Northwestern India was a key setting for the emergence of South Asia's earliest complex society, the Indus Civilization. Indus cities arose around 2600 B.C. across extensive and ecologically diverse areas of western South Asia, and concentrations of archaeological sites have been reported in the modern states of Rajasthan, Haryana, and Punjab in India. It has frequently been noted that the density of settlements across the alluvial plains of northwestern India appears to increase after ca. 1900 B.C. Climate change appears to have played a role in this shift, as changes in settlement density seem to have favored the variability of local environmental conditions in northwestern India in the face of a weakening in the Indian Summer Monsoon around 2200–2100 B.C.

The increase in settlement density in northwestern India may have been due to the strong possibility that this region received more reliable rainfall from a weakened monsoon. As people left Indus cities, they appear to have populated particular areas, establishing new small-scale settlements and re-occupying mounds that had been abandoned in earlier periods. This apparent shift both resulted from and contributed to a process of de-urbanization, wherein smaller and more dispersed settlements replaced larger population aggregations. Much attention has been given to the process of urbanization that brought together multiple groups of specialized artisans and agro-pastoralists. However, it is unclear how de-urbanization transformed these social relations, as it was a dispersed process that unfolded at a large number of sites across an extensive area, thus necessitating a large-scale approach that incorporates the results of multiple projects.

To utilize multiple datasets in aggregate studies, it is necessary to compare the approaches, questions, and methods that contributed to each researcher's agenda. It has been noted that site location reports from northwestern India vary in their intensity of survey coverage, adherence to modern administrative boundaries, and assumptions about the locations of past watercourses. To address these challenges, this paper describes the assembly of a pilot database that integrates all site location data from a sample region that encompasses two major surveys carried out by the "Land, Water and Settlement" project. The data were then analyzed using Geographic Information Systems (GIS) analyses; this was the first stage of a larger effort to integrate site locations from northwestern India into a single relational database, which is being carried out for the "TwoRains" project. This approach is informed by Kintigh, who has advocated increasing the scale of archaeological investigations without compromising the detail recorded in specific reports. It allows the analysis of site location data at different levels of certainty. The pilot database facilitated a test of the following hypotheses: first, that in northwestern India, the Mature Harappan period saw the nucleation of settled population; second, that the Late Harappan period saw an increase in settlement density. Our results support these hypotheses and enhance the interpretation that site density increased in particular locations with the decline of Indus cities. It follows that the landscapes of urbanization and de-urbanization created by Indus populations integrated a range of varied environments to produce and sustain social complexity.

Ancient Towns: There are number of towns in India having historical background spanning over 2000 years. Most of them developed as religious and cultural centres. Varanasi is one of the important towns among these. Prayag (Allahabad), Pataliputra (Patna), Madurai are some other examples of ancient towns in the country.

4.2.2 Landscape Archaeology and the Indus Civilization

Landscape archaeology provides the approaches necessary to frame research on past social processes. It has been foundational to modeling social complexity in ancient Mesopotamia, and has also been critical to the study of complex societies across the globe. Large scale analyses are necessary for outlining the interaction between emerging complex societies and their varied local settings, revealing patterns that are difficult to explain in reference to their local settings alone and thus must result from processes of greater regional integration. By incorporating data from locations across broad and varied environments, landscape approaches have the potential to challenge traditional models of complexity and urbanism. Such approaches have revealed processes such as the heterarchical clustering of settlements and alternative political trajectories.

Wilkinson argued that relationships between archaeological remains and their environmental contexts result in "signature landscapes" that exemplify the prevailing configurations of social, cultural, and economic processes within specific environmental settings and chronological periods. Signature landscapes can be compared to one another to investigate social change. Site locations are key to this approach, but to address large-scale processes that take place throughout a landscape typically requires aggregating data built up by many projects. A framework for integrating heterogeneous survey datasets has been set out by Lawrence and Bradbury (2012), who characterize site locations using factors such as boundary certainty, geographical precision, and archaeological significance, ascertaining different levels of certainty in archaeological datasets. Boundary certainty addresses the size of archaeological sites and lies beyond the scope of this paper, but site location reports from northwestern India can be used to establish a basic level of certainty based on geographical precision (locations) and archaeological significance (approximate chronology). Linking multiple datasets has become essential to investigating shifts in settlement density that illustrate how populations engage with and retreat from local ecologies as social relations transform. This approach is particularly applicable to northwestern India, where integrating a wide range of site location reports has the potential to cast the Indus Civilization's signature landscapes, and interrelationships between varied local environments and social complexity, into high relief.

4.2.3 The Indus Civilization in Northwestern India

After a protracted period of village-based occupation, the first cities in South Asia appeared during the Mature Harappan period of the Indus Civilization (ca. 2600–1900 B.C.), which were the largest of thousands of settlements across areas that today lie in western India and Pakistan. Five Indus sites are typically considered cities, and their locations in contrasting environments support the interpretation that they were to some degree politically discrete. At the same time, the aspects of Indus material culture that were shared across such a vast and varied extent suggest that the Indus Civilization's political organization resulted in signature landscapes that were distinct from those materialized by other early complex societies. Excavations at Indus sites have produced evidence of a broad range of sophisticated technologies, including copper metallurgy, standardized weights and measures, and engraved stamp seals. Indus settlements also present examples of civic coordination and planning, though they lack direct evidence for the extreme forms of social differentiation and political hierarchy reported in other complex societies.

Landscape approaches and archaeological surveys have been essential to challenging past narratives that suggest that the Indus Civilization was socio-culturally uniform and homogeneous. Initial surveys highlighted its great extent, and subsequent studies identified local variation in material culture. The increase in fieldwork in India between 1960 and 1980, predominantly recorded in Indian Archaeology: A Review, has been used by multiple researchers to generate site location lists. One such study by Joshi and colleagues suggested that the distribution of site locations revealed "economic pockets" during the Mature Harappan period, which were apparent concentrations of settlements that were closely knit and perhaps economically self-sufficient. As features of the Urban Phase, economic pockets were thought to support one or more large settlements.

Smaller settlements, which have many of the same characteristics as the cities themselves, comprise the majority of Indus sites. Surveys of the settlement distribution along the Beas River in Pakistan's Punjab revealed that the economic diversification and intensification apparent in assemblages from the city of Harappa is also apparent in the material assemblages of nearby smaller settlements. Other studies have used survey data to clarify site distribution patterns in other Indus regions, including Sindh in Pakistan, and Gujarat in India.

The plains of northwestern India are characterized by a range of alluvial environments, an absence of mineral resources, extensive irrigation farming, and numerous archaeological sites from all periods. Some site locations were initially reported as early as 1832, and relatively informal excavations at Indus sites in this region began in the early twentieth century. Field methods and recording improved with the reinvigoration of the Archaeological Survey of India under Sir John Marshall, but remained rudimentary by modern standards. Parts of what is now northwestern India were later explored by Stein (1942) and Ghosh (1952), who assumed that settlement densities in the region resulted from proximity to now-dry watercourses. Further surveys through the 1970s and 1980s brought to light many important Indus sites, including Mitathal and Rakhigarhi, and there were several attempts to collate these data.

Unfortunately, the majority of these studies predate the use of Global Positioning Systems (GPS), so there is a degree of imprecision in the reported site location coordinates.

During the same period, excavations were also undertaken at the sites of Kalibangan, Banawali, and Mitathal. These excavations were essential to developing ceramic typologies for northwestern India, which typically include pottery vessel types and styles like those found at the cities of Harappa and Mohenjo-daro along with other types and styles with local characteristics. Subsequently, excavations were carried out at Rakhigarhi, which appears to have been urban in scale and complexity, and the smaller sites of Bhirrana and Kunal. More recent excavations at Farmana have unearthed large mud-brick houses, a coordinated street plan, and an extensive cemetery, highlighting additional associations between elements of material culture found at other major Indus cities and local artifact styles. Material culture assemblages from these sites are believed to correspond to the periods nested within the overarching chronology of the Indus Civilization, such as those employed by Meadow and Kenoyer (1997, 2003), Possehl (2002), and Wright (2010, 2012). These periods include the Early Harappan, Mature Harappan, and Late Harappan periods. Following the Indus Civilization comes a sequence of phases marked by distinctive pottery types, such as Painted Gray Ware. This framework is widely utilized in South Asian archaeology, though the attribution of many types and styles to specific periods is not straightforward.

Since 2000 there have been many surveys conducted in several states in northwestern India, including Haryana, Rajasthan, and Punjab. Most archaeological surveys in northwestern India have employed a "village-to-village" methodology, wherein a survey team visits the contemporary villages within an administrative unit and asks local informants where archaeological materials can be found. The number of villages and intensity of agricultural land use therefore impact the results of these surveys. Many site locations are only readily accessible through secondary studies, which combine the primary results of published and unpublished survey projects, and which reinforce the notion that the region was home to several dynamic settlement concentrations, though they differ on specific interpretations. For example, Kumar argued that settlement density in northwestern India increased markedly during the Late Harappan period, while Chakrabarti and Saini suggested that the change in population between the Mature and Late Harappan periods was less dramatic, indicating that that migration from the declining cities may be unlikely.

It has been clear for some time that a high-resolution evaluation of these site location data will improve scholarly understanding of the processes of urbanization and de-urbanization that created and transformed the Indus Civilization's signature landscapes. The "Land, Water and Settlement" (hereafter LWS) project produced two complementary site location datasets that can anchor data assembly projects: the Rakhigarhi Hinterland Survey and the Ghaggar Hinterland Survey. LWS focused on rural life in northwestern India, and expanded and refined a subset of site location datasets from this region. The LWS surveys demonstrated that during the Mature Harappan period there was an overall reduction in settlement density that sustained the emergence of larger urban settlements like Rakhigarhi. During the Late Harappan period, the number of sites in northwestern India appears to increase, but these settlements are typically small in size. This transformation is likely associated with climate change, and it has been suggested that a weakening summer monsoon prompted communities in northwestern India to diversify their agricultural practices. However, it is clear that this diversity emerged well before cities and may have provided the risk buffering and mitigation necessary to maintain food surpluses in the face of climate change.

New landscape approaches to the Indus Civilization have the potential to reveal how social complexity integrates vast and varied environments in the face of dramatic changes in social scale. However, the environmental and socio-cultural diversity and variation across the extensive region occupied by Indus populations inhibit the understanding Indus landscapes if site location reports remain confined to the spatial silos of individual studies. Assembling Indus site location reports into larger integrated databases creates an opportunity to critically assess settlement densities and identify research strategies that will increase certainty by revealing areas where data need to be reviewed and re-examined and locations that will benefit from additional survey.

More research on the diverse range of social processes that unfolded in early complex societies is needed. It is particularly critical to determine when transformations in past landscapes reinforce current models of social complexity, and when they demand the revision of traditional models, and the Indus Civilization is particularly important in the regard. Investigating the Indus Civilization's signature landscapes may reveal how particular environments, and variation within them at smaller scale, interact with heterarchical social processes, such as those outlined by Crumley (1995) and McIntosh (2005). Moreover, most classic studies of site location data tend to emphasize the relationship between an early complex society and a particular environment. The Indus offers a fundamentally different challenge: an example of an extensive early complex society that encompassed a great range of different environments.

4.3 THE CHARACTERISTIC OF URBANIZATION DURING THE MEDIEVAL PERIOD IN INDIA

Characteristic of urbanization during the medieval period in India were:

- In the early medieval period, the rate of urbanisation was slow due to unstable political conditions. Only fortress towns under the patronage of chieftains and petty rulers could grow.
- Those towns also developed which were along the main routes of travel, and by the river-side had trade in food grains, cloth, swords, carpets, perfumes and several other handicraft articles.

- It was only under the rule of Akbar that urban life was reconstituted and redeveloped. All centres – 'dasturs' (districts) as well as 'parganas' (tehsils) beside capitals in nature were also 'garrison towns' where armies were invariably stationed for protection.
- Urban centres in the medieval period were surrounded by agricultural land. Farmers and labourers usually resided near or outside the town limit. The areas within the walls of a town were inhabited by artisans of different castes engaged in handicrafts and other works.

4.3.1 Modern India

The contribution of the agricultural sector to the GDP of India started to decline and the percentage contribution from secondary sector increased. The period after 1941, witnessed rapid growth of four metropolitan cities in India, which were Kolkata, Delhi, Mumbai, and Chennai.

The nation's economy saw a rise due to industrial revolution and the invention of new technologies increased the standard of living of people living in urban areas. The growth of public sector resulted in development of public transport, roads, water supply, electricity, and hence the infrastructure of urban areas. Some of the cities like Three tier also increasing population.

Maharashtra was the most urbanized major state in India till 1991, stood behind Tamil Nadu in 2001 and third after it in 2011, with Kerala being second, with the urban-total state population ratio. However, Maharashtra's urban population of 41 million, far exceeds that of Tamil Nadu which is at 27 million, as per the 2001 census. The spatial distribution of large cities in India is uneven as out of 100 most populous cities in the country more than 50 are confined to only 5 states namely, Uttar Pradesh, Maharashtra, Tamil Nadu, Kerala and Andhra Pradesh. Other than state capitals and major industrial centre, large cities are mainly concentrated in the National Capital Region (NCR), the western and southern part of India.

4.3.2 Causes of Urbanization in India

The main causes of urbanisation in India are:

- Expansion in government services, as a result of the Second World War
- Migration of people during the partition of India
- The Industrial Revolution
- Eleventh five-year plan that aimed at urbanisation for the economic development of India
- Economic opportunities are just one reason people move into cities
- Infrastructure facilities in the urban areas

- Growth of private sector after 1990.
- Growth of employment in cities is attracting people from rural areas as well as smaller cities to large towns. According to Mckinsey India's urban population will grow from 340 million in 2008 to 590 million in 2030.
- Therefore, it is being driven by economic compulsions where people move out for economic advancements to areas offering better job opportunities.
- It is also driven by land fragmentations, villages being erased due to roads and highway constructions, dam constructions and other activities.
- Agriculture is the primary source of livelihood, but it's no more profitable: Indian rural economy is primarily based on agriculture. Indian agriculture sector accounts for 18 percent of India's Gross Domestic Product (GDP) and it is estimated that it provides employment to 50% of the countries workforce, but ground reality differs. Many farmers in different states of India are leaving farming, primarily because of high input cost and low income from agriculture. Also on the other hand with usage of fertilizers, chemicals and hybrid seeds, land fertility is declining. This encourages many farmers to commit suicide. In 2014, the National Crime Records Bureau of India reported 5,650 farmer suicides. As per the figures given by central government in 2015, there were 12602 farmer suicides. This includes 8,007 farmers-cultivators and 4,595 agricultural labourers. The farmer's suicide rate in India has ranged between 1.4 and 1.8 per 100,000 total populations, over a 10-year period through 2005. Thus people (including farmers) are migrating to cities.

Giving children a better future by educating them in a great educational institute is now an alarming reason of urbanization.

4.3.3 Consequences of Urbanization in India

Rapid rise in urban population, in India, is leading to many problems like increasing slums, decrease in standard of living in urban areas, also causing environmental damage.

The Industrial Revolution of the 18th century caused countries like the United States and the United Kingdom to become superpowers, but conditions elsewhere are worsening. India's urban growth rate is 2.07%; seemingly insignificant compared to Rwanda, with 7.6%. India has around 300 million people living in metropolitan areas. This has greatly increased housing issues: with overcrowded cities, many people are forced to live in unsafe conditions, such as illegal buildings. Water lines, roads and electricity are lacking in quality, resulting in a decline in living standards. It is also contributing to the issues presented by pollution.

Urbanization also results in a disparity in the market, owing to the large demands of the growing population and the primary sector struggling to cope.

It could be argued that urbanization impacts the migrants themselves on multiple levels. Networks of friends and family become support systems during the initial transformation phase and the struggle to find work in a fast-paced environment. Their struggles may take months, or even years, to adjust to the new surrounds in order to find a stable job. Migrants are responsible for supporting both themselves in the city and the family left at home.

Some of the positive effects resulting from rural to urban migration occur in the agrarian communities from which migrants came. Family members left at home, usually the elderly and young, are eased out of financial pressures as their relatives work to provide higher standards of living for their dependants. Their quality of life is often additionally improved by the provisions that the migrant sends back.

On the other hand, rural to urban migration poses a big challenge for developing cities due to migrant populations flocking in. How will cities support it in terms of resources, land and space?

Cities offer solutions in terms of high rise buildings (affordable housing), metros (affordable transport), affordable schooling, established local clinics, water ATMs and many new initiatives. However, the problems include:

1. National Institution for Transforming India [NITI Ayog], has released report 'Composite Water Management Index' in June 2018 and stated that 21 cities (including Delhi, Mumbai, and Chennai) in India would run out of groundwater by 2020.
2. The latest database of Numbeo lists three Indian cities among the top 10 cities of the world for having worst traffic conditions. These cities include Mumbai, Pune and Kolkata.
3. Population growth and rapid urbanization are combining to create huge challenges for Indian cities. According to McKinsey, the country's cities are expected to grow from 340 million people in 2008 to a whopping 590 million in 2030 and this growth will be very rapid. Meeting demand for urban services in these cities will require US$1.1 trillion in capital investment over the next 20 years. Without the right design and planning, this massive urban growth could exacerbate existing problems of congestion, pollution, and traffic safety.
4. Delhi is now considered the most polluted city in the world, according to the Brookings Institution, while at least two-and-a-half million premature deaths are blamed on poor air quality across the country as a whole.
5. The Economic Survey Report of India 2017-18 estimated that percentage of agricultural workers of total work force would drop to 25.7 percent by 2050 from 58.2 percent in 2001. What is the cause of this decline? People migrating to cities and adding to the unemployment in cities. It is interesting to know that because

of migration from rural pockets to urban (cities), unemployment in cities is increasing and in rural pockets it's decreasing. As per the Centre for Monitoring Indian Economy (CMIE) overall unemployment rate of India as on 18 October 2018 is 6.9%, whereas for urban India its 7.5% and for rural India its 6.6%.

4.4 URBAN CHALLENGES

The built environment: the supply of both land and infrastructure is falling behind demand.

- Out of date and static master plans immobilize the supply of serviced land and buildable space, blocking efficient and productive development. While important laws have been revoked to remove land restrictions and rent controls, many of the legal instruments to operationalize those reforms are still lacking;
- Inappropriate FSI, zoning and development control regulations inhibit development and trunk infrastructure;
- Fractured planning structures are incapable of integrated planning for land use, infrastructure planning, and finance across metropolitan areas;
- The legal, regulatory and institutional basis for land management is generally lacking, hindering the capacity of the private sector to be competitive and the ability of the public sector to use land based financial instruments to finance infrastructure investments.
- Urban services, including water & wastewater, and solid waste, do not reach many residents, and those they do serve receive sporadic, unreliable services. And, in the case of non-notified slums, service providers are prohibited from serving residents.
- Public transport services provide inadequate services, and non-motorized transit for pedestrians, bicycles, and handcarts is limited, forcing this traffic to compete with cars, trucks and motorcycles for space on the streets.
- The proliferation of slums is largely the result of failures in land markets and regulations, compounded by limited access to housing finance.
- Rigid master plans and restrictive zoning regulations limit the land available for building, constricting cities' abilities to grow in accordance with changing needs.

Weak and unpredictable financing frameworks limit the ability of ULBs to manage their resources effectively.

- Overlapping institutions across three tiers of government diffuse accountability across agencies, parastatal bodies and elected governments. This situation undermines India's robust democracy, clouds issues of responsibility, and blocks the development of coherent regulatory frameworks and sustainable service delivery models.

- Dominance of state governments: India's states are often on the scale of countries. These state governments cannot effectively provide service or good governance at the local level. Even India's mega-cities do not have control over their own policies, planning, finances, assets, or institutions.
- ULBs and local service entities have neither clear responsibility nor the fiscal and operational autonomy to deliver adequate urban services;
- Weakness of local governments: ULBs lack capacity, i.e. systems and trained human resources in areas such as financial and organizational management; land use and infrastructure planning; asset management; and project identification, design and management.
- This leaves them trapped in a low level equilibrium, dependent on the State and Centre. Citizens have limited opportunities to hold City managers (who are appointed by the states and are not elected) and service providers accountable for the quantity or quality of services provided..

4.5 GOVERNMENT PRIORITIES AND PROGRAMS

Government's urban development strategy. There two urban related ministries at the national (GoI) level- the Ministry of Urban Development (MoUD) and Ministry of Housing and Urban Poverty Alleviation (MoHUPA). The Government of India's overarching urbandevelopment objectives is to create economically productive, efficient, inclusive and responsive ULBs, by focusing on strategic outcomes: (i) universal access to a minimum level of services; (ii) establishment of city wide frameworks for planning and governance; (iii) modern and transparent budgeting, accounting and FM; (iv) financial sustainability for ULBs and service delivery institutions; (v) utilization of e-governance; (vi) transparency and accountability in urban service delivery and management; (vii) Slum-free cities.

In pursuance of these goals, the Government of India (GoI) launched a flagship urban development program called the Jawaharlal Nehru National Urban Renewal Mission (JnNURM), in December 2005. The Mission targets 65 ULBs (7 with populations greater than 4 million, 28 greater than 1 million and 30 other ULBs of religious, historic or tourist importance). JnNURM is reform and incentive based - in return for a commitment to adopt the obligatory reforms over a period of seven years, ULBs may access funds for investment and capacity building. The investment component of the Mission consists of two sub-missions: (i) Urban Infrastructure and Governance (UIG), implemented by the Ministry of Urban Development (MoUD), with investments including (a) water, sanitation, sewerage and drainage; (b) solid waste management (SWM); (c) urban transport; (d) street lighting; and (e) environmental protection; and (ii) Basic Services to the Urban Poor (BSUP), implemented by the Ministry of Housing and Urban Poverty Alleviation (MoHUPA), with investments supporting integrated development of slums. More recently GoI launched the Slum-free City program of Rajiv Awas Yojana (RAY) a scheme working towards the goal of a slum-free India.

CHAPTER 5
URBAN PLANNING AND DESIGN

5.1 INTRODUCTION

Urban planners direct the development of cities and towns. A well-developed, effective urban plan requires thorouresearch and input from numerous stakeholders, including citizens, landowners, and government staff. Urban planners should think about their plans from the perspectives of all who will be impacted by their efforts. And since implementing the plan uses taxpayer dollars, the plan should be as practical and cost-effective as possible.

From a local government perspective, urban planning touches on numerous city-life elements—new and pre-existing land, buildings, roads, communal spaces, transportation, economic development, infrastructure, and the environment, among others. We've broadly categorized these aspects into different conceptual areas. These areas are commonly referred to as types of urban planning, but it's important to understand that they are not mutually exclusive—for an urban plan to be cohesive, it should include many or all of the areas.

Urban planning, design and regulation of the uses of space that focus on the physical form, economic functions, and social impacts of the urban environment and on the location of different activities within it. Because urban planning draws upon engineering, architectural, and social and political concerns, it is variously a technical profession, an endeavour involving political will and public participation, and an academic discipline. Urban planning concerns itself with both the development of open land ("greenfields sites") and the revitalization of existing parts of the city, thereby involving goal setting, data collection and analysis, forecasting, design, strategic thinking, and public consultation. Increasingly, the technology of Geographic Information Systems (GIS) has been used to map the existing urban system and to project the consequences of changes. In the late 20th century the term sustainable development came to represent an ideal outcome in the sum of all planning goals. As advocated by the United Nations-sponsored World Commission on Environment and Development in Our Common Future (1987), sustainability refers to "development that meets the needs of the present without compromising the ability of future generations to meet their own needs." While there is widespread consensus on this general goal, most major planning decisions involve trade-offs between subsidiary objectives and thus frequently involve conflict.

The modern origins of urban planning lie in a social movement for urban reform that arose in the latter part of the 19th century as a reaction against the disorder of the industrial city. Many visionaries of the period sought an ideal city, yet practical considerations of adequate sanitation, movement of goods and people, and provision of amenities also drove the desire for planning. Contemporary planners seek to balance the conflicting demands of social equity, economic growth, environmental sensitivity, and aesthetic appeal. The result of the planning process may be a formal master plan for an entire city or metropolitan area, a neighbourhood plan, a project plan, or a set of policy alternatives. Successful implementation of a plan usually requires entrepreneurship and political astuteness on the part of planners and their sponsors, despite efforts to insulate planning from politics. While based in government, planning increasingly involves private-sector participation in "public-private partnerships."

Urban planning emerged as a scholarly discipline in the 1900s. In Great Britain the first academic planning program began at the University of Liverpool in 1909, and the first North American program was established at Harvard University in 1924. It is primarily taught at the postgraduate level, and its curriculum varies widely from one university to another. Some programs maintain the traditional emphasis on physical design and land use; others, especially those that grant doctoral degrees, are oriented toward the social sciences. The discipline's theoretical core, being somewhat amorphous, is better defined by the issues it addresses than by any dominant paradigm or prescriptive approach. Representative issues especially concern the recognition of a public interest and how it should be determined, the physical and social character of the ideal city, the possibility of achieving change in accordance with consciously determined goals, the extent to which consensus on goals is attainable through communication, the role of citizens versus public officials and private investors in shaping the city, and, on a methodological level, the appropriateness of quantitative analysis and the "rational model" of decision making. Most degree programs in urban planning consist principally of applied courses on topics ranging from environmental policy to transportation planning to housing and community economic development.

5.2 HISTORY OF URBAN PLANNING

Evidence of planning has been unearthed in the ruins of cities in China, India, Egypt, Asia Minor, the Mediterranean world, and South and Central America. Early examples of efforts toward planned urban development include orderly street systems that are rectilinear and sometimes radial; division of a city into specialized functional quarters; development of commanding central sites for palaces, temples, and civic buildings; and advanced systems of fortification, water supply, and drainage. Most of the evidence is in smaller cities that were built in comparatively short periods as colonies. Often the central cities of ancient states grew to substantial size before they achieved governments capable of imposing controls.

For several centuries during the Middle Ages, there was little building of cities in Europe. Eventually towns grew up as centres of church or feudal authority, of marketing or trade. As the urban population grew, the constriction caused by walls and fortifications led to overcrowding, the blocking out of air and light, and very poor sanitation. Certain quarters of the cities, either by custom or fiat, were restricted to different nationalities, classes, or trades, as still occurs in many contemporary cities of the developing world.

In the Neolithic period, agriculture and other techniques facilitated larger populations than the very small communities of the Paleolithic, which probably led to the stronger, more coercive governments emerging at that time. The pre-Classical and Classical periods saw a number of cities laid out according to fixed plans, though many tended to develop organically. Designed cities were characteristic of the Mesopotamian,Harrapan, and Egyptian civilizations of the third millennium BCE (see Urban planning in ancient Egypt).

Distinct characteristics of urban planning from remains of the cities of Harappa, Lothal, andMohenjo-daro in the Indus Valley Civilization(in modern-day northwestern India and Pakistan) lead archeologists to conclude that they are the earliest examples of deliberately planned and managed cities. The streets of many of these early cities were paved and laid out at right angles in a grid pattern, with a hierarchy of streets from major boulevards to residential alleys. Archaeological evidence suggests that many Harrapan houses were laid out to protect from noise and enhance residential privacy; many also had their own water wells, probably for both sanitary and ritual purposes. These ancient cities were unique in that they often had drainage systems, seemingly tied to a well-developed ideal of urban sanitation.

The Greek Hippodamus (c. 407 BC) has been dubbed the "Father of City Planning" for his design of Miletus; Alexander commissioned him to lay out his new city of Alexandria, the grandest example of idealized urban planning of the ancient Mediterranean world, where the city's regularity was facilitated by its level site near a mouth of the Nile. The Hippodamian, or grid plan, was the basis for subsequent Greek and Roman cities.

The ancient Romans used a consolidated scheme for city planning, developed for military defense and civil convenience. The basic plan consisted of a central forum with city services, surrounded by a compact, rectilinear grid of streets, and wrapped in a wall for defense. To reduce travel times, two diagonal streets crossed the square grid, passing through the central square. A river usually flowed through the city, providing water, transport, and sewage disposal. Many European towns, such asTurin, preserve the remains of these schemes, which show the very logical way the Romans designed their cities. They would lay out the streets at right angles, in the form of a square grid. All roads were equal in width and length, except for two, which were slightly wider than the others. One of these ran east-west, the other, north-south, and intersected in the middle to form the center of the grid. All roads were made of carefully fitted flag stones and filled in with smaller, hard-packed rocks and pebbles. Bridges were constructed where needed. Each square marked by four roads was called aninsula, the Roman equivalent of a modern city block.

Each insula was 80 yards (73 m) square, with the land within it divided. As the city developed, each insula would eventually be filled with buildings of various shapes and sizes and crisscrossed with back roads and alleys. Most insulae were given to the first settlers of a Roman city, but each person had to pay to construct his own house.

The city was surrounded by a wall to protect it from invaders and to mark the city limits. Areas outside city limits were left open as farmland. At the end of each main road was a large gateway with watchtowers. A portculliscovered the opening when the city was under siege, and additional watchtowers were constructed along the city walls. An aqueduct was built outside the city walls.

The collapse of Roman civilization saw the end of Roman urban planning, among other arts. Urban development in the Middle Ages, characteristically focused on a fortress, a fortified abbey, or a (sometimes abandoned) Roman nucleus, occurred "like the annular rings of a tree",whether in an extended village or the center of a larger city. Since the new center was often on high, defensible ground, the city plan took on an organic character, following the irregularities of elevation contours like the shapes that result from agricultural terracing.

The ideal of wide streets and orderly cities was not lost, however. A few medieval cities were admired for their wide thoroughfares and orderly arrangements, but the juridical chaos of medieval cities (where the administration of streets was sometimes passed down through noble families), and the characteristic tenacity of medieval Europeans in legal matters prevented frequent or large-scale urban planning until the Renaissance and the early-modern strengthening of central government administration, as European (and soon after, North American) society transited from city-states to what we would recognize as a more modern concept of a nation-state.

Florence was an early model of the new urban planning, which took on a star-shaped layout adapted from the new star fort, designed to resist cannon fire. This model was widely imitated, reflecting the enormous cultural power of Florence in this age; "the Renaissance was hypnotized by one city type which for a century and a half- from Filarete to Scamozzi- was impressed upon utopian schemes: this is the star-shaped city". Radial streets extend outward from a defined center of military, communal or spiritual power.

Only in ideal cities did a centrally planned structure stand at the heart, as in Raphael's Sposalizio Illustration of 1504. As built, the unique example of a rationally planned quattrocento new city center, that of Vigevano (1493-95), resembles a closed space instead, surrounded by arcading.

Filarete's ideal city, building on Leone Battista Alberti's De re aedificatoria, was named "Sforzinda" in compliment to his patron; its twelve-pointed shape, circumscribable by a "perfect" Pythagorean figure, the circle, took no heed of its undulating terrain in Filarete's manuscript. This process occurred in cities, but ordinarily not in the industrial suburbs characteristic of this era, which remained disorderly and characterized by crowding and organic growth.

Following the 1695 bombardment of Brussels by the French troops of King Louis XIV, in which a large part of the city center was destroyed, Governor Max Emanuel proposed using the reconstruction to completely change the layout and architectural style of the city. His plan was to transform the medieval city into a city of the new baroque style, modeled on Turin, with a logical street layout, with straight avenues offering long, uninterrupted views flanked by buildings of a uniform size. This plan was opposed by residents and municipal authorities, who wanted a rapid reconstruction, did not have the resources for grandiose proposals, and resented what they considered the imposition of a new, foreign, architectural style. In the actual reconstruction, the general layout of the city was conserved, but it was not identical to that before the cataclysm. Despite the necessity of rapid reconstruction and the lack of financial means, authorities did take several measures to improve traffic flow, sanitation, and the aesthetics of the city. Many streets were made as wide as possible to improve traffic flow.

Many Central American civilizations also planned their cities, including sewage systems and running water. InMexico, Tenochtitlan was the capital of the Aztec empire, built on an island in Lake Texcoco in what is now the Federal District in central Mexico. At its height, Tenochtitlan was one of the largest cities in the world, with over 200,000 inhabitants.

Modern urban planning dates from the 1850s and the contrasting projects to update Paris and extend Barcelona. In 1852, Baron Georges-Eugène Haussmann was commissioned to remodel the Medieval street plan of Paris by demolishing swathes of the old city and laying out wide boulevards, extending outwards beyond the old city limits. Haussmann's project encompassed all aspects of urban planning, both in the centre of Paris and in the surrounding districts, with regulations imposed on building facades, public parks, sewers and water works, city facilities, and public monuments. Beyond aesthetic and sanitary considerations, the wide thoroughfares facilitated troop movement and policing.

The plan chosen to extend Barcelona was a rigorous project based on a scientific analysis of the city and its modern requirements. It was drawn up by the Catalan engineer Ildefons Cerdà to fill the space beyond the city walls after they were demolished from 1854. He is credited with inventing the term 'urbanization' and his approach was codified in his General Theory of Urbanization (1867). Cerdà's Example (Catalan for 'extension') consisted of 550 regular blocks with chamfered corners to facilitate the movement of trams, crossed by three wider avenues. His objectives were to improve the health of the inhabitants, towards which the blocks were built around central gardens and orientated NW-SE to maximize the sunlight they received, and assist social integration.

In the developed countries of Western Europe, North America, Japan, and Australasia, planning and architecture can be said to have gone through various paradigms or stages of consensus in the last 200 years. Firstly, there was the industrialised city of the 19th century, where building was largely controlled by businesses and wealthy elites. Around 1900, a movement began for providing citizens, especially factory workers, with healthier environments. The concept of the garden city arose and several model towns were built, such as Letchworth andWelwyn Garden City in Hertfordshire, UK, the world's first garden cities. These were small in size, typically providing for a few thousand residents.

In the 1920s, the ideas of modernism began to surface in urban planning. Based on the ideas of Le Corbusier and using new skyscraper-building techniques, the modernist city stood for the elimination of disorder, congestion, and the small scale, replacing them with preplanned and widely spaced freeways and tower blocks set within gardens. There were plans for large-scale rebuilding of cities in this era, such as the Plan Voisin (based on Le Corbusier's Ville Contemporaine), which proposed clearing and rebuilding most of central Paris. No large-scale plans were implemented until after World War II, however. Throughout the late 1940s and 1950s, housing shortages caused by wartime destruction led many cities to subsidize housing blocks. Planners used the opportunity to implement the modernist ideal of towers surrounded by gardens. The most prominent example of an entire modernist city isBrasilia in Brazil, constructed between 1956 and 1960.

5.2.1 Reaction

By the late 1960s and early 1970s, many planners felt that modernism's clean lines and lack of human scale sapped vitality from the community, blaming them for high crime rates and social problems.

Modernist planning fell into decline in the 1970s when the construction of cheap, uniform tower blocks ended in most countries, such as Britain and France. Since then many have been demolished and replaced by other housing types. Rather than attempting to eliminate all disorder, planning now concentrates on individualism and diversity in society and the economy; this is the post-modernist era.

Minimally planned cities still exist. Houston is a large city (with a metropolitan population of 5.5 million) in a developed country without a comprehensive zoning ordinance. Houston does, however, restrict development densities and mandate parking, even though specific land uses are not regulated. Also, private-sector developers in Houston use subdivision covenants and deed restrictions to effect land-use restrictions resembling zoning laws. Houston voters have rejected comprehensive zoning ordinances three times since 1948. Even without traditional zoning, metropolitan Houston displays large-scale land-use patterns resembling zoned regions comparable in age and population, such as Dallas. This suggests that non-regulatory factors such as urban infrastructure and financing may be as important as zoning laws in shaping urban form.

5.3 URBAN PLANNING EXAMPLES

5.3.1 Raleigh, North Carolina Aims to Improve Quality of Life

Raleigh's city planning department has a lofty goal: Create a city where the quality of life is second to none. This goal ties directly to the city's overall strategic plan, making strategic urban planning a priority.

To achieve this goal and prepare for likely changes across the city, Raleigh leadership is planning a number of different tactics, including conducting studies to evaluate the impact of population growth. These studies inform future strategic planning, helping the city to prioritize initiatives and allocate resources appropriately.

Many of the key focus areas of the city's strategic plan rely on urban planning, but there are three that stand out over the others:

- Growth & Natural Resources
- Safe, Vibrant & Healthy Community
- Transportation & Transit

The city planning department is a key player in the success of these focus areas and the initiatives that support them. Having a clear urban plan has allowed the city to make improvements and be able to spread the message that Raleigh is a "great place to live, work, and play."

5.3.2 New York Elevates Nature and Pedestrians

New York's High Line experiment transformed a 1930s elevated railroad into gardens that stretch a mile and a half. Instead of tearing down the tracks, city leadership gave it new life through careful urban planning efforts. This experiment was so successful that it's now a top visitor attraction—with popularity that's been said to dwarf the famed Statue of Liberty. Its success has also inspired other cities to recreate a similar experience for their own citizens.

New York has included green space in all of its planning going forward, and there are many other examples of great outdoor spaces that are giving the city a feeling of openness in one of the most densely populated areas in the country. Waterfront parks, the Downtown Boathouse, and Fort Tryon Park are just a few of the many spaces available to enjoy the outdoors.

5.3.3 Eugene, Oregon Goes Green

Urban planning tends to be environmentally focused in Eugene, which was once named the fifth greenest city in America and the best city for urban farming. What makes the city so green? For one, it boasts an innovative public power grid that draws 85% of its energy from renewables. In addition, the city is very accessible by bike and hybrid transport. Eugene also has an ambitious goal it's been working toward for the past decade: being carbon neutral by the end of 2020.

This isn't Eugene just being trendy. It was known as the Emerald City since before Earth Day was created, and part of its brand is to be green all year round. Thus, to live and execute on that strategy, it has put in place some ambitious goals and modeled its planning process around these goals.

5.4 SUSTAINABLE DEVELOPMENT AND SUSTAINABILITY

Sustainable development and sustainability influence today's urban planners. Some planners argue that modern lifestyles use too many natural resources, polluting or destroying ecosystems, increasing social inequality, creating urban heat islands, and causing climate change. Many urban planners, therefore, advocate sustainable cities.

However, sustainable development is a recent, controversial concept. Wheeler, in his 1998 article, defines sustainable urban development as "development that improves the long-term social and ecological health of cities and towns." He sketches a 'sustainable' city's features: compact, efficient land use; less automobile use, yet better access; efficient resource use; less pollution and waste; the restoration of natural systems; good housing and living environments; a healthy social ecology; a sustainable economy; community participation and involvement; and preservation of local culture and wisdom.

Because of political and governance structures in most jurisdictions, sustainable planning measures must be widely supported before they can affect institutions and regions. Actual implementation is often a complex compromise.

5.4.1 Collaborative Strategic Goal Oriented Programming

It is a collaborative and communicative way of strategic programming, decision making, implementation, and monitoring oriented towards defined and specific goals. It is based on sound analysis of available information, emphasizes stakeholder participation, works to create awareness among actors, and is oriented towards managing development processes. It was adopted as a theoretical framework for analyzing redevelopment processes in large urban distressed areas in European cities.

Background of CoSGOP

CoSGOP is derived from goal-oriented planning, which was oriented towards the elaboration and implementation of projects based on a logical framework, which was useful for embedding a specific project in a wider development frame and defining its major elements. This approach had weaknesses: its logical rules were strictly applied and the expert language did not encourage participation. CoSGOP introduced a new approach characterized by communication with and active involvement of stakeholders and those to be affected by the program; strategic planning based on the identification of strengths and weakness, opportunities and threats, as well as on scenario-building and visioning; the definition of goals as the basis for action; and long-term, flexible programming of interventions by stakeholders.

Elements of CoSGOP

CoSGOP is not a planning method but a process model. It provides a framework for communication and joint decision-making, in a structured process characterized by feedback loops. It also facilitates stakeholder learning. The essential elements of CoSGOP are analysis of stakeholders; analysis of problems and potentials; development of goals, improvement priorities, and alternatives; specification of an improvement program and its main activities; assessment of possible impacts of the improvement program; definition and detailed specification of key projects and their implementation; continuous monitoring of improvement activities, feedback, and adjustment of the programme.

Application

CoSGOP has been applied in European cross-border policy programming, as well in local and regional development programming. In 2004, the CoSGOP model was applied in the LUDA Project, starting with an analysis of the European experience of urban regeneration projects.

5.4.2 Collaborative Planning in the United States

Collaborative planning arose in the US in response to the inadequacy of traditional public participation techniques to provide real opportunities for the public to make decisions affecting their communities. Collaborative planning is a method designed to empower stakeholders by elevating them to the level of decision-makers through direct engagement and dialogue between stakeholders and public agencies, to solicit ideas, active involvement, and participation in the community planning process. Active public involvement can help planners achieve better outcomes by making them aware of the public's needs and preferences and by using local knowledge to inform projects. When properly administered, collaboration can result in more meaningful participation and better, more creative outcomes to persistent problems than can traditional participation methods. It enables planners to make decisions that reflect community needs and values, it fosters faith in the wisdom and utility of the resulting project, and the community is given a personal stake in its success.

Experiences in Portland and Seattle have demonstrated that successful collaborative planning depends on a number of interrelated factors: the process must be truly inclusive, with all stakeholders and affected groups invited to the table; the community must have final decision-making authority; full government commitment must be manifest; participants should be given clear objectives by planning staff, who facilitate the process by providing guidance, consultancy, expert opinions, and research; and facilitators should be trained in conflict resolution and community organization.

5.5 TYPES OF URBAN PLANNING

5.5.1 Strategic Urban Planning

Strategic urban planning focuses on setting high-level goals and determining desired areas of growth for a city or metropolitan area. The result of the planning process is a strategic plan—also called the development plan, core strategy, or comprehensive plan. The strategic plan's goals may include easing transportation throughout the city, creating more community spaces, improving citizens' quality of life, or encouraging people to visit or move to the city. This is generally the highest level of the planning process and other components of planning typically will fit into this type of plan.

5.5.2 Land-Use Planning

Land-use planning largely concerns legislation and policy, adopting planning instruments like governmental statutes, regulations, rules, codes, and policies to influence land use.

On a broad level, these planning instruments deal with the type, location, and amount of land needed to carry out different functions of the city. They also serve to zone or reserve land for certain purposes such as:

- Residential, for buildings like apartment homes, single-family residences, and condominiums
- Commercial, for buildings like retail shops and office buildings
- Industrial, for structures like manufacturing plants and warehouses
- Municipal, for structures like police stations and courthouses

As with subsequent types of urban planning, consulting with the community and relevant stakeholders is an important part of land-use planning to ensure transparency, and incorporate a wide range of interests into the overall plan. If you communicate your strategic plan well, then transportation, commercial and industrial planning should flow right into your plans.

5.5.3 Master Planning

Master planning is typically used for greenfield development projects, or building on undeveloped land—instead of modifying pre-existing structures or spaces, you're starting from scratch.

This type of urban planning envisions a future state for a given space, and what it will take to achieve that vision. Urban planners must consider the required zoning and infrastructure to make the project possible, such as residential and commercial land, transportation considerations, road locations, etc. They must also plan the location of urban amenities such as community facilities, schools, parks, and the like.

Again, consulting with landowners and government agencies impacted by the plan is an essential task here. Additionally, you may need to bring in professional consultants to gather important expertise and insights, ensure the plan considers all potential angles, and set the completed space up for success for years to come.

5.5.4 Urban Revitalization

In contrast to master planning, urban revitalization focuses on improving areas that are in a state of decline. The exact definition of a declining area will differ from city to city—for example, areas that have a troubling number of failing businesses or a stagnant or decreasing population growth. The improvement tactics city leaders use for revitalization will depend on the root cause of decline, and may include things like repairing roads, developing infrastructure, cleaning up pollution, and adding to parks and other public spaces, etc.

Community interaction is especially important with this urban planning concept, as local residents and business owners often have insights that can help inform and tailor planning efforts. You may need to change land use from industrial to residential to get the loft apartments you want, or you may need to involve environmental planning to clean up any messes from previous use cases.

5.5.5 Economic Development

Economic development is about identifying areas of growth to foster greater financial prosperity within the city, specifically by enticing companies to build or move offices there. Subsequently, those companies then hire local talent and drive commuter traffic to the new office. More workers dining at local restaurants for lunch, getting gas at nearby gas stations, and stopping by local grocery stores on the way home will boost visibility and spend in the area.

Sometimes an economic development department lives outside of the planning department of a municipality, so it is important to help that group navigate Land Use Plans, Master Plans, and Infrastructure Plans to ensure that any development projects are workable. Of course it will be important to coordinate with environmental plans as well.

5.5.6 Environmental Planning

Environmental planning is a type of strategic development that emphasizes sustainability. Considerations for this type of urban planning include air pollution, noise pollution, wetlands, habitats of endangered species, flood zone susceptibility, and coastal zone erosion, along with a host of other environmental factors dealing with the relationship between natural and human systems.

Environmental plans need to be filed alongside master, revitalization, and infrastructure plans. If it seems like there are a lot of steps and a lot of requirements, don't get discouraged. While it seems complicated, it will be best in the long run if your plans all interact well.

5.5.7 Infrastructure Planning

Infrastructure planning deals with the fundamental facilities and systems that serve a city and its people, and how those facilities can support goals laid out in the strategic plan. This type of urban planning covers:

- Public works infrastructure such as water supply, sewage, electricity, and telecommunications
- Community infrastructure such as schools, hospitals, and parks
- Safety and transportation such as roads, police, and fire facilities

As you can see from the above urban planning concepts, good planning takes a lot of work. But when done correctly, planning at the city, county, and state levels can have a positive, lasting impact on your community.

5.6 ASPECTS OF URBAN PLANNING

5.6.1 Aesthetics

In developed countries, there has been a backlash against excessive human-made clutter in the visual environment, such as signposts, signs, and hoardings. Other issues that generate strong debate among urban designers are tensions between peripheral growth, housing density and new settlements. There are also debates about the mixing tenures and land uses, versus distinguishing geographic zones where different uses dominate. Regardless, all successful urban planning considers urban character, local identity, respects heritage, pedestrians, traffic, utilities and natural hazards.

Planners can help manage the growth of cities, applying tools like zoning and growth management to manage the uses of land. Historically, many of the cities now thought the most beautiful are the result of dense, long lasting systems of prohibitions and guidance about building sizes, uses and features. These allowed substantial freedoms, yet enforce styles, safety, and often materials in practical ways. Many conventional planning techniques are being repackaged using the contemporary term smart growth.

There are some cities that have been planned from conception, and while the results often don't turn out quite as planned, evidence of the initial plan often remains.

5.6.2 Safety and Security

Historically within the Middle East, Europe and the rest of the Old World, settlements were located on higher ground (for defense) and close to fresh water sources. Cities have often grown onto coastal and flood plains at risk of floods and storm surges. Urban planners must consider these threats. If the dangers can be localised then the affected regions can be made into parkland or green belt, often with the added benefit of open space provision.

Extreme weather, flood, or other emergencies can often be greatly mitigated with secure emergency evacuation routes and emergency operations centres. These are relatively inexpensive and unintrusive, and many consider them a reasonable precaution for any urban space. Many cities will also have planned, built safety features, such as levees, retaining walls, and shelters.

In recent years, practitioners have also been expected to maximize the accessibility of an area to people with different abilities, practicing the notion of "inclusive design," to anticipate criminal behaviour and consequently to "design-out crime" and to consider "traffic calming" or "pedestrianisation" as ways of making urban life more pleasant.

Some city planners try to control criminality with structures designed from theories such as socio-architecture or architectural determinism a subset of environmental determinism. These theories say that an urban environment can influence individuals' obedience to social rules and level of power. Refer to Foucault and the Encyclopedia of the Prison System for more details. The theories often say that psychological pressure develops in more densely developed, unadorned areas. This stress causes some crimes and some use of illegal drugs. The antidote is usually more individual space and better, more beautiful design in place of functionalism.

Oscar Newman's defensible space theory cites the modernist housing projects of the 1960s as an example of environmental determinism, where large blocks of flats are surrounded by shared and disassociated public areas, which are hard for residents to identify with. As those on lower incomes cannot hire others to maintain public space such as security guards or grounds keepers, and because no individual feels personally responsible, there was a general deterioration of public space leading to a sense of alienation and social disorder.

Jane Jacobs is another notable environmental determinist and is associated with the "eyes on the street" concept. By improving 'natural surveillance' of shared land and facilities of nearby residents by literally increasing the number of people who can see it, and increasing the familiarity of residents, as a collective, residents can more easily detect undesirable or criminal behavior. However, this is not a new concept. This was prevalent throughout the middle eastern world during the time of Mohamad. It was not only reflected in the general structure of the outside of the home but also the inside.

Jacobs went further, though, in emphasizing the details in how to achieve this 'natural surveillance', in stressing the necessity of multiple uses on city streets, so that different people co-mingle with different stores and parks in a condensed part of city space. By doing this, as well as by making city streets interesting, she theorized a continuous animation of social actions during an average city day, which would keep city streets interesting and well occupied throughout a 24 hour period. She presented the North End in Boston, Massachusetts, as an idealization of this persistent occupation and tasking in a condensed city space, as a model for criminal control.

The "broken-windows" theory argues that small indicators of neglect, such as broken windows and unkempt lawns, promote a feeling that an area is in a state of decay. Anticipating decay, people likewise fail to maintain their own properties. The theory suggests that abandonment causes crime, rather than crime causing abandonment.

Some planning methods might help an elite group to control ordinary citizens. Haussmann's renovation of Paris created a system of wide boulevards which prevented the construction of barricades in the streets and eased the movement of military troops. In Rome, the Fascists in the 1930s created ex novo many new suburbs in order to concentrate criminals and poorer classes away from the elegant town.

Other social theories point out that in Britain and most countries since the 18th century, the transformation of societies from rural agriculture to industry caused a difficult adaptation to urban living. These theories emphasize that many planning policies ignore personal tensions, forcing individuals to live in a condition of perpetual extraneity to their cities. Many people therefore lack the comfort of feeling "at home" when at home. Often these theorists seek a reconsideration of commonly used "standards" that rationalize the outcomes of a free (relatively unregulated) market.

5.6.3 Slums

The rapid urbanization of the last century caused more slums in the major cities of the world, particularly in developing countries. Planning resources and strategies are needed to address the problems of slum development. Many planners are calling for slum improvement, particularly the Commonwealth Association of Planners. When urban planners work on slums, they must cope with racial and cultural differences to ensure that racial steering does not occur.

According to the United Nations, the percentage of urban dwellers living in slums decreased from 47 percent to 37 percent in the developing world between 1990 and 2005. However, due to rising population, and the rise especially in urban populations, the number of slum dwellers is rising. One billion people worldwide live in slums and the figure will likely grow to 2 billion by 2030.

The term slum has traditionally referred to housing areas that were once relatively affluent but which deteriorated as the original dwellers moved on to newer and better parts of the city, but has come to include the vast informal settlements found in cities in the developing world.

Although their characteristics vary between geographic regions, they are usually inhabited by the very poor or socially disadvantaged. Slum buildings vary from simple shacks to permanent and well-maintained structures. Most slums lack clean water, electricity, sanitation and other basic services.

Slums were often "fixed" by clearance. However, more creative solutions are beginning to emerge such as Nairobi's "Camp of Fire" program, where established slum-dwellers promise to build proper houses, schools, and community centers without government money, in return for land on which they have been illegally squatting on for 30 years. The "Camp of Fire" program is one of many similar projects initiated by Slum Dwellers International, which has programs in Africa, Asia, and South America.

The characteristics and politics associated with slums vary from place to place. Slums are usually characterized byurban decay, high rates of poverty,illiteracy, andunemployment. They are commonly seen as "breeding grounds" for social problems such as crime, drug addiction,alcoholism, high rates of mental illness, and suicide. In many poor countries they exhibit high rates of disease due to unsanitary conditions, malnutrition, and lack of basic health care. However, some like Dharavi, Mumbai, are a hive of business activity such as leather work, cottage industries, etc. Rural depopulation with thousands arriving daily into the cities makes slum clearance an uphill struggle.

5.6.4 Decay

Urban decay is a process by which a city, or a part of a city, falls into a state of disrepair and neglect. It is characterized by depopulation, economic restructuring, property abandonment, high unemployment, fragmented families, political disenfranchisement, crime, and desolate urban landscapes.

During the 1970s and 1980s, urban decay was often associated with central areas of cities in North America and Europe. During this time, changes in global economies, demographics, transportation, and policies fostered urban decay. Many planners spoke of "white flight" during this time. This pattern was different than the pattern of "outlying slums" and "suburban ghettos" found in many cities outside of North America and Western Europe, where central urban areas actually had higher real estate values.

Starting in the 1990s, many of the central urban areas in North America have been experiencing a reversal of the urban decay, with rising real estate values, smarter development, demolition of obsolete social housing and a wider variety of housing choices.

5.6.5 Reconstruction and Renewal

Areas devastated by war or invasion challenge urban planners. Resources are scarce. The existing population has needs. Buildings, roads, services and basic infrastructure like power, water and sewerage are often damaged, but with salvageable parts. Historic, religious or social centers also need to be preserved and re-integrated into the new city plan. A prime example of this is the capital city of Kabul, Afghanistan, which, after decades of civil war and occupation, has regions of rubble and desolation. Despite this, the indigenous population continues to live in the area, constructing makeshift homes and shops out of salvaged materials. Any reconstruction plan, such as Hisham Ashkouri's City of Light Development, needs to be sensitive to the needs of this community and its existing culture and businesses.

Urban Reconstruction Development plans must also work with government agencies as well as private interests to develop workable designs.

Urban renewal is a program of land redevelopment in areas of moderate to high density urban land use. Renewal has had both successes and failures. Its modern incarnation began in the late 19th century in developed nations and experienced an intense phase in the late 1940s - under the rubric of reconstruction. The process has had a major impact on many urban landscapes, and has played an important role in the history and demographics of cities around the world.

Urban renewal may involve relocation of businesses, the demolition of structures, the relocation of people, and the use of eminent domain (government purchase of property for public purpose) as a legal instrument to take private property for city-initiated development projects.

In some cases, renewal may result in urban sprawl and less congestion when areas of cities receive freeways and expressways.

Urban renewal has been seen by proponents as an economic engine and a reform mechanism, and by critics as a mechanism for control. It may enhance existing communities, and in some cases result in the demolition of neighborhoods.

Many cities link the revitalization of the central business district and gentrification of residential neighborhoods to earlier urban renewal programs. Over time, urban renewal evolved into a policy based less on destruction and more on renovation and investment, and today is an integral part of many local governments, often combined with small and big business incentives.

5.6.6 Transport

Transport within urbanized areas presents unique problems. The density of an urban environment increases traffic, which can harm businesses and increase pollution unless properly managed. Parking space for private vehicles requires the construction of large parking garages in high density areas. This space could often be more valuable for other development.

Good planning uses transit oriented development, which attempts to place higher densities of jobs or residents near high-volume transportation. For example, some cities permit commerce and multi-story apartment buildings only within one block of train stations and multilane boulevards, and accept single-family dwellings and parks farther away.

Floor area ratio is often used to measure density. This is the floor area of buildings divided by the land area. Ratios below 1.5 are low density. Ratios above five constitute very high density. Most exurbs are below two, while most city centres are well above five. Walk-up apartments with basement garages can easily achieve a density of three. Skyscrapers easily achieve densities of thirty or more.

Transportation planning is a field involved with the evaluation, assessment, design and siting of transportation facilities.

City authorities may try to encourage higher densities to reduce per-capita infrastructure costs. In the UK, recent years have seen a concerted effort to increase the density of residential development in order to better achieve sustainable development. Increasing development density has the advantage of making mass transport systems, district heating and other community facilities (schools, health centres, etc.) more viable. However critics of this approach dub the densification of development as 'town cramming' and claim that it lowers quality of life and restricts market-led choice.

Problems can often occur at residential densities between about two and five. These densities can cause traffic jams for automobiles, yet are too low to be commercially served by trains or light rail systems. The conventional solution is to use buses, but these and light rail systems may fail where automobiles and excess road network capacity are both available, achieving less than 2% ridership.

The Lewis-Mogridge Position claims that increasing road space is not an effective way of relieving traffic jams as latent or induced demand invariably emerges to restore a socially tolerable level of congestion.

Transportation planning historically has followed the rational planning model of defining goals and objectives, identifying problems, generating alternatives, evaluating alternatives, and developing plans. Other models for planning includerational actor, transit oriented development, satisficing,incremental planning, organizational process, and political bargaining.

However, planners are increasingly expected to adopt a multi-disciplinary approach, especially due to the rising importance of environmentalism.

5.6.7 Economics

Urban economics is broadly the economic study of urban areas; as such, it involves using the tools of economics to analyze urban issues such as crime, education, public transit, housing, and local government finance. More narrowly, it is a branch of microeconomics that studies urban spatial structure and the location of households and firms.

Much urban economic analysis relies on a particular model of urban spatial structure, the monocentric city model pioneered in the 1960s by William Alonso, Richard Muth, and Edwin Mills. While most other forms of neoclassical economics do not account for spatial relationships between individuals and organizations, urban economics focuses on these spatial relationships to understand the economic motivations underlying the formation, functioning, and development of cities.

Urban economics is rooted in the location theories of von Thünen, Alonso, Christaller, and Lösch that began the process of spatial economic analysis. Economics is the study of the allocation of scarce resources, and as all economic phenomena take place within a geographical space, urban economics focuses of the allocation of resources across space in relation to urban areas. Other branches of economics ignore the spatial aspects of decision making but urban economics focuses not only on the location decisions of firms, but also of cities themselves as cities themselves represent centers of economic activity.

Many spatial economic topics can be analyzed within either an urban or regional economics framework as some economic phenomena primarily affect localized urban areas while others are felt over much larger regional areas. Arthur O'Sullivan believes urban economics is divided into six related themes: market forces in the development of cities, land use within cities, urban transportation, urban problems and public policy, housing and public policy, and local government expenditures and taxes.

5.6.8 Suburbanization

In some countries, declining satisfaction with the urban environment is held to blame for continuing migration to smaller towns and rural areas. Successful urban planning supported Regional planning can bring benefits to a much larger hinterland or city region and help to reduce both congestion along transport routes and the wastage of energy implied by excessive commuting.

5.6.9 Environmental Factors

Environmental protection and conservation are of utmost importance to many planning systems across the world. Not only are the specific effects of development to be mitigated, but attempts are made to minimize the overall effect of development on the local and global environment. This is commonly done through the assessment of Sustainable urban infrastructure and microclimate. In Europe this process is known as a Sustainability Appraisal.

In most advanced urban or village planning models, local context is critical. In many, gardening and other outdoor activities assumes a central role in the daily life of citizens. Environmental planners focus now on smaller and larger systems of resource extraction and consumption, energy production, and waste disposal. A practice known as Arcology seeks to unify the fields of ecology andarchitecture, using principles of landscape architecture to achieve a harmonious environment for all living things. On a small scale, the eco-village theory has become popular, as it emphasizes a traditional 100-140 person scale for communities.

An urban planner can use a number of quantitative tools to forecast impacts of development on the environmental, including roadway air dispersion models to predict air quality impacts of urban highways and roadway noise models to predict noise pollution effects of urban highways. As early as the 1960s, noise pollution was addressed in the design of urban highways as well as noise barriers.The Phase I Environmental Site Assessment can be an important tool to the urban planner by identifying early in the planning process any geographic areas or parcels which have toxic constraints.

Tall buildings in particular can have a substantial effect in channelling winds and shading large areas. The microclimate around the building will typically be assessed as part of the environmental impact assessment for the building. The placement and design of buildings may also be affected by the land on which they are placed. Soil and rock considerations such as depth to bedrock may influence the height of very tall structures, as in Manhattan, though there is less impact than previously supposed, and geological conditions such as fault lines may affect building requirements.

5.6.10 Light and Sound

The urban canyon effect is a colloquial, non-scientific term referring to street space bordered by very high buildings. This type of environment may shade the sidewalk level from direct sunlight during most daylight hours. While an oft-decried phenomenon, it is rare except in very dense, hyper-tall urban environments, such as those found in Lower and Midtown Manhattan, Chicago's Loop and Hong Kong's Kowloon and Central.

In urban planning, sound is usually measured as a source of pollution. Another perspective on urban sounds is developed in Soundscape studies emphasising that sound aesthetics involves more than noise abatement and decibel measurements. Hedfors coined 'Sonotope' as a useful concept in urban planning to relate typical sounds to a specific place.

Light pollution has become a problem in urban residential areas, not only as it relates to its effects on the night sky, but as some lighting is so intrusive as to cause conflict in the residential areas and paradoxically intense improperly installed security lighting may pose a danger to the public, producing excessive glare. The development of the full cutoff fixture, properly installed, has reduced this problem considerably.

5.7 THEORIES OF PLANNING

Planning theory is generally called procedural because it generally concerns itself with the process through which planning occurs and whether or not that process is valid. Lane (2005) traces the intellectual history through its different procedural approaches, especially as they relate to public participation. He calls the first type of blueprint planning. What Lane calls blueprint planning is that which is associated with the early planning thinkers like Ebenezer Howard and Patrick Geddes. Basically, it involved planning for outcomes. The planner has an end-state in mind (i.e., Howard's garden city) and seeks to achieve that end-state through high levels of codification and control. This model provides essentially no outlet for public participation, except possibly voting against those implementing this planning strategy when they come up for election. This type of planning has left two important legacies on contemporary practice: the idea that planning is an apolitical activity and the idea that there is a single, unified public interest.

After, the "fall" of blueprint planning in the late 1950s and early 1960s, the synoptic model began to emerge as a dominant force in planning. Lane (2005) describes synoptic planning as having four central elements:

Indented line "(1) an enhanced emphasis on the specification of goals and targets; (2) an emphasis on quantitative analysis and predication of the environment; (3) a concern to identify and evaluation alternative policy options; and (4) the evaluation of means against ends."

Public participation was first introduced into this model and it was generally integrated into the system process described above. However, the problem was that the idea of a single public interest still dominated attitudes, effectively devaluing the importance of participation because it suggests the idea that the public interest is relatively easy to find and only requires the most minimal form of participation.

Blueprint and synoptic planning both employ what is called the rational paradigm of planning. The rational model is perhaps the most widely accepted model among planning practitioners and scholars, and is considered by many to be the orthodox view of planning. As its name clearly suggests, the goal of the rational model is make planning as rational and systematic as possible. Proponents of this paradigm would generally come up with a list of steps that the planning process can be at least relatively neatly sorted out into and that planning practitioners should go through in order when setting out to plan in virtually any area. As noted above, this paradigm has clear implications for public involvement in planning decisions.

Beginning in the late 1950s and early 1960s, critiques of the rational paradigm began to emerge and formed into several different schools of planning thought. The first of these schools is Linblom'sincrementalism. Lindblom describes planning as "muddling through" and thought that practical planning required decisions to be made incrementally. This incremental approach meant choosing from small number of policy approaches that can only have a small number consequences and are firmly bounded by reality, constantly adjusting the objectives of the planning process and using multiple analyses and evaluations. Lane (2005) explains the public involvement implications of this philosophy. Though this perspective of planning could be considered a large step forward in that it recognizes that there are number of "public interests" and because it provides room for the planning process to de less centralized and incorporate the voices other than those of planners, it in practice would only allow for the public to involved in a minimal, more reactive rather than proactive way.

The mixed scanning model, developed by Etzioni, takes a similar, but slightly different approach. Etzioni (1968) suggested that organizations plan on two different levels: the tactical and the strategic. He posited that organizations could accomplish this by essentially scanning the environment on multiple levels and then choose different strategies and tactics to address what they found there. While Lindblom's approach only operated on the functional level Etzioni argued, the mixed scanning approach would allow planning organizations to work on both the functional and more big-picture oriented levels. Lane explains though, that this model does not do much more at improving public involvement since the planner or planning organization is still at its focus and since its goal is not necessarily to achieve consensus or reconcile differing points of view on a particular subject.

By the late 1960s and early 1970s, planners began to look for new approaches because as happened nearly a decade before, it was realized that the current models were not necessarily sufficient. As had happened before, a number of different models emerged. Lane (2005) notes that it is most useful to think of these model as emerging from a social transformation planning tradition as opposed to a social guidance one, so the emphasis is more bottom-up in nature than it is top-down.

Transactive planning was a radical break from previous models. Instead of considering public participation as method that would be used in addition to the normal training planning process, participation was a central goal. For the first time, the public was encouraged to take an active role in the policy setting process, while the planner took on the role of the as a distributor of information and a feedback source. Transactive planning focuses on interpersonal dialogue that develops ideas, which will be turned into action. One of the central goals is mutual learning where the planner gets more information on the community and citizens become more educated about planning issues.

Advocacy planning is another radical departure from past theoretical models. This model takes the perspective that there are large inequalities in the political system and in the bargaining process between groups that result in large numbers of people unorganized and unrepresented in the process. It concerns itself with ensuring that all people are equally represented in the planning process by advocating for the interests of the underprivileged and seeking social change. Again, public participation is a central tenet of this model. A plurality of public interests is assumed, and the role of planner is essentially as facilitator who either advocates directly for underrepresented groups directly or encourages them to become part of the process.

The bargaining model views planning as the result of give and take on the part of a number of interests who are all involved in the process. It argues that this bargaining is the best way to conduct planning within the bounds of legal and political institutions. Like the advocacy model, this model recognizes that there are inherent inequalities in society, but it asserts that each group or individual in our unequal society has a chance to influence planning decisions, even if they are unable to dominate it or win the benefits that they are seeking. The most interesting part of this theory of planning is that makes public participation the central dynamic in the decision-making process. Decisions are made first and foremost by the public, and the planner plays a more minor role.

The communicative approach to planning is perhaps the most difficult to explain. It focuses on using communication to help different interests in the process understand each other. The idea is that each individual will approach a conversation with his or her own subjective experience in mind and that from that conservation shared goals and possibilities will emerge. Again, participation plays a central role under this model. The model seeks to include as a broad range of voice toe enhance the debate and negotiation that is supposed to form the core of actual plan making. In this model, participation is actually fundamental to the planning process happening. Without the involvement of concerned interests there is no planning.

Looking at each of these models it becomes clear that participation is not only shaped by the public in a given area or by the attitude of the planning organization or planners that work for it. In fact, public participation is largely influences by how planning is defined, how planning problems are defined, the kinds of knowledge that planners choose to employ and how the planning context is set. Though some might argue that is too difficult to involve the public through transactive, advocacy, bargaining and communicative models because transportation is some ways more technical than other fields, it is important to note that transportation is perhaps unique among planning fields in that its systems depend on the interaction of a number of individuals and organizations.

5.7.1 The Scope of Planning

Throughout the first half of the 20th century, the influence of planning broadened within Europe as various national and local statutes increasingly guided new development. European governments became directly involved with housing provision for the working class, and decisions concerning the siting of housing construction shaped urban growth. In the United States, local planning in the form of zoning began with the 1916 New York City zoning law, but it was not until the Great Depression of the 1930s that the federal government intervened in matters of housing and land use. During World War II, military mobilization and the need to coordinate defense production caused the development of the most extensive planning frameworks ever seen in the United States and Britain. Although the wartime agencies were demobilized after hostilities ended, they set a precedent for national economic and demographic planning, which, however, was much more extensive in Britain than in the United States.

5.7.2 Postwar Approaches

During the postwar period European governments mounted massive housing and rebuilding programs within their devastated cities. These programs were guided by the principles of modernist planning promulgated through the Congrès International Architecture Moderne (CIAM), based on the ideas of art and architectural historian Siegfried Giedion, Swiss architect Le Corbusier, and the International school rooted in Germany's Bauhaus. High-rise structures separated by green spaces prevailed in the developments built during this period. Their form reflected both the need to produce large-scale, relatively inexpensive projects and the architects' preference for models that exploited new materials and technologies and could be replicated universally. Government involvement in housing development gave the public sector a more direct means of controlling the pattern of urban growth through its investments, rather than relying on regulatory devices as a means of restricting private developers.

Within Britain the Greater London Plan of Leslie Patrick Abercrombie called for surrounding the metropolitan area with an inviolate greenbelt, construction of new towns beyond the greenbelt that would allow for lowering of population densities in the inner city, and the building of circumferential highways to divert traffic from the core. The concept of the sharp separation of city from country prevailed also throughout the rest of Britain and was widely adopted in the Scandinavian countries, Germany, and the Netherlands as well. In the United States the burgeoning demand for housing stimulated the construction of huge suburban subdivisions. Construction was privately planned and financed, but the federal government encouraged it through tax relief for homeowners and government-guaranteed mortgages. Suburban planning took place at the municipal level in the form of zoning and subdivision approval, public development of sewerage and water systems, and schools. The lack of metropolitan-wide planning jurisdictions resulted in largely unplanned growth and consequent urban sprawl. Within central cities, however, the federal government subsidized land clearance by local urban renewal authorities and the construction of public housing (i.e., publicly owned housing for low-income people). Local government restricted its own reconstruction activities to public facilities such as schools, police stations, and recreation centres. It relied on private investors for the bulk of new construction, simply indicating what would be desirable. Consequently, many cleared sites lay vacant for decades when the private market did not respond.

5.8 PLANNING AND GOVERNMENT

The place of the city-planning function in the structure of urban government developed in different ways in different countries. In many countries today, private developers must obtain governmental permission in order to build. In the United States, however, they may build "as of right" if their plans conform to the municipality's zoning code. On the European continent, where municipal administration is strongly centralized, city planning occurs within the sphere of an executive department with substantial authority. In the United Kingdom the local planning authority is the elected local council, while a planning department acts in an executive and advisory capacity. Developers denied permission to build can appeal the verdict to the central government.

Although the mayor and council have final decision-making power in U.S. cities, an independent planning commission of appointed members usually takes primary responsibility for routine planning functions. Planning activity primarily consists of the approval or disapproval of private development proposals. In larger cities the commission has a staff reporting to it. During the period of a large, federally financed urban renewal program in place from 1949 to 1974, most American cities had powerful semi-independent urban renewal authorities that were responsible for redevelopment planning. Some of these still exist, but in most places they either became subordinate to the mayor or combined with economic development agencies, which are often quasi-autonomous corporations. While they are appointed by the mayor and council, these agencies usually report to an independent board of directors drawn primarily from the business community. Especially as city government became preoccupied with economic development planning, the agencies were authorized to enter into development agreements with private investors.

In some countries, most notably in northern Europe, national governments made city planning part of their overall effort to deal with issues of growth and social welfare. Even in the United States—where the initiative remained with local governments and where metropolitan government never gained a significant foothold—the federal government became involved with local planning issues through the creation and execution of national housing and urban renewal legislation and through the supervisory role of the federal Department of Housing and Urban Development, established in 1965. As developing countries gained independence from colonial powers in the 1960s and '70s, planning structures became highly centralized within the new national governments, which typically laid down the framework for city planning.

5.8.1 Competing Models

Starting in the 20th century, a number of urban planning theories came into prominence and, depending on their popularity and longevity, influenced the appearance and experience of the urban landscape. The primary goal of city planning in the mid-20th century was comprehensiveness. An increasing recognition of the interdependence of various aspects of the city led to the realization that land use, transport, and housing needed to be designed in relation to each other. Developments in other disciplines, particularly management science and operations research, influenced academic planners who sought to elaborate a universal method—also known as "the rational model"—whereby experts would evaluate alternatives in relation to a specified set of goals and then choose the optimum solution. The rational model was briefly hegemonic, but this scientific approach to public-policy making was quickly challenged by critics who argued that the human consequences of planning decisions could not be neatly quantified and added up.

The modernist model, involving wholesale demolition and reconstruction under the direction of planning officials isolated from public opinion, came under fierce attack both intellectually and on the ground. Most important in undermining support for the modernist approach was urbanologist Jane Jacobs. In her book The Death and Life of Great American Cities (1961), she sarcastically described redeveloped downtowns and housing projects as comprising the "radiant garden city"—a sly reference to the influence of Le Corbusier's "towers in the park" and Ebenezer Howard's antiurban garden city. Jacobs criticized large-scale clearance operations for destroying the complex social fabric of cities and imposing an inhuman orderliness. Rather than seeing high population density as an evil, she regarded it as an important factor in urban vitality. She considered that a lively street life made cities attractive, and she promoted diversity of uses and population groups as a principal value in governing urban development. According to Jacobs, urban diversity contributes to sustainable growth, whereas undifferentiated urban settings tend to depend upon unsustainable exploitation, exhibited in the extreme form by lumber or mining towns that collapse after the valuable resources have been removed. Jacobs was not alone in her criticism. Beginning in the 1960s, urban social movements, at times amounting to insurrection, opposed the displacements caused by large-scale modernist planning. In cities throughout the United States and Europe, efforts at demolishing occupied housing provoked fierce opposition. Within developing countries, governmental attempts to destroy squatter settlements stimulated similar counteroffensives.

By the end of the 20th century, planning orthodoxy in the United States and Europe began to take Jacobs's arguments into account. New emphasis was placed on the rehabilitation of existing buildings, historical preservation, adaptive reuse of obsolete structures, mixed-use development, and the "24-hour city"—i.e., districts where a variety of functions would create around-the-clock activity. Major new projects, while still sometimes involving demolition of occupied housing or commercial structures, increasingly came to be built on vacant or "brownfields" sites such as disused railroad yards, outmoded port facilities, and abandoned factory districts. Within developing countries, however, the modernist concepts of the earlier period still retained a significant hold. Thus, for example, China, in preparation for the Beijing Olympics of 2008, engaged in major displacement of its urban population to construct roads and sports facilities, and it likewise developed new commercial districts by building high-rise structures along the functionalist Corbusian model.

5.8.2 Contemporary Planning

The ways in which planning operated at the beginning of the 21st century did not conform to a single model of either a replicable process or a desirable outcome. Within Europe and the United States, calls for a participatory mode—one that involved residents most likely to be affected by change in the planning process for their locales—came to be honoured in some cities but not in others. The concept of participatory planning has spread to the rest of the world, although it remains limited in its adoption. Generally, the extent to which planning involves public participation reflects the degree of democracy enjoyed in each location. Where government is authoritarian, so is planning. Within a more participatory framework, the role of planner changes from that of expert to that of mediator between different groups, or "stakeholders." This changed role has been endorsed by theorists supporting a concept of "communicative rationality." Critics of this viewpoint, however, argue that the process may suppress innovation or simply promote the wishes of those who have the most power, resulting in outcomes contrary to the public interest. They are also concerned that the response of "not in my backyard" ("NIMBYism") precludes building affordable housing and needed public facilities if neighborhood residents are able to veto any construction that they fear will lower their property values.

In sum, the enormous variety of types of projects on which planners work, the lack of consensus over processes and goals, and the varying approaches taken in different cities and countries have produced great variation within contemporary urban planning. Nevertheless, although the original principle of strict segregation of uses continues to prevail in many places, there is an observable trend toward mixed-use development—particularly of complementary activities such as retail, entertainment, and housing—within urban centres.

5.8.3 Changing Objectives

Although certain goals of planning, such as protection of the environment, remain important, emphases among the various objectives have changed. In particular, economic development planning, especially in old cities that have suffered from the decline of manufacturing, has come to the fore. Planners responsible for economic development behave much like business executives engaged in marketing: they promote their cities to potential investors and evaluate physical development in terms of its attractiveness to capital and its potential to create jobs, rather than by its healthfulness or conformity to a master plan. Such planners work to achieve development agreements with builders and firms that will contribute to local commerce. Especially in the United States and the United Kingdom, planning agencies have concerned themselves with promoting economic development and have become involved in negotiating deals with private developers. In the United Kingdom these can include the trading of planning permission for "planning gain" or other community benefits; in other words, developers may be allowed to build in return for providing funds, facilities, or other benefits to the community. In the United States, where special permission is not required if the building fits into the zoning ordinance, deals usually involve some kind of public subsidy. Typical development agreements involve offering land, tax forgiveness, or regulatory relief to property developers in return for a commitment to invest in an area or to provide amenities. An agreement may also be struck between the city and a private firm in which the firm agrees to move into or remain in an area in return for various concessions. Many such arrangements generate controversy, especially if a municipality exercises the right of eminent domain and takes privately owned land for development projects.

A late 20th-century movement in planning, variously called new urbanism, smart growth, or neotraditionalism, has attracted popular attention through its alternative views of suburban development. Reflecting considerable revulsion against urban sprawl, suburban traffic congestion, and long commuting times, this movement has endorsed new construction that brings home, work, and shopping into proximity, encourages pedestrian traffic, promotes development around mass-transit nodes, and mixes types of housing. Within the United Kingdom, Prince Charles became a strong proponent of neotraditional planning through his sponsorship of Poundbury, a new town of traditional appearance in Dorset. Similar efforts in the United States, where growth on the metropolitan periphery continued unabated, chiefly arose as limited areas of planned development amid ongoing dispersal and sprawl. Although the movement's primary influence has been in new suburban development, it has also been applied to the redevelopment of older areas within the United Kingdom and the United States. Paternoster Square in London, adjacent to Saint Paul's Cathedral, and a number of HOPE VI schemes in the United States have been erected in accordance with neotraditional or new urbanist ideas.

5.9 PROCESS OF URBAN PLANNING

Prior to the 1950, Urban Planning was seldom considered a unique profession.Planning focused on top-down processes by which the urban planner created the plans. The planner would know architecture, surveying, or engineering, bringing to the town planning process ideals based on these disciplines. They typically worked for national or local governments.

Changes to the planning process Strategic Urban Planning over past decades have witnessed the metamorphosis of the role of the urban planner in the planning process. More citizens calling for democratic planning & development processes have played a huge role in allowing the public to make important decisions as part of the planning process. Community organizers and social workers are now very involved in planning from the grassroots level. The term advocacy planning was coined by Paul Davidoff in his influential 1965 paper, "Advocacy and Pluralism in Planning" which acknowledged the political nature of planning and urged planners to acknowledge that their actions are not value-neutral and encouraged minority and under represented voices to be part of planning decisions.

Ozawa and Seltzer (1999) advocate a communicative planning model in education to teach planners to work within the social and political context of the planning process. In their paper "Taking Our Bearings: Mapping a Relationship among Planning Practice, Theory, and Education," the authors demonstrate the importance of educating planners beyond the rational planning model in which planners make supposedly value-neutral recommendations based on science and reason. Through a survey of employers, it was found that the most highly rated skills in entry-level professional hiring are communication-based. The results suggest this view of planning as a communicative discourse as a possible bridge between theory and practice, and indicate that the education of planners needs to incorporate synthesis and communication across the curriculum.

Developers have also played huge roles in development, particularly by planning projects. Many recent developments were results of large and small-scale developers who purchased land, designed the district and constructed the development from scratch. The Melbourne Docklands, for example, was largely an initiative pushed by private developers to redevelop the waterfront into a high-end residential and commercial district.

Recent theories of urban planning, espoused, for example by Salingaros see the city as a adaptive system that grows according to process similar to those of plants. They say that urban planning should thus take its cues from such natural processes. Such theories also advocate participation by inhabitants in the design of the urban environment, as opposed to simply leaving all development to large-scale construction firms.

In the process of creating an urban plan or urban design, carrier-infill is one mechanism of spatial organization in which the city's figure and ground components are considered separately. The urban figure, namely buildings, are represented as total possible building volumes, which are left to be designed by architects in following stages. The urban ground, namely in-between spaces and open areas, are designed to a higher level of detail.

The carrier-infill approach is defined by an urban design performing as the carrying structure that creates the shape and scale of the spaces, including future building volumes that are then infilled by architects' designs. The contents of the carrier structure may include street pattern, landscape architecture, open space, waterways, and other infrastructure. The infill structure may contain zoning, building codes, quality guidelines, and Solar Access based upon asolar envelope. Carrier-Infill urban design is differentiated from complete urban design, such as in the monumental axis of Brasília, in which the urban design and architecture were created together.

In carrier-infill urban design or urban planning, the negative space of the city, including landscape, open space, and infrastructure is designed in detail. The positive space, typically building site for future construction, are only represented as unresolved volumes. The volumes are representative of the total possible building envelope, which can then be infilled by individual architects.

5.10 PRINCIPLES OF URBAN DESIGN

Public spaces are frequently subject to overlapping management responsibilities of multiple public agencies or authorities and the interests of nearby property owners, as well as the requirements of multiple and sometimes competing users. The design, construction and management of public spaces therefore typically demands consultation and negotiation across a variety of spheres. Urban designers rarely have the degree of artistic liberty or control sometimes offered in design professions such as architecture. It also typically requires interdisciplinary input with balanced representation of multiple fields including engineering, ecology, local history, and transport planning.

The scale and degree of detail considered varies depending on context and needs. It ranges from the layout of entire cities, as with l'Enfant's plan for Washington DC, Griffin and Mahony's plan for Canberra and Doxiadis' plan for Islamabad (although such opportunities are obviously rare), through 'managing the sense of a region' as described by Kevin Lynch, to the design of street furniture.

Urban design may encompass the preparation of design guidelines and regulatory frameworks, or even legislation to control development, advertising, etc. and in this sense overlaps with urban planning. It may encompass the design of particular spaces and structures and in this sense overlaps with architecture, landscape architecture, highway engineering and industrial design. It may also deal with 'place management' to guide and assist the use and maintenance of urban areas and public spaces.

Much urban design work is undertaken by urban planners, landscape architects and architects but there are professionals who identify themselves specifically as urban designers. Many architecture, landscape and planning programs incorporate urban design theory and design subjects into their curricula and there are an increasing number of university programs offering degrees in urban design, usually at post-graduate level.

Urban design considers:

- **Urban structure** - How a place is put together and how its parts relate to each other
- **Urban typology, density and sustainability** - Spatial types and morphologies related to intensity of use, consumption of resources and production and maintenance of viable communities

- **Accessibility** - Providing for ease, safety and choice when moving to and through places
- **Legibility and wayfinding** - Helping people to find their way around and understand how a place works
- **Animation** - Designing places to stimulate public activity
- **Function and fit** - Shaping places to support their varied intended uses
- **Complementary mixed uses** - Locating activities to allow constructive interaction between them
- **Character and meaning** - Recognizing and valuing the differences between one place and another
- **Order and incident** - Balancing consistency and variety in the urban environment in the interests of appreciating both
- **Continuity and change** - Locating people in time and place, including respect for heritage and support for contemporary culture
- **Civil society** - Making places where people are free to encounter each other as civic equals, an important component in building social capital

5.10.1 Equality Issues in Urban Design

Until the 1970s, urban designers had taken little account of the needs of people with disabilities. At that time, disabled people began to form movements demanding recognition of their potential contribution if social obstacles were removed. Disabled people challenged the 'medical model' of disability which saw physical and mental problems as an individual 'tragedy' and people with disabilities as 'brave' for enduring them. They proposed instead a 'social model' which said that barriers to disabled people result from the design of the built environment and attitudes of able-bodied people. 'Access Groups' were established composed of people with disabilities who audited their local areas, checked planning applications and made representations for improvements. The new profession of 'access officer' was established around that time to produce guidelines based on the recommendations of access groups and to oversee adaptations to existing buildings as well as to check on the accessibility of new proposals. Many local authorities now employ access officers who are regulated by the Access Association. A new chapter of the Building Regulations (Part M) was introduced in 1992. Although it was beneficial to have legislation on this issue the requirements were fairly minimal but continue to be improved with ongoing amendments. The Disability Discrimination Act 1995 continues to raise awareness and enforce action on disability issues in the urban environment.

5.10.2 Urban Density

Urban density is a term used in urban planning and urban design to refer to the number of people inhabiting a given urbanized area. As such it is to be distinguished from other measures of population density. Urban density is considered an important factor in understanding how cities function. Research related to urban density occurs across diverse areas, including economics, health, innovation, psychology and geography as well as sustainability.

It is commonly asserted that higher density cities are more sustainable than low density cities. Much planning theory, particularly in North America, the UK, Australia and New Zealand has been developed premised on raising urban densities, such as New Urbanism, Transit-oriented development, and Smart growth.

However, the link between urban density and aspects ofsustainability remains a contested area of planning theory. Many experts on sustainable urbanism, including prominent urban designer Jan Gehl, argue that low-density, dispersed cities are unsustainable as they are automobile dependent. A minority, such as Randy O'Toole, author of Gridlock and a senior fellow at the Cato Institute, counter that raising densities results in more expensive real estate, greater road congestion and more air pollution. At a broader level though, there is evidence to indicate a strong negative correlation between the total energy consumption of a city and its overall urban density, i.e. the lower the density, the more energy consumed.

5.10.3 Measuring Urban Density

Urban density is a very specific measurement of the population of an urbanized area, excluding non-urban land-uses. Non-urban uses include regional open space, agriculture and water-bodies.

There are a variety of other ways of measuring the density of urban areas:

- **Floor area ratio** - The total floor area of buildings divided by land area of the lot they are built on
- **Residential density** - The number of dwelling units in any given area
- **Population density** - The number of people in any given area
- **Employment density** - The number of jobs in any given area
- **Gross density** - Any density figure for a given area of land that includes uses not necessarily directly relevant to the figure (usually roads and other transport infrastructure)
- **Net density** - A density figure for a given area of land that excludes land not directly related to the figure.
- **Weighted density** - A density metric which measures the density at which the average citizen lives. It is determined by calculating the standard density of each census tract, assigning each a weight equal to its share of the total population, and then adding the segments.

5.11 URBAN CONSOLIDATION

Urban consolidation refers to a diverse set of planning policies intended to make better use of existing urban infrastructure by encouraging development within existing urbanised areas (so-called 'brownfield sites') rather than on non-urbanised land (aka 'greenfield sites'), thus limiting urban sprawl. Urban Consolidation involves increasing the number of houses or units within existing areas so they can have more efficient use of services and reduce the overall impact on the environment. Urban Consolidation is touted as a means to reduce the total amount of land needed to house the population.

There are broadly three kinds of urban consolidation: Market-led consolidation of existing residential areas involves residential redevelopment of established dwellings as well as non-residential land and buildings at higher densities than the metropolitan average. Transit-oriented development (or TOD) involves high-density residential and mixed-use buildings within walkable precincts around public transport nodes, often referred to as Activity centres. The third approach is to require that all new development on the urban fringe of existing metropolitan areas is at higher densities than the current average for those cities.

CHAPTER

ENVIRONMENTAL DESIGN AND PLANNING

6

6.1 INTRODUCTION

Environmental design and planning is the moniker used by several Ph.D. programs that take a multidisciplinary approach to the built environment. Typically environmental design and planning programs address architectural history or design (interior or exterior), city or regional planning, landscape architecture history or design, environmental planning, construction science, cultural geography, or historic preservation. Social science methods are frequently employed; aspects of sociology or psychology can be part of a research program.

The concept of "environmental" in these programs is quite broad and can encompass aspects of the natural, built, work, or social environments.

6.2 ENVIRONMENTAL PLANNING

Environmental Planning is the process of facilitating decision making to carry out development with due consideration given to the natural environmental, social, political, economic and governance factors and provides a holistic frame work to achieve sustainable outcomes.

6.2.1 Elements of Environmental Planning

Environmental Planning concerns itself with the decision making processes where they are required for managing relationships that exist within and between natural systems and human systems. Environmental Planning endeavours to manage these processes in an effective, orderly, transparent and equitable manner for the benefit of all constituents within such systems for the present and for the future. Present day Environmental Planning practices are the result of continuous refinement and expansion of the scope of such decision making processes. Some of the main elements of present day environmental planning are:

- Social & economic development
- Urban development
- Regional development
- Natural resource management & integrated land use
- Infrastructure systems
- Governance frameworks

The environmental planning assessments encompass areas such as land use, socio-economics,transportation, economic and housing characteristics, air pollution, noise pollution, the wetlands, habitat of the endangered species, flood zones susceptibility, coastal zones erosion, and visual studies among others, and is referred to as an Integrated environmental planning assessment.

6.2.2 Environmental Planning in the United States

In the United States, for any project, environmental planners deal with a full range of environmental regulations from federal to state and city levels, administered federally by the Environmental Protection Agency. A rigorous environmental process has to be undertaken to examine the impacts and possible mitigation of any construction project. Depending on the scale and impact of the project, an extensive environmental review is known as an Environmental Impact Statement (EIS), and the less extensive version is Environmental Assessment (EA). Procedures follow guidelines from National Environmental Policy Act (NEPA), State Environmental Quality Review Act (SEQRA) and/or City Environmental Quality Review (CEQR), and other related federal or state agencies published regulations.

The Association of Environmental Professionals (AEP) is a non-profit organization of interdisciplinary professionals including environmental science, resource management, environmental planning and other professions contributing to this field. AEP is the first organization of its kind in the USA, and its influence and model have spawned numerous other regional organizations throughout the United States. Its mission is to improve the technical skills of members, and the organization is dedicated to "the enhancement, maintenance and protection of the natural and human environment". From inception in the mid 1970s the organization has been closely linked with the maintenance of theCalifornia Environmental Quality Act (CEQA), due to California being one of the first states to adopt a comprehensive legal framework to govern the environmental review of public policy and project review.

6.3 ENVIRONMENTAL DESIGN

Environmental design is the process of addressing surrounding environmental parameters when devising plans, programs, policies, buildings, or products. Classical prudent design may have always considered environmental factors; however, the environmental movement beginning in the 1940s has made the concept more explicit.

Environmental design can also refer to the applied arts and sciences dealing with creating the human-designed environment. These fields include architecture, geography, urban planning, landscape architecture, and interior design. Environmental design can also encompass interdisciplinary areas such as historical preservation and lighting design. In terms of a larger scope, environmental design has implications for the industrial design of products: innovative automobiles, wind-electricity generators, solar-electric equipment, and other kinds of equipment could serve as examples. Currently, the term has expanded to apply to ecological and sustainability issues.

Examples of the environmental design process include use of roadway noise computer models in design of noise barriers and use of roadway air dispersion models in analyzing and designing urban highways. Designers consciously working within this more recent framework of philosophy and practice seek a blending of nature and technology, regarding ecology as the basis for design. Some believe that strategies of conservation, stewardship, and regeneration can be applied at all levels of scale from the individual building to the community, with benefit to the human individual and local and planetary ecosystems.

Specific examples of large scale environmental design projects include:

- Boston Transportation Planning Review
- BART - Bay Area Rapid Transit System Daly City Turnback project and airport extension.
- Metropolitan Portland, Oregon light rail system

The theory and practice of Environmental impact assessment developed from the appreciation that development projects can have negative impacts - externalities which harm the environment. This led to the enactment of the US National Environmental Policy Act of 1969.

6.3.1 Environmental Impact Design

Environmental impact design is concerned with modifications to the design of development projects to achieve positive impacts - externalities which benefit the environment and raise the stock of public goods. Examples of positive impacts include:

- Habitat creation as a result of afforestation projects
- Coastal management projects which contribute to ecological and recreational objectives

- Flood defense projects which create greenways
- Public open space projects which contribute to surface water management objectives
- Bridge designs which enhance the landscape and contribute to non-transportation objectives

6.3.2 Environmental Impact Assessment

An environmental impact assessment is an assessment of the possible positive or negative impact that a proposed project may have on the environment, together consisting of the environmental, social and economic aspects.

The purpose of the assessment is to ensure that decision makers consider the ensuing environmental impacts when deciding whether to proceed with a project. The International Association for Impact Assessment (IAIA) defines an environmental impact assessment as "the process of identifying, predicting, evaluating and mitigating the biophysical, social, and other relevant effects of development proposals prior to major decisions being taken and commitments made." EIAs are unique in that they do not require adherence to a predetermined environmental outcome, but rather they require decisionmakers to account for environmental values in their decisions and to justify those decisions in light of detailed environmental studies and public comments on the potential environmental impacts of the proposal.

EIAs began to be used in the 1960s as part of a rational decision making process. It involved a technical evaluation that would lead to objective decision making. EIA was made legislation in the US in the National Environmental Policy Act (NEPA) 1969. It has since evolved as it has been used increasingly in many countries around the world. As per Jay et al.(2006), EIA as it is practiced today, is being used as a decision aiding tool rather than decision making tool. There is growing dissent on the use of EIA as its influence on development decisions is limited and there is a view it is falling short of its full potential. There is a need for stronger foundation of EIA practice through training for practitioners, guidance on EIA practice and continuing research.

EIAs have often been criticized for having too narrow spatial and temporal scope. At present no procedure has been specified for determining a system boundary for the assessment. The system boundary refers to 'the spatial and temporal boundary of the proposal's effects'. This boundary is determined by the applicant and the lead assessor, but in practice, almost all EIAs address the direct, on-site effects alone.

However, as well as direct effects, developments cause a multitude of indirect effects through consumption of goods and services, production of building materials and machinery, additional land use for activities of various manufacturing and industrial services, mining of resources etc. The indirect effects of developments are often an order of magnitude higher than the direct effects assessed by EIA. Large proposals such as airports or ship yards cause wide ranging national as well as international environmental effects, which should be taken into consideration during the decision-making process.

Broadening the scope of EIA can also benefit threatened species conservation. Instead of concentrating on the direct effects of a proposed project on its local environment some EIAs used a landscape approach which focused on much broader relationships between the entire population of a species in question. As a result, an alternative that would cause least amount of negative effects to the population of that species as a whole, rather than the local subpopulation, can be identified and recommended by EIA.

There are various methods available to carry out EIAs, some are industry specific and some general methods:

- **Industrial products** - Product environmental life cycle analysis (LCA) is used for identifying and measuring the impact on the environment of industrial products. These EIAs consider technological activities used for various stages of the product: extraction of raw material for the product and for ancillary materials and equipment, through the production and use of the product, right up to the disposal of the product, the ancillary equipment and material.
- **Genetically modified plants** - There are specific methods available to perform EIAs of genetically modified plants. Some of the methods are GMP-RAM, INOVA etc.
- **Fuzzy Arithmetic** - EIA methods need specific parameters and variables to be measured to estimate values of impact indicators. However many of the environment impact properties cannot be measured on a scale e.g. landscape quality, lifestyle quality, social acceptance etc. and moreover these indicators are very subjective. Thus to assess the impacts we may need to take the help of information from similar EIAs, expert criteria, sensitivity of affected population etc. To treat this information, which is generally inaccurate, systematically, fuzzy arithmetic and approximate reasoning methods can be utilised. This is called as a fuzzy logic approach.

At the end of the project, an EIA should be followed by an audit. An EIA audit evaluates the performance of an EIA by comparing actual impacts to those that were predicted. The main objective of these audits is to make future EIAs more valid and effective. The two main considerations are:

- **Scientific** - To check the accuracy of predictions and explain errors.
- **Management**- To assess the success of mitigation in reducing impacts.

Some people believe that audits be performed as a rigorous scientific testing of the null hypotheses. While some believe in a simpler approach where you compare what actually occurred against the predictions in the EIA document.

After an EIA, the precautionary and polluter pays principles may be applied to prevent, limit, or require strict liability or insurance coverage to a project, based on its likely harms. Environmental impact assessments are sometimes controversial.

6.3.3 Applications of EIA

Australia

The history of EIA in Australia could be linked to the enactment of the U.S. National Environment Policy Act (NEPA) in 1970, which made the preparation of environmental impact statements a requirement. In Australia, one might say that the EIA procedures were introduced at a State Level prior to that of the Commonwealth (Federal), with a majority of the states having divergent views to the Commonwealth. One of the pioneering states was New South Wales, whose State Pollution Control Commission issued EIA guidelines in 1974. At a Commonwealth (Federal) level, this was followed by passing of the Environment Protection (Impact of Proposals) Act in 1974. The Environment Protection and Biodiversity Conservation Act 1999 (EPBC) superseded the Environment Protection (Impact of Proposals) Act 1974 and is the current central piece for EIA in Australia on a Commonwealth (Federal) level. An important point to note is that this Commonwealth Act does not affect the validity of the States and Territories environmental and development assessments and approvals; rather the EPBC runs as a parallel to the State/Territory Systems. Overlap between federal and state requirements is addressed via bilateral agreements or one off accredition of state processes, as provided for in the EPBC Act.

6.3.4 The Commonwealth Level

The EPBC Act provides a legal framework to protect and manage nationally and internationally important flora, fauna, ecological communities and heritage places-defined in the EPBC Act as matters of 'national environmental significance'. Following are the eight matters of 'national environmental significance' to which the EPBC ACT applies :

- World Heritage sites;
- National Heritage places;
- RAMSAR wetlands of international significance;
- Listed threatened species and ecological communities;
- Migratory species protected under international agreements;
- The Commonwealth marine environment;
- Nuclear actions (including uranium mining); and
- National Heritage.

In addition to this, the EPBC Act aims at providing a streamlined national assessment and approval process for activities. These activities could be by the Commonwealth, or its agents, anywhere in the world or activities on Commonwealth land; and activities that are listed as having a 'significant impact' on matters of 'national environment significance'.

The EPBC Act comes into play when a person (a 'proponent') wants an action (often called a 'proposal' or 'project') assessed for environmental impacts under the EPBC Act, he or she must refer the project to the Department of Environment, Water, Heritage and the Arts (Australia). This 'referral' is then released to the public, as well as relevant state, territory and Commonwealth ministers, for comment on whether the project is likely to have a significant impact on matters of national environmental significance. The Department of Environment, Water, Heritage and the Arts assess the process and makes recommendation to the minister or the delegate for the feasibility. The final discretion on the decision remains of the minister, which is not solely based on matters of 'national environmental significance' but also the consideration of social and economic impact of the project.

The Australian Government environment minister cannot intervene in a proposal if it has no significant impact on one of the eight matters of 'national environmental significance' despite the fact that there may be other undesirable environmental impacts. This is primarily due to the division of powers between the States and the Federal government and due to which the Australian Government environment minister cannot overturn a state decision.

There are strict civil and criminal penalties for the breach of EPBC Act. Depending on the kind of breach, civil penalty (maximum) may go up to $550,000 for an individual and $5.5 million for a body corporate, or for criminal penalty (maximum) of seven years imprisonment and/or penalty of $46,200.

6.3.5 The State and Territory Level

Australian Capital Territory (ACT)

EIA in ACT is administered with the help of Part 4 of the Land (Planning and Environment) Act 1991 (Land Act) and Territory Plan (plan for land-use).

New South Wales (NSW)

In New South Wales, the Environment Planning Assessment Act 1979 (EPA) establishes three pathways for EIA. The first is under Part 3A of the Act where Major Projects require an environmental assessment. The second is under Part 4 of the Act dealing with development control. If a project does not require approval under Part 3A or Part 4 it is then potentially captured by the third pathway, Part 5 dealing with environment impact assessment.

Northern Territory (NT)

The EIA process in Northern Territory is chiefly administered under the Environmental Assessment Act (EEA). Although EEA is the primary tool for EIA in Northern Territory, there are further provisions for proposals in the Inquiries Act 1985 (NT).

Queensland (QLD)

There are four main EIA processes in Queensland. Firstly, under the Integrated Planning Act 1997 (IPA) for development projects other than mining. Secondly, under the Environmental Protection Act 1994 (EP Act) for some mining and petroleum activities. Thirdly, under the State Development and Public Works Organization Act 1971 (State Development Act) for 'significant projects'. Finally, Environment Protection and Biodiversity Conservation Act 1999 for 'controlled actions'.

South Australia (SA)

The local governing tool for EIA in South Australia is the Development Act 1993. There are three levels of assessment possible under the Act in the form of an environment impact statement (EIS), a Public Environmental Report (PER) or a Development Report (DR).

Tasmania (TAS)

In Tasmania, an integrated system of legislation is used to govern development and approval process, this system is a mixture of the Environmental Management and Pollution Control Act 1994 (EMPCA), Land Use Planning and Approvals Act 1993 (LUPAA), State Policies and Projects Act 1993 (SPPA), and Resource Management and Planning Appeals Tribunal Act 1993.

Victoria (VIC)

The EIA process in Victoria is intertwined with the Environment Effects Act 1978 and the Ministerial Guidelines for Assessment of Environmental Effects.

Western Australia (WA)

The Environmental Protection Act 1986 (Part 4) provides the legislative framework for the EIA process in Western Australia. The EPA Act oversees the planning and development proposals and assesses their likely impacts on the environment.

Canada

The Canadian Environmental Assessment Act (CEAA) is the legal basis for the federal Environmental Assessment (EA) process. CEAA came into force in 1995. Legislative amendments were introduced in 2001 and came into force on October 30, 2003. EA is defined as a planning tool to identify, understand, assess and mitigate, where possible, the environmental effects of a project. Under the CEAA, all federal government departments and agencies are required to undertake an EA for projects relating to a physical work and for any proposed physical activity listed in the Inclusion List Regulations where it exercises one or more of the following CEAA triggers:

- Proposes or undertakes a project
- Grants money or any other form of financial assistance to a project
- Grants an interest in the land to enable a project to be carried out
- Exercises a regulatory duty in relation to a project, such as issuing a permit or license that is included in the Law List Regulations.

If a federal government department or agency exercises one or more of the above-mentioned triggers, it becomes a Responsible Authority (RA) under CEAA. As an RA, the federal department or agency in question must ensure that an EA is carried out in accordance with the CEAA and must consider the EA findings before a decision is made that could allow the project to proceed.

China

The Environmental Impact Assessment Law (EIA Law) requires an environmental impact assessment to be completed prior to project construction. However, if a developer completely ignores this requirement and builds a project without submitting an environmental impact statement, the only penalty is that the Environmental Protection Bureau (EPB) may require the developer to do a make-up environmental assessment. If the developer does not complete this make-up assessment within the designated time, only then is the EPB authorized to fine the developer. Even so, the possible fine is capped at a maximum of about US$25,000, a fraction of the overall cost of most major projects. The lack of more stringent enforcement mechanisms has resulted in a significant percentage of projects not completing legally required environmental impact assessments prior to construction.

China's State Environmental Protection Administration (SEPA) used the legislation to halt 30 projects in 2004, including three hydro-power plants under the Three Gorges Project Company. Although one month later, most of the 30 halted projects resumed their construction, reportedly having passed the environmental assessment, the fact that these key projects' construction was ever suspended was notable.

A joint investigation by SEPA and the Ministry of Land and Resources in 2004 showed that 30-40% of the mining construction projects went through the procedure of environment impact assessment as required, while in some areas only 6-7% did so. This partly explains why China has witnessed so many mining accidents in recent years.

SEPA alone cannot guarantee the full enforcement of environmental laws and regulations, observed Professor Wang Canfa, director of the centre to help environmental victims at China University of Political Science and Law. In fact, according to Wang, the rate of China's environmental laws and regulations that are actually enforced is estimated to be barely 10%.

Egypt

Environmental Impact Assessment (EIA) EIA is implemented in Egypt under the umbrella of the Ministry of state for environmental affairs. The Egyptian Environmental Affairs Agency (EEAA) is responsible for the EIA services.

In June 1997, the responsibility of Egypt's first full time Minister of State for Environmental Affairs was assigned as stated in the Presidential Decree no.275/1997. From thereon, the new ministry has focused, in close collaboration with the national and international development partners, on defining environmental policies, setting priorities and implementing initiatives within a context of sustainable development.

According to the Law 4/1994 for the Protection of the Environment, the Egyptian Environmental Affairs Agency (EEAA) was restructured with the new mandate to substitute the institution initially established in 1982. At the central level, EEAA represents the executive arm of the Ministry.

The purpose of EIA is to ensure the protection and conservation of the environment and natural resources including human health aspects against uncontrolled development. The long-term objective is to ensure a sustainable economic development that meets present needs without compromising future generations ability to meet their own needs. EIA is an important tool in the integrated environmental management approach.

EIA must be performed for new establishments or projects and for expansions or renovations of existing establishments according to the Law for the Environment.

European Union

The European Union has established a mix of mandatory and discretionary procedures to assess environmental impacts. European Union Directive (85/337/EEC) on Environmental Impact Assessments (known as the EIA Directive) was first introduced in 1985 and was amended in 1997. The directive was amended again in 2003, following EU signature of the 1998 Aarhus Convention. In 2001, the issue was enlarged to the assessment of plans and programmes by the so called Strategic Environmental Assessment (SEA) Directive (2001/42/EC), which is now in force.Under the EU directive, an EIA must provide certain information to comply. There are seven key areas that are required:

Description of the Project

- Description of actual project and site description
- Break the project down into its key components, i.e. construction, operations, decommissioning
- For each component list all of the sources of environmental disturbance
- For each component all the inputs and outputs must be listed, e.g., air pollution, noise, hydrology

Alternatives that have been Considered

- Examine alternatives that have been considered
- Example: in a biomass power station, will the fuel be sourced locally or nationally?

Description of the Environment

- List of all aspects of the environment that may be affected by the development
- Example: populations, fauna, flora, air, soil, water, humans, landscape, cultural heritage
- This section is best carried out with the help of local experts, e.g. the RSPB in the UK

Description of the Significant Effects on the Environment

- The word significant is crucial here as the definition can vary
- 'Significant' needs to be defined
- The most frequent method used here is use of the Leopold matrix
- The matrix is a tool used in the systematic examination of potential interactions
- Example: in a windfarm development a significant impact may be collisions with birds

Mitigation

- This is where EIA is most useful
- Once section 4 has been completed it will be obvious where the impacts will be greatest
- Using this information ways to avoid negative impacts should be developed
- Best working with the developer with this section as they know the project best
- Using the windfarm example again construction could be out of bird nesting seasons

Non-Technical Summary

- The EIA will be in the public domain and be used in the decision making process
- It is important that the information is available to the public
- This section is a summary that does not include jargon or complicated diagrams
- It should be understood by the informed lay-person

Lack of Know-How/Technical Difficulties

- This section is to advise any areas of weakness in knowledge

- It can be used to focus areas of future research
- Some developers see the EIA as a starting block for poor environmental management

The Netherlands

EIA was implemented in Dutch legislation on September 1, 1987. The categories of projects that require an EIA are summarised in Dutch legislation, the Wet milieubeheer. The use of thresholds for activities makes sure that EIA is obligatory for those activities that may have considerable impacts on the environment.

For projects and plans that fit these criteria, a EIA report is required. The EIA report defines a.o. the proposed initiative, it makes clear the impact of that initiative on the environment and compares this with the impact of possible alternatives with less a negative impact.

Hong Kong

EIA in Hong Kong, since 1998, is regulated by the Environmental Impact Assessment Ordinance 1997.

The original proposal to construct the Lok Ma Chau Spur Line overground across the Long Valley failed to get through EIA, and the KCR Corporation had to change its plan and build the railway underground. In April 2011, the EIA of the Hong Kong section of the Hong Kong-Zhuhai-Macau Bridge was found to have breached the ordinance, and was declared unlawful. The appeal by the government was allowed in September 2011. However, it was estimated that this EIA court case had increased the construction cost of the Hong Kong section of the bridge by HK$6.5 billion in money-of-the-day prices.

India

The Ministry of Environment and Forests (MoEF) of India have been in a great effort in Environmental Impact Assessment in India. The main laws in nation are Water Act(1974), The Indian Wildlife (Protection) Act (1972), The Air (Prevention and Control of Pollution) Act (1981) and The Environment (Protection) Act (1986). The responsible body for this is Central Pollution Control Board. Environmental Impact Assessment (EIA) studies need a significant amount of primary and secondary environmental data. The primary data are those which need to be collected in the field to define the status of environment (like air quality data, water quality data etc.). The secondary data are those data which have been collected over the years and can be used to understand the existing environmental scenario of the study area. The environmental impact assessment (EIA) studies are conducted over a short period of time and therefore the understanding the environmental trends based on few months of primary data has its own limitations. Ideally, the primary data has to be considered along with the secondary data for complete understanding of the existing environmental status of the area. In many EIA studies, the secondary data needs could be as high as 80% of the total data requirement. EIC is the repository of one stop secondary data source for environmental impact assessment in India.

The Environmental Impact Assessment (EIA) experience in India indicates that the lack of timely availability of reliable and authentic environmental data has been a major bottle neck in achieving the full benefits of EIA. The environment being a multi-disciplinary subject, a multitude of agencies is involved in collection of environmental data. However, there is no single organization in India which tracks the data available amongst these agencies and makes it available in one place, in a form and manner required by practitioners in the field of environmental impact assessment in India. Further, the environmental data is not available in value added forms that can enhance the quality of the EIA. This in turn adversely affects the time and efforts required for conducting the environmental impact assessments (EIAs) by project proponents and also timely environmental clearances by the regulators. With this background, Environmental Information Centre (EIC) has been set up to serve as a professionally managed clearing house of environmental information that can be used by MoEF, project proponents, consultants, NGOs and other stakeholders involved in the process of environmental impact assessment in India. EIC caters to the need of creating and disseminating of organized environmental data for various developmental initiatives all over the country.

EIC stores data in GIS format and makes it available to all environmental impact assessment studies and to EIA stakeholders in a cost effective and timely manner.

Malaysia

In Malaysia, Section 34A, Environmental Quality Act, 1974 requires developments that have significant impact to the environment are required to conduct the Environmental impact assessment.

Nepal

In Nepal, EIA has been integrated in major development projects since the early 1980s. In the planning history of Nepal, the sixth plan (1980-85), for the first time, recognized the need for EIA with the establishment of Environmental Impact Study Project (EISP) under the Department of Soil Conservation in 1982 to develop necessary instruments for integration of EIA in infrastructure development projects. However, the government of Nepal enunciated environment conservation related policies in the seventh plan (NPC, 1985-1990). In order to enforce this policy and make necessary arrangements, a series of guidelines were developed, thereby incorporating the elements of environmental factors right from the project formulation stage of the development plans and projects and to avoid or minimize adverse effects on the ecological system. In addition, it has also emphasized that EIAs of industry, tourism, water resources, transportation, urbanization, agriculture, forest and other developmental projects be conducted.

In Nepal, the government's Environmental Impact Assessment Guideline of 1993 inspired the enactment of the Environment Protection Act (EPA) of 1997 and the Environment Protection Rules (EPR) of 1997(EPA and EPR have been enforced since 24 and 26 June 1997 respectively in Nepal) to internalizing the environmental assessment system. The process institutionalized the EIA process in development proposals and enactment, which makes the integration of IEE and EIA legally binding to the prescribed projects. The projects, requiring EIA or IEE, are included in Schedules 1 and 2 of the EPR, 1997. Progresses were made in the Environmental protection issue during the 8th five year plan (1992-1997). The following development in Environmental protection were achieved during that time:

- Formulation of Environmental Protection Act 1997
- Establishment of Ministry of Environment
- Development of National Environmental Policies and Action Plan, EIA guidelines developed
- Consideration of environmental concerns in hydropower projects
- Development of industrial, irrigation and agricultural policies that undertook environmental concerns

New Zealand

In New Zealand, EIA is usually referred to as Assessment of Environmental Effects (AEE). The first use of EIA's dates back to a Cabinet minute passed in 1974 called Environmental Protection and Enhancement Procedures. This had no legal force and only related to the activities of government departments. When the Resource Management Act was passed in 1991, an EIA was required as part of a resource consent application. Section 88 of the Act specifies that the AEE must include "such detail as corresponds with the scale and significance of the effects that the activity may have on the environment". While there is no duty to consult any person when making a resource consent application (Sections 36A and Schedule 4), proof of consultation is almost certain to be required by local councils when making a decision about whether or not to publicly notifiy the consent application under Section 93.

Russian Federation

Russia holds the world's largest natural gas reserves, the second largest coal reserves, and the eighth largest oil reserves. Russia is also the world's largest exporter of natural gas, the second largest oil exporter and the third largest energy consumer.

As of 2004, the state authority responsible for conducting the State EIA in Russia has been split between two Federal bodies: 1) Federal service for monitoring the use of natural resources - a part of the Russian Ministry for Natural Resources and Environment and 2) Federal Service for Ecological, Technological and Nuclear Control. The two main pieces of environmental legislation in Russia are: The Federal Law 'On Ecological Expertise, 1995 and the 'Regulations on Assessment of Impact from Intended Business and Other Activity on Environment in the Russian Federation, 2000.

Federal Service for Monitoring the Use of Natural Resources

In 2006, the parliament committee on ecology in conjunction with the Ministry for Natural Resources and Environment, created a working group to prepare a number of amendments to existing legislation in order to cover such topics as stringent project documentation for building of potentially environmentally damaging objects as well as building of projects on the territory of protected areas. There has been some success in this area, as evidenced from abandonment of plans to construct a gas pipe-line through the only remaining habitat of the critically endangered Amur leopard in the Russian Far East.

Federal Service for Ecological, Technological and Nuclear Control

The government's decision to hand over control over several important procedures, including state EIA in the field of all types of energy projects, to the Federal Service for Ecological, Technological and Nuclear Control had caused a major controversy and criticism from environmental groups that blamed the government for giving nuclear power industry control over the state EIA.

Not surprisingly the main problem concerning State EIA in Russia is the clear differentiation of jurisdiction between the two above-mentioned Federal bodies.

Sri Lanka

Environmental Impact Assessments is one popular approach to assist in smart growth in democratic countries is for law-makers to require prospective developers to prepare environmental impact assessments of their plans as a condition for state and/or local governments to go for Environmental Impact Assessments

These reports often indicate how significant impacts generated by the development will be mitigated - the cost of which is usually paid by the developer. These assessments are frequently controversial. Conservationists, neighborhood advocacy groups and NIMBYs are often skeptical about such impact reports, even when they are prepared by independent agencies and subsequently approved by the decision makers rather than the promoters. Conversely, developers will sometimes strongly resist being required to implement the mitigation measures required by the local government as they may be quite costly.

These assessments are frequently controversial. Conservationists, neighborhood advocacy groups and NIMBYs are often skeptical about such impact reports, even when they are prepared by independent agencies and subsequently approved by the decision makers rather than the promoters. Conversely, developers will sometimes strongly resist being required to implement the mitigation measures required by the local government as they may be quite costly.

The importance of the Environmental Impact Assessment as an effective tool for the purpose of integrating environmental considerations with development planning is highly recognized in Sri Lanka. The application of this technique is considered as a means of ensuring that the likely effects of new development projects on the environment are fully understood and taken into account before development is allowed to proceed. The importance of this management tool to foresee potential environmental impacts and problems caused by proposed projects and its use as a mean to make project more suitable to the environment are highly appreciated.

United States

Under United States environmental law an Environmental Assessment (EA) is compiled to determine the need for an Environmental Impact Statement (EIS), and originated in the National Environmental Policy Act (NEPA), enacted in 1969. Certain actions of federal agencies must be preceded by an EA or EIS. Contrary to a widespread misconception, NEPA does not prohibit the federal government or its licensees/permittees from harming the environment, nor does it specify any penalty if the EA or EIS turns out to be inaccurate, intentionally or otherwise. NEPA requires that plausible statements as to the prospective impacts be disclosed in advance. The purpose of NEPA process is to ensure that the decision maker is fully informed of the environmental aspects and consequences prior to making the final decision.

6.3.6 Environmental Assessment (EA)

An Environmental Assessment (EA) is an environmental analysis prepared pursuant to the National Environmental Policy Act to determine whether a federal action would significantly affect the environment and thus require a more detailed Environmental Impact Statement (EIS). The release of an Environmental Assessment results in either a Finding of No Significant Impact (FONSI) or an Environmental Impact Statement (EIS).

The Environmental Assessment is a concise public document prepared by the federal action agency that serves to:

1. briefly provide sufficient evidence and analysis for determining whether to prepare an EIS or a Finding of No Significant Impact (FONSI)
2. Demonstrate compliance with the act when no EIS is required
3. facilitate the preparation of a EIS when a FONSI cannot be demonstrated

The Environmental Assessment includes a brief discussion of the purpose and need of the proposal and of its alternatives as required by CFR 102(2)(E), of the environmental impacts of the proposed actions and alternatives, as well as a listing of agencies and stakeholders consulted. The action agency must approve an EA before it is made available to the public. The EA is made public through notices of availability by local, state, or regional clearing houses, newspapers, etc.

There is a 15-30 day review period required for an Environmental Assessment, while the document is made available for public commentary.

An agency will release either a Draft Environmental Assessment (Draft EA) or a Draft Environmental Impact Statement (DEIS) for comment. Interested parties and the general public have the opportunity to comment on the draft, after which the agency will address all comments received and prepare a decision document, either a FONSI, Notice of Intent (NOI) to prepare a EIS or a Record of Decision for an EIS. The agency will then approve the " Final Environmental Assessment" (Final EA) or Final Environmental Impact Statement (FEIS). Commenting on the Draft EA is typically done in writing or email, submitted to the lead agency as defined in the Notice of Availability. Draft EIS's require public hearings, so comments can be made in person, as well as in writing. Occasionally, the agency will later release a "Supplemental Environmental Assessment or a Supplemental Environmental Impact Statement (SEIS), if the project parameters or environmental conditions change substantially after the issuance of the FONSI or ROD.

The adequacy of an EIS can be challenged in federal court. Major proposed projects have been blocked because of an agency's failure to prepare an acceptable EIS. One prominent example was the Westway landfill and highway development in and along the Hudson River in New York City.Another prominent case involved the Sierra Club suing the Nevada Department of Transportation over its denial of Sierra Club's request to issue a supplemental EIS addressing air emissions of particulate matter and hazardous air pollutants in the case of widening US Highway 95 through Las Vegas.The case reached the United States Court of Appeals for the Ninth Circuit, which led to construction on the highway being halted until the court's final decision. The case was settled prior to the court's final decision.

Several state governments that have adopted "little NEPAs," state laws imposing EIS requirements for particular state actions. Some those state laws such as the California Environmental Quality Actrefer to the required environmental impact studies as environmental impact reports.

The structure of a generic Environmental Assessment is as follows:

1. Summary
2. Introduction
    - Structure
    - Background
    - Purpose and Need for Action
    - Proposed Action
    - Decision Framework
    - Public Involvement

    - Issues
3. Alternatives, including the Proposed Action
    - Alternatives
    - Mitigation Common to All Alternatives
    - Comparison of Alternatives
4. Environmental Consequences
5. Consultation and Coordination

These variety of state requirements are yielding voluminous data not just upon impacts of individual projects, but also to elucidate scientific areas that had not been sufficiently researched. For example, in a seemingly routine Environmental Impact Report for the city of Monterey, California, information came to light that led to the official federal endangered species listing of Hickman's potentilla, a rare coastal wildflower.

6.4 ENVIRONMENTAL IMPACT STATEMENT (EIS)

An Environmental Impact Statement (EIS), under United States environmental law, is a document required by the National Environmental Policy Act (NEPA) for certain actions "significantly affecting the quality of the human environment". An EIS is a tool for decision making. It describes the positive and negative environmental effects of a proposed action, and it usually also lists one or more alternative actions that may be chosen instead of the action described in the EIS. Several US state governments require that a document similar to an EIS be submitted to the state for certain actions. For example, in California, an Environmental Impact Report (EIR) must be submitted to the state for certain actions, as described in the California Environmental Quality Act (CEQA).

The purpose of the NEPA is to promote informed decision-making by federal agencies by making "detailed information concerning significant environmental impacts" available to both agency leaders and the public. The NEPA was the first piece of legislation that created a comprehensive method to assess potential and existing environmental risks at once. One of the primary authors of the act was Lynton K. Caldwell. It also encourages communication and cooperation between all the actors involved in environmental decisions, including government officials, private businesses, and citizens.

In particular, an EIS acts as an enforcement mechanism to ensure that the federal government adheres to the goals and policies outlined in the NEPA. An EIS should be created in a timely manner as soon as the agency is planning development or is presented with a proposal for development. The statement should use an interdisciplinary approach so that it accurately assesses both the physical and social impacts of the proposed development.

Not all federal actions require a full EIS. If the action may or may not cause a significant impact the agency can first prepare a smaller, shorter document called an Environmental

Assessment (EA). The finding of the EA determines whether an EIS is required. If the EA indicates that no significant impact is likely, then the agency can release a finding of no significant impact (FONSI) and carry on with the proposed action. Otherwise, the agency must then conduct a full-scale EIS. Most EAs result in a FONSI. A limited number of federal actions may avoid the EA and EIS requirements under NEPA if they meet the criteria for a categorical exclusion (CATEX). A CATEX is usually permitted when a course of action is identical or very similar to a past course of action and the impacts on the environment from the previous action can be assumed for the proposed action, or for building a structure within the footprint of an existing, larger facility or complex. For example, two proposed sections of Interstate 69 in Kentucky were granted a CATEX from NEPA requirements as these portions of I-69 will be routed over existing freeways requiring little more than minor spot improvements and a change of highway signage. Additionally, a CATEX can be issued during an emergency situation when time does not permit the preparation of an EA or EIS. An example of the latter is when the Federal Highway Administration issued a CATEX to construct the replacement bridge in the wake of the I-35W Mississippi River Bridge Collapse.

Contrary to a widespread misconception, NEPA does not prohibit the federal government or its licensees/permittees from harming the environment, but merely requires that the prospective impacts be understood and disclosed in advance. The intent of NEPA is to help key decisionmakers and stakeholders balance the need to implement an action with its impacts on the surrounding human and natural environment, and provide opportunities for mitigating those impacts while keeping the cost and schedule for implementing the action under control. However, many activities require various federal permits to comply with other environmental legislation, such as the Clean Air Act, the Clean Water Act, Endangered Species Act and Section 4(f) of the Federal Highway Act to name a few. Similarly, many states and local jurisdictions have enacted environmental laws and ordinances, requiring additional state and local permits before the action can proceed. Obtaining these permits typically requires the lead agency to implement the Least Environmentally Damaging Practicable Alternative (LEDPA) to comply with federal, state, and local environmental laws that are ancillary to NEPA.

6.4.1 Layout

An EIS typically has four sections:

- An Introduction including a statement of the Purpose and Need of the Proposed Action.
- A description of the Affected Environment.
- A Range of Alternatives to the proposed action. Alternatives are considered the "heart" of the EIS.
- An analysis of the environmental impacts of each of the possible alternatives. This section covers topics such as:

- Impacts to threatened or endangered species
- Air and water quality impacts
- Impacts to historic and cultural sites, (particularly sites of significant importance to Native American tribes)
- Social and Economic impacts to local communities
- Cost analysis for each alternative, including costs to mitigate expected impacts, to determine if the proposed action is a prudent use of taxpayer dollars

While not required in the EIS, the following subjects may be included as part of the EIS or as separate documents based on agency policy.

- Financial Plan for the proposed action identifying the sources of secured funding for the action. For example, the Federal Highway Administration has started requiring states to include a financial plan showing that funding has been secured for major highway projects before it will approve an EIS and issue a Record of Decision.
- An Environmental Mitigation Plan is often requested by the Environmental Protection Agency (EPA) if substantial environmental impacts are expected from the preferred alternative.
- Additional documentation to comply with state and local environmental policy laws and secure required federal, state, and local permits before the action can proceed.

Every EIS is required to analyze a No Action Alternative, in addition to the range of alternatives presented for study. The No Action Alternative identifies the expected environmental impacts in the future if existing conditions were left as is with no action taken by the lead agency. Analysis of the No Action Alternative is used to establish a baseline upon which to compare the proposed "Action" alternatives.

6.4.2 NEPA Process

The NEPA process is designed to involve the public and gather the best available information in a single place so that decision makers can be fully informed when they make their choices.

The process has the following steps:

- **Proposal:** In this stage, the needs and objectives of a project have been decided, but the project has not been financed.
- **Categorical Exclusion (CATEX):** As discussed above, the government may exempt an agency from the process. The agency can then proceed with the project and skip the remaining steps.
- **Environmental Assessment (EA):** The proposal is analyzed in addition to the local environment with the aim to reduce the negative impacts of the development on the area.

- **Finding of No Significant Impact (FONSI):** Occurs when no significant impacts are identified in an EA. A FONSI typically allows the lead agency to proceed without having to complete an EIS.

6.4.3 Environmental Impact

Scoping: The first meetings are held to discuss existing laws, the available information, and the research needed. The tasks are divided up and a lead group is selected. Decision makers and all those involved with the project can attend the meetings.

Notice: The public is notified that the agency is preparing an EIS. The agency also provides the public with information regarding how they can become involved in the process. The agency announces its project proposal with a notice in the Federal Register, notices in local media, and letters to citizens and groups that it knows are likely to be interested. Citizens and groups are welcome to send in comments helping the agency identify the issues it must address in the EIS (or EA).

Draft EIS (DEIS): Based on both agency expertise and issues raised by the public the agency prepares a Draft EIS with a full description of the affected environment, a reasonable range of alternatives, and an analysis of the impacts of each alternative.

Comment: Affected individuals then have the opportunity to provide feedback through written and public hearing statements.

Final EIS (FEIS) and Proposed Action: Based on the comments on the Draft EIS, the agency writes a Final EIS, and announces its Proposed Action. The public is not invited to comment on this, but if they are still unhappy, or feel that the agency has missed a major issue, they may protest the EIS to the Director of the agency. The Director may either ask the agency to revise the EIS, or explain to the protester why their complaints are not actually taken care of.

Re-evaluation: Prepared following an approved FEIS or ROD when unforeseen changes to the proposed action or its impacts occurs, or when a substantial period of time has passed between approval of an action and the planned start of said action. Based on the significance of the changes, three outcomes may result from a re-evaluation report: (1) the action may proceed with no substantive changes to the FEIS, (2) significant impacts are expected with the change that can be adequately addressed in a Supplemental EIS (SEIS), or (3) the circumstances force a complete change in the nature and scope of the proposed action, thereby voiding the pre-existing FEIS (and ROD, if applicable), requiring the lead agency to restart the NEPA process and prepare a new EIS to encompass the changes.

Supplemental EIS (SEIS): Typically prepared after either a Final EIS or Record of Decision has been issued and new environmental impacts that were not considered in the original EIS are discovered, requiring the lead agency to re-evaluate its initial decision and consider new alternatives to avoid or mitigate the new impacts. Supplemental EISs are also prepared when the size and scope of a federal action changes, when a significant period of time has lapsed since the FEIS was completed to account for changes in the surrounding environment during that time, or when all of the proposed alternatives in an EIS are deemed to have unacceptable environmental impacts and new alternatives are proposed.

Record of Decision (ROD): Once all the protests are resolved the agency issues a Record of Decision which is its final action prior to implementation. If members of the public are still dissatisfied with the outcome, they may sue the agency in Federal court.

Often, the agencies responsible for preparing an EA or EIS do not compile the document directly, but outsource this work to private-sector consulting firms with expertise in the proposed action and its anticipated effects on the environment. Because of the intense level of detail required in analyzing the alternatives presented in an EIS or EA, such documents may take years or even decades to compile, and often compose of multiple volumes that can be thousands to tens of thousands of pages in length.

To avoid potential conflicts in securing required permits and approvals after the ROD is issued, the lead agency will often coordinate with stakeholders at all levels, and resolve any conflicts to the greatest extent possible during the EIS process. Proceeding in ths fashion helps avoid interagency conflicts and potential lawsuits after the lead agency reaches its decision.

6.4.4 Tiering

On exceptionally large projects, especially proposed highway and railroad corridors that cross long distances, the lead agency may use a two-tiered process prior to implemeting the proposed action. In such cases, the Tier I EIS would analyze the potential socio-environmental impacts along a general corridor, but would not identify the exact location of where the action would occur. A Tier I ROD would be issued approving the general area where the action would be implemented. Following the Tier I ROD, the approved Tier I area is further broken down into subareas, and a Tier II EIS is then prepared for each subarea, that identifies the exact location of where the proposed action will take place. The preparation of Tier II EISs for each subarea proceeds at its own pace, independent from the other subareas within the Tier I area. For example, parts of the proposed Interstate extension in Indiana and Texas are being studied through a two-tiered process.

6.4.5 Strengths

By requiring agencies to complete an EIS, the act encourages them to consider the environmental costs of a project and introduces new information into the decision-making process. The NEPA has increased the influence of environmental analysts and agencies in the federal government by increasing their involvement in the development process. Because an EIS requires expert skill and knowledge, agencies must hire environmental analysts. Unlike agencies who may have other priorities, analysts are often sympathetic to environmental issues. In addition, this feature introduces scientific procedures into the political process.

6.4.6 Limitations

The differences that exist between science and politics limit the accuracy of an EIS. Although analysts are members of the scientific community, they are affected by the political atmosphere. Analysts do not have the luxury of an unlimited time for research. They are also affected by the different motives behind the research of the EIS and by different perspectives of what constitutes a good analysis. In addition, government officials do not want to reveal an environmental problem from within their own agency.

Citizens often misunderstand the environmental assessment process. The public does not realize that the process is only meant to gather information relevant to the decision. Even if the statement predicts negative impacts of the project, decision makers can still proceed with the proposal.

Environmental impact statements presented to citizens and government officials frequently include very precise data. However, the quality and context of the data, such as the margin of error and the range, is omitted.

6.5 ENVIRONMENTAL INDICATORS

Environmental indicators are simple measures that tell us what is happening in the environment. Since the environment is very complex, indicators provide a more practical and economical way to track the state of the environment than if we attempted to record every possible variable in the environment. For example, concentrations of Ozone Depleting Substances (ODS) in the atmosphere, tracked over time, is a good indicator with respect to the environmental issue of stratospheric ozone depletion..

6.5.1 Kemi's Idea of Definitions and Use

Environmental indicators have been defined in different ways but common themes exist.

"An environmental indicator is a numerical value that helps provide insight into the state of the environment or human health. Indicators are developed based on quantitative measurements or statistics of environmental condition that are tracked over time. Environmental indicators can be developed and used at a wide variety of geographic scales, from local to regional to national levels."

"A parameter or a value derived from parameters that describe the state of the environment and its impact on human beings, ecosystems and materials, the pressures on the environment, the driving forces and the responses steering that system. An indicator has gone through a selection and/or aggregation process to enable it to steer action."

Environmental indicator criteria and frameworks have been used to help in their selection and presentation.

It can be considered, for example, that there are major subsets of environmental indicators in-line with the Pressure-State-Response model developed by the OECD. One subset of environmental indicators is the collection of ecological indicators which can include physical, biological and chemical measures such as atmospheric temperature, the concentration of ozone in the stratosphere or the number of breeding bird pairs in an area. These are also referred to as "state" indicators as their focus is on the state of the environment or conditions in the environment. A second subset is the collection of indicators that measure human activities or anthropogenic pressures, such as greenhouse gasemissions. These are also referred to as "pressure" indicators. Finally, there are indicators, such as the number of people serviced by sewage treatment, which track societal responses to environmental issues.

Environmental indicators, in turn, should be considered as a subset of sustainable development indicators which are meant to track the overall sustainability of a society with respect to its environmental, social and economic integrity and health.

A common framework spearheaded by the European Environment Agency is the "DPSIR" or "drivers, pressures, state, impact, response" framework. Drivers and pressures are indicators of the human activities and resulting pressures on the environment in the form of pollution or land-use change, for example. State and impact indicators are the resulting conditions in the environment and the implications for the health of ecosystems and humans. The response indicators measure the reaction of human society to the environmental issue. Criteria tend to focus on three key areas - scientific credibility, policy/social relevance and practical monitoring and data requirements.

Environmental indicators are used by governments, non-government organizations, community groups and research institutions to see if environmental objectives are being met, to communicate the state of the environment to the general public and decision makers and as a diagnostic tool through detecting trends in the environment.

Environmental indicators can be measured and reported at different scales. For example, a town may track air quality along with water quality and count the number of rare species of birds to estimate the health of the environment in their area. Indicators are developed for specific ecosystems, such as the Great-Lakes in North America. National governments use environmental indicators to show status and trends with respect to environmental issues of importance to their citizens. Others have attempted to monitor and assess the state of the planet using indicators.

The type of indicators selected or developed should be partially based on who will be using the information from the indicators. There are generally three possible audiences to consider, each with different information needs. These audiences are:

1) technical experts and science advisors, 2) policy-makers, decision makers and resource managers, and 3) general public and media.

The technical experts and scientists will be interested in detailed and complex indicators. These indicators should have scientific validity, sensitivity, responsiveness and have data available on past conditions. The audience that includes policy-makers and resource managers will be concerned with using indicators that are directly related to evaluating policies and objectives. They require their indicators to be sensitive, responsive and have historical data available like the technical audience, but they are also looking for indicators that are cost-effective and have meaning for public awareness. Finally, the general public responds to indicators that have clear and simple messages and are meaningful to them, such as the UV index and the air quality index.

6.6 URBAN AREAS

An urban area is characterized by higher population density and vast human features in comparison to areas surrounding it. Urban areas may be cities, towns or conurbations, but the term is not commonly extended to rural settlements such as villages and hamlets.

Urban areas are created and further developed by the process of urbanization. Measuring the extent of an urban area helps in analyzing population density and urban sprawl, and in determining urban and rural populations.

Unlike an urban area, a metropolitan area includes not only the urban area, but also satellite cities plus intervening ruralland that is socio-economically connected to the urban core city, typically by employment ties through commuting, with the urban core city being the primary labor market. In fact, urbanized areas agglomerate and grow as the core population/economic activity center within a larger metropolitan area or envelope.

In the US, Metropolitan areas tend to be defined using counties or county sized political units as building blocks of much larger, albeit more condensed population units. Counties tend to be stable political boundaries; economists prefer to work with economic and social statistics based on metropolitan areas. Urbanized areas are a more relevant statistic for determining per capita land usage and densities.

It is difficult to compare countries based on the percentage of urban population since many countries have different definitions of what size population is necessary to make a community "urban."

In Sweden and Denmark, a village of 200 people is counted as an "urban" population but it takes a city of 30,000 in Japan. Most other countries fall somewhere in between. Australia and Canada use 1000, Israel and France use 2000 and the United States and Mexico call a town of 2500 residents urban.

Due to these differences, we have a problem with comparisons. Let us assume that in Japan and in Denmark there are 100 villages of 250 people each. In Denmark, all of these 25,000 people are counted as "urban" residents but in Japan, the residents of these 100 villages are all "rural" populations. Similarly, a single city with a population of 25,000 would be an urban area in Denmark but not in Japan.

Japan is 78% and Denmark is 85% urbanized. Unless we are aware of what size of a population makes an area urban we can not simply compare the two percentages and say "Denmark is more urbanized than Japan."

Below is a table with some countries, the minimum population which makes an area urban and the percent of the country which is "urbanized." Notice that some countries with a higher minimum population have a lower percentage of urbanized population.

Country	Min. Pop.	Urban Pop.
Sweden	200	83%
Denmark	200	85%
South Africa	500	57%
Australia	1000	85%
Canada	1000	77%
Israel	2000	90%
France	2000	74%
United States	2500	75%
Mexico	2500	71%
Belgium	5000	97%
Iran	5000	58%
Nigeria	5000	16%
Spain	10,000	64%
Turkey	10,000	63%
Japan	30,000	78%

Hartshorn, Truman A. Interpreting the City: An Urban Geography. 1992.

Famighetti, Robert (ed.). The World Almanac and Book of Facts. 1997.

6.7 URBAN DECAY

Urban decay (or urban rot) is the process whereby a previously functioning city, or part of a city, falls into disrepair and decrepitude. It may feature deindustrialization, depopulation or changing population, economic restructuring, abandoned buildings, high local unemployment, fragmented families, political disenfranchisement, crime, and a desolate, inhospitable city landscape.

Since the 1970s and 1980s, urban decay has been associated with Western cities, especially in North America and parts of Europe. Since then, major structural changes in global economies, transportation, and government policy created the economic and then the social conditions resulting in urban decay.

The effects counter the development of most of Europe and North America; in countries beyond, urban decay is manifested in the peripheral slums at the outskirts of a metropolis, while the city center and the inner city retain high real estate values and sustain a steadily increasing populace. In contrast, North American and British cities often experience population flights to the suburbs and exurbcommuter towns, i.e., white flight. Another characteristic of urban decay is blight-the visual, psychological, and physical effects of living among empty lots, buildings and condemned houses. Such desolate properties are socially dangerous to the community because they attract criminals and street gangs, contributing to the volume of crime.

Urban decay has no single cause; it results from combinations of inter-related socio-economic conditions-including the city's urban planning decisions, the poverty of the local populace, the construction of freeway roads and rail road lines that bypass the area, depopulation by suburbanization of peripheral lands, real estate neighborhood redlining, and xenophobic immigration restrictions.

6.7.1 Background of Urban Decay

During the Industrial Revolution, from the late eighteenth century to the early nineteenth century, rural people moved from the country to the cities for employment in manufacturing industry, thus causing the urban population boom. However, subsequent economic change left many cities economically vulnerable. Studies such as the Urban Task Force, the Urban White Paper, and a study of Scottish cities posit that areas suffering industrial decline-high unemployment, poverty, and a decaying physical environment-prove "highly resistant to improvement".

Changes in means of transport, from the public to the private - specifically, the private motor car - eliminated some of the cities' public transport service advantages, e.g., fixed-route buses and trains. In particular, at the end of World War II, many political decisions favored suburban development and encouraged suburbanization, by drawing city taxes from the cities to build new infrastructure for remote, racially-restricted suburban towns. That was the context of racial discrimination exercised as "white flight", the middle- and upper-class abandonment of U.S. cities, and the start of urban sprawl; only the non-white and the poor inhabited the cities.

After World War II, Western economies lifted tariffs and outsourced most of their manufacturing industries and businesses overseas, where foreign labor is cheaper than domestic. During the change from a manufacturing to a services economy, buying an automobile became economically feasible for most people. In the U.S., the federal government legislated discriminatory lending practices for theFederal Housing Administration (FHA) via redlining.

Later, under president Dwight D. Eisenhower, urban centers were drained further through the building of the Interstate Highway System. In North America this shift manifested itself in strip malls, suburban retail and employment centers, and very low-density housing estates. Large areas of many northern cities in the United States experienced population decreases and a degradation of urban areas.Inner-city property values declined and economically disadvantaged populations moved in. In the U.S., the new inner-city poor were often African-Americans that migrated from the South in the 1920s and 1930s. As they moved into traditional white European-American neighborhoods, ethnic frictions served to accelerate flight to the suburbs. In Western Europe the experience differs, in that the effect was often unknowingly assisted by public sector policies designed to clear 18th- and 19th-century slum areas and movements of people out into state-subsidized, lower-density suburban housing.

On continental Europe and Oceania, the historical core of major cities has usually remained relatively affluent; it is generally the inner-city districts and the edge-of-town suburbs made up of single-class state-subsidised housing, such as the French "cités" or "banlieues" and British council estates, which suffer the worst decay and blight. Due to higher population densities in Europe, economics dictates that extremely low-density housing would be impractical.

6.7.2 Examples of Decay

The car manufacturing sector was the base for Detroit's prosperity, and employed the majority of its residents. When the industry began relocating outside of the city, it experienced massive population loss with associated urban decay, particularly after the 1967 riots. To ensure he had a stable electoral base Mayor Coleman Young cut back on police and fire services, while lobbying for federally funded public housing. He increased the city's commuter and income taxes, encouraging those residents of Detroit who could to move out. According to the U.S. Census, in 1950 the city's population was around 1.85 million; by 2010, this had declined to 714,000, a loss of 1,136,000 people (61%). In addition, the homeless population has grown, and there are many abandoned structures in Detroit.

Britain experienced severe urban decay in the 1970s and 1980s. Major cities like Glasgow, the towns of the South Wales valleys, and some of the major industrial cities like Birmingham, Manchester,Liverpool, Newcastle, and east London, all experienced population decreases, with large areas of 19th-century housing experiencing market price collapse.

Large French cities are often surrounded by decayed areas. While city centers tend to be occupied mainly by middle- and upper-class residents, cities are often surrounded by large mid- to high-rise housing projects. The concentration of poverty and crime radiating from the developments often causes the entire suburb to fall into a state of urban decay, as more affluent citizens seek housing in the city or further out in semi-rural areas. In November 2005, the decaying northern suburbs of Paris were the scene of severe riots sparked in part by the substandard living conditions in public housing projects.

6.7.3 Response to Urban Decay

The main responses to urban decay have been through positive public intervention and policy, through a plethora of initiatives, funding streams, and agencies, using the principles of New Urbanism. Gentrification has also had a significant effect, and remains the primary means of a "natural" remedy.

In the United States, early government policies included "urban renewal" and building of large scale housing projects for the poor. Urban renewal demolished entire neighborhoods in many inner cities; in many ways, it was a cause of urban decay rather than a remedy. Housing projects became crime-infested mistakes. These government efforts are now thought by many to have been misguided. For multiple reasons, some cities have rebounded from these policy mistakes. Meanwhile, some of the inner suburbs built in the 1950s and 60s are beginning the process of decay, as those who are living in the inner city are pushed out due to gentrification.

In Western Europe and Asia, where land is much less in supply and urban areas are generally recognised as the drivers of the new information and service economies, urban regeneration has become an industry in itself, with hundreds of agencies and charities set up to tackle the issue. European cities have the benefit of historical organic development patterns already concurrent to the New Urbanist model, and although derelict, most cities have attractive historical quarters and buildings ripe for redevelopment.

In the inner-city estates and suburban cités, the solution is often more drastic, with 1960s and 70s state housing projects being totally demolished and rebuilt in a more traditional European urban style, with a mix of housing types, sizes, prices, and tenures, as well as a mix of other uses such as retailor commercial. One of the best examples of this is in Hulme, Manchester, which was cleared of 19th-century housing in the 1950s to make way for a large estate of high-rise flats. During the 1990s, it was cleared again to make way for new development built along new urbanist lines.

6.7.4 Contributing Factors to Urban Decay

Homelessness

Homelessness describes the condition of people without a regular dwelling. People who are homeless are unable or unwilling to acquire and maintain regular, safe, and adequate housing, or lack "fixed, regular, and adequate night-time residence." The legal definition of "homeless" varies from country to country, or among different entities or institutions in the same country or region.

The term homeless may also include people whose primary night-time residence is in a homeless shelter, a warming center, a domestic violence shelter or other ad hoc housing situation. American Government homeless enumeration studies also include persons who sleep in a public or private place not designed for use as a regular sleeping accommodation for human beings.

In western countries the large majority of homeless are men (75-80%), with single males particularly overrepresented.

The basic problem of homelessness is the need for personal shelter, warmth and safety. Other difficulties include:

- Personal security, quiet, and privacy, especially for sleeping
- Safekeeping of bedding, clothing and possessions, which may have to be carried at all times
- Hygiene and sanitary facilities
- Cleaning and drying of clothes
- Obtaining, preparing and storing food in quantities
- Keeping contacts, without a permanent location or mailing address
- Hostility and legal powers against urban vagrancy.

Homeless people face many problems beyond the lack of a safe and suitable home. They are often faced with many social disadvantages also, reduced access to private and public services and reduced access to vital necessities:

- Reduced access to health care and dental services.
- Limited access to education.
- Increased risk of suffering from violence and abuse.
- General rejection or discrimination from other people.
- Loss of usual relationships with the mainstream
- Not being seen as suitable for employment.
- Reduced access to banking services
- Reduced access to communications technology

There is sometimes corruption and theft by the employees of a shelter as evidenced by a 2011 investigative report by FOX 25 TV in Boston wherein a number of Boston public shelter employees were found stealing large amounts of food over a period of time from the shelter's kitchen for their private use and catering.

Modern Ruins

Modern ruins is a neologism referring to ruins ofarchitecture constructed in the recent past, generally in the most recent century, or since the 19th century.

The term is most frequently used by people performingurban exploration of man-made architecture that is abandoned or no longer accessible to the generalpublic, such as structures abandoned through the process of urban decay. Enough documentation on these sites may have been lost over time that this unscientific exploration resembles archaeology of ancient ruins in the methods used to collect information.

Public Housing

Public housing is a form of housing tenure in which the property is owned by a government authority, which may be central or local. Social housing is an umbrella term referring to rental housing which may be owned and managed by the state, by non-profit organizations, or by a combination of the two, usually with the aim of providingaffordable housing. Social housing can also be seen as a potential remedy to housing inequality.

Although the common goal of public housing is to provide affordable housing, the details, terminology, definitions of poverty and other criteria for allocation vary within different contexts.

Slum

A slum, as defined by United Nations agency UN-HABITAT, is a run-down area of a city characterized by substandard housing and squalor and lacking in tenure security.

Shrinking Cities

Shrinking cities are cities that are experiencing acute population loss. Deindustrialisation and out-migration are some of the common reasons that cities shrink. In the United States, this problem is most commonly associated with the Rust Belt, while parts of Eastern Europe also experience similar problems. Since the infrastructure of such cities was built to support a larger population, its maintenance can become a serious concern. A related phenomenon is counter urbanization.

Global Context

The current population of the world is 6.5 billion people. Currently about 3 billion live in cities. By 2030, population in cities is expected to be 4.1 billion. Though some cities are continuing to attract residents, others are losing residents at varying rates.[vague] Saskia Sassen's "global cities" theory forecasts these urban "winners" and "losers": the winners being those cities with agglomerated financial and specialized services and the losers being those with outdated industrial infrastructure and manufacture economies.

In the last 50 years, about 370 cities with more than 100,000 residents have undergone population losses of more than 10%. More than 25 percent of the depopulating cities are in the United States, and most of those are in the midwest.

6.8 URBAN EXPLORATION

Urban exploration (often shortened asurbex or UE) is the examination of the normally unseen or off-limits parts of urban areas or industrial facilities. Urban exploration is also commonly referred to as infiltration, although some people consider infiltration to be more closely associated with the exploration of active or inhabited sites. It may also be referred to as draining urban spelunking, urban caving, or building hacking.

The nature of this activity presents various risks, including both physical danger and the possibility of arrest and punishment. Many, but not all, of the activities associated with urban exploration could be considered trespassing or other violations of local or regional laws, including-but not limited to-invasion of privacy and certain broadly-interpreted anti-terrorism laws.

6.8.1 Targets of Exploration

Urban explorers often attempt some or all of these subsets of urban exploration.

Abandonments

"Haikyo" redirects here.

Ventures into abandoned structures are perhaps the most common example of urban exploration. At times, sites are entered first by locals, and may sport large amounts of graffiti and other acts of vandalism. Explorers face various risks in abandoned structures including collapsing roofs and floors, broken glass, guard dogs, the presence of chemicals and other harmful substances, most notably asbestos, hostile squatters and motion detectors. Some explorers wear respirators to protect their airways and proper attire to protect their bodies.

Although targets of exploration vary from one country to another, high-profile abandonments include amusement parks, grain elevators, factories, power plants, missile silos, fallout shelters, hospitals, asylums, schools, poor houses, and sanatoriums.

In Japan, ruins are known as Haikyo, (literally "abandoned place") but the term is synonymous with the practice of urban exploration. Haikyo are particularly common in Japan because of its rapid industrialization (e.g., Hashima Island), damage during World War II, the 1980s real estate bubble and the 2011 earthquake and sunami.

Many explorers find decay of uninhabited space to be profoundly beautiful and some are also proficient freelance photographers. Abandoned locations can be, at times, heavily guarded with motion sensors and active security. Others are more easily accessible and carry less risk of discovery. Abandoned sites are also popular among historians, preservationists, architects, archaeologists, industrial archaeologists, and ghost hunters.

Active Buildings

Another aspect of urban exploration is the practice of exploring active or in use buildings which includes gaining access to secured or "member-only" areas, mechanical rooms, roofs, elevator rooms, abandoned floors and other normally unseen parts of working buildings. The term "infiltration" is often associated with the exploration of active structures. People entering restricted areas may be committing trespass and civil prosecution may result.

Catacombs

Catacombs such as those found in Paris, Rome, Odessaand Naples have been investigated by urban explorers. The Mines of Paris, comprising much of the underground tunnels that are not open to public tourism like the catacombs, have been considered the "Holy Grail" by some due to their extensive nature and history. Explorers of these are known as cataphiles.

Sewers and Storm Drains

Entry into storm drains, or draining, is another common form of urban exploration. Groups devoted to the task have arisen, such as the Cave Clan in Australia. Draining has a specialized set of guidelines, the foremost of which is "When it rains, no drains!" The dangers of becoming entrapped, washed away, or killed increase dramatically during a heavy rainfall.

A small subset of explorers enter sanitary sewers. Sometimes they are the only connection to caves or other subterranean features. Sewers are among the most dangerous locations to explore owing to risk of poisoning by buildups of toxic gas (commonly methane and hydrogen sulfide). There have been numerous fatalities around the world where people are overcome by toxic gas from sewers.

Transit Tunnels

Exploring active and abandoned subway and underground railway tunnels, bores and stations is often considered to be trespassing and can result in civil prosecution. As a result, this type of exploration is rarely publicized. One important exception to this is the abandoned subway of Rochester, NY, the only American city to have an abandoned, formerly used, subway system.

Utility Tunnels

Universities and other large institutions, such as hospitals, often distribute steam for heating buildings and autoclaves from a central heating plant. These high pressure steam pipes are generally run through utility tunnels, which are often accessible solely for the purposes of maintenance. Many of these steam tunnels, such as those on college campuses, often also have a tradition of exploration by its students. This was once called vadding at MIT, though students there now refer to it as roof and tunnel hacking.

Steam tunnels, in general, have been secured heavily in recent years, due to their use for carrying network backbones and perceived risk of their use in terrorist activities, safety and liability.

Some steam tunnels have dirt floors, no efficient lighting and have temperatures upwards of 45°C(113°F). Others have concrete floors, bright light, and feature a cool low-grade temperature. Most steam tunnels have large intake fans to bring in fresh air and push the hot air out the back. Most active steam tunnels do not contain airborne asbestos but proper breathing protection may be required for other hazards. It is wise to take proper care inside active utility tunnels, since pipes can spew boiling hot water from leaky valves, puddles pool at your feet, and forceful steam may leak inward resulting in burns and slips.

6.8.2 Popularity of Urban Exploration

The rise in the popularity of urban exploration can be attributed to its increased media attention. Recent television shows, such as Urban Explorers on the Discovery Channel, MTV's Fear, and the Ghost Hunting exploits of The Atlantic Paranormal Society have packaged the hobby for a popular audience. Talks and exhibits on urban exploration have appeared at the 5th and 6th Hackers on Planet Earth Conference, complementing numerous newspaper articles and interviews.

With the rise in the relative popularity of the hobby due to this increased focus, there has been increasing discussion on whether the extra attention has been beneficial to urban exploration as a whole. The unspoken rule of urban exploring is "take nothing but photographs, leave nothing but footprints", but because of the rising popularity, many individuals who may have other intentions are creating a concern among many property owners.

6.8.3 Safety and Legality

Urban exploration is a hobby that comes with a number of inherent dangers. Storm water drains are not designed with human access as their primary use. They can be subject to flash flooding and bad air. There have been a number of deaths in storm water drains, but these are usually during floods, and are normally not Urban Explorers.

Many old abandoned structures feature hazards such as unstable structures, unsafe floors, asbestos, carbon monoxide, carbon dioxide, exposed electrical wires and entrapment hazards.

Asbestos is a long term health risk for urban explorers, along with breathing in contaminants from dried bird feces, otherwise known as pigeon lung, a form of hypersensitivity pneumonitis. Urban explorers may use dust masks and respirators to alleviate this danger. Some sites are occasionally used by substance abusers for either recreation or disposal, and there may be used and/or infected syringe needles route, such as those commonly used with heroin.

The growing popularity of the activity has resulted not just in increased attention from explorers, but also from vandals and Law Enforcement. The illicit aspects of urban exploring, which may include trespassing and breaking and entering, have brought along with them critical articles in mainstream newspapers.

In Australia, the web-site of the Sydney Cave Clan was shut down by lawyers for the Roads and Traffic Authority of New South Wales, after they raised concerns that the portal could "risk human safety and threaten the security of its infrastructure." Another web-site belonging to the Bangor Explorers Guild was criticized by the Maine State Police for potentially encouraging behavior that "could get someone hurt or killed." Likewise, the Toronto Transit Commission has also used the Internet to crimp Subway Tunnel Explorations, going as far as to send Investigators to various Explorers' homes.

Jeff Chapman, who authored Infiltration, stated that genuine urban explorers "never vandalize, steal or damage anything." The thrill comes from that of "discovery and a few nice pictures." Some Explorers will also request permission for entry.

6.8.4 Urban Warfare

Urban warfare is combat conducted in urban areas such as towns and cities. Urban combat is very different from combat in the open at both the operational and tactical level. Complicating factors in urban warfare include the presence of civilians and the complexity of the urban terrain.

Some civilians may be difficult to distinguish from combatants such as armed militias and gangs, and particularly individuals who are simply trying to protect their homes from attackers. Tactics are complicated by a three-dimensional environment, limited fields of view and fire because of buildings, enhanced concealment and cover for defenders, below ground infrastructure, and the ease of placement of booby traps and snipers.

6.8.5 Urban Operations

Urban military operations in World War II often relied on large quantities of artillery fire and air support varying from ground attack fighters to heavy bombers. In some particularly vicious urban warfare operations such as Stalingrad and Warsaw, all weapons were used irrespective of their consequences.

However, when liberating occupied territory some restraint was often applied, particularly in urban settings. For example, Canadian operations in both Ortona and Groningen avoided the use of artillery altogether to spare civilians and buildings, and during the Battle of Manila in 1945, General MacArthur initially placed a ban on artillery and air strikes to save civilian lives.

Armies are bound by laws of war governing military necessity to the amount of force which can be applied when attacking an area where there are known to be civilians. Until the 1970s this was covered by customary law and IV Hague Convention "The Laws and Customs of War on Land" of 1907 and specifically articles 25-29. This has since been supplemented by the "Additional Protocols to the Geneva Conventions of 12 August 1949, and relating to the Protection of Victims of International and Non-International Armed Conflicts."

Sometimes distinction and proportionality, as in the case of the Canadians in Ortona, causes the attacking force to restrain from using all the force they could when attacking a city. In other cases, such as the Battle of Stalingrad and the Battle of Berlin, both armies considered evacuating civilians only to find it impractical.

When Russian forces attacked Grozny in 1999, large amounts of artillery fire were used. The Russian Army handled the issue of civilian casualties by warning the inhabitants that they were going to launch an all-out assault on Grozny and requested that all civilians leave the city before the start of the artillery bombardment.

Fighting in an urban environment can offer some advantages to a weaker defending force or to guerrilla fighters through ambush-induced attrition losses. The attacking army must account for three dimensions more often, and consequently expend greater amounts of manpower in order to secure a myriad of structures, and mountains of rubble.

Ferroconcrete structures will be ruined by heavy bombardment, but it is very difficult to demolish such a building totally when it is well defended. Soviet forces had to fight room by room; while defending the Red October Steel Factory during the Battle of Stalingrad, and in 1945, during the race to capture the Reichstag; despite heavy bombardment with artillery at point blank range.

It is also difficult to destroy underground or heavily fortified structures such as bunkers and utility tunnels; during the Battle of Budapest in 1944 fighting broke out in the sewers, as both Axis and Soviet troops used them for troops movement.

6.8.6 Urban Warfare Tactics

The characteristics of an average city include tall buildings, narrow alleys, sewage tunnels and possibly a subway system. Defenders may have the advantage of detailed local knowledge of the area, right down to the layout inside of buildings and means of travel not shown on maps.

The buildings can provide excellent sniping posts while alleys and rubble-filled streets are ideal for planting booby traps. Defenders can move from one part of the city to another undetected using underground tunnels and springambushes.

Meanwhile, the attackers tend to become more exposed than the defender as they must use the open streets more often, unfamiliar with the defenders' secret and hidden routes. During a house to house search the attacker is often also exposed on the streets.

Battle of Monterrey, Mexico

The Battle of Monterrey was the US Army's first major encounter with urban warfare. It occurred in September 1846 when the US Army under Zachary Taylor invaded the town. The US Army had no prior training in urban warfare and the Mexican defenders hid on rooftops, shot through loopholes, and stationed cannons in the middle of the city's streets. The houses at Monterrey were made of thick adobe, with strong double doors and few windows. The rooftops were lined with a two foot tall wall that acted as a parapet for the defending soldiers. Each home was a fort unto itself.

On September 21, 1846 the US Army which included some of its best soldiers, recent West Point graduates, marched down the city's streets and were cut down by the Mexican defenders. They could not see the men hidden behind walls, loopholes, or rooftops. They tried to march straight down the street until the intense fire drove them to hide in adjacent buildings. Taylor tried to move artillery into the city but it could not hit the well-hidden defenders any better than the US soldiers could. Two days later the US again assaulted the city from two sides and this time they fought differently.

Not wanting to repeat the mistakes of the 21st, General William Jenkins Worth listened to his Texan advisers. These men had fought in Mexican cities before at the Battle of Mier in 1842 and the Battle of Bexar in 1835. They understood that the army needed to "mouse hole" through each house and root out the defenders in close combat.

Worth's men used pick axes to chip holes in the adobe walls of the homes, or sometimes in the roof of the house from where the soldiers could drop in. Or they used ladders to climb to the top of a rooftop and assault the Mexican defenders in hand to hand combat. The typical assault on a home would include one man who would run to the door of the house and chip the door away with a pick axe while under covering fire. Once the door showed signs of weakening, 3-4 other soldiers would run to the door and barge in with the revolvers blazing. Worth lost few men on the 23rd using these new urban warfare techniques.

Battle of Berlin

A Soviet combat group was a mixed arms unit of about eighty men, divided into assault groups of six to eight men, closely supported by field artillery. These were tactical units which were able to apply the tactics of house to house fighting that the Soviets had been forced to develop and refine at each Festungsstadt (fortress city) they had encountered from Stalingrad to Berlin.

The German tactics in the battle of Berlin were dictated by three considerations: the experience that the Germans had gained during five years of war; the physical characteristics of Berlin; and the tactics used by the Soviets.

Most of the central districts of Berlin consisted of city blocks with straight wide roads, intersected by several waterways, parks and large railway marshalling yards. The terrain was predominantly flat but there were some low hills like that of Kreuzberg that is 66m above sea level.

Much of the housing stock consisted of apartments blocks built in the second half of the 19th century. Most of those, thanks to housing regulations and few elevators, were five stories high, built around a courtyard which could be reached from the street through a corridor large enough to take a horse and cart or small trucks used to deliver coal.

In many places these apartment blocks were built around several courtyards, one behind the other, each one reached through the outer courtyards by a ground-level tunnel similar to that between the first courtyard and the road. The larger, more expensive flats faced the street and the smaller, less expensive ones were found around the inner courtyards.

Just as the Soviets had learned a lot about urban warfare, so had the Germans. The Waffen-SS did not use the makeshift barricades erected close to street corners, because these could be raked by artillery fire from guns firing over open sights further along the straight streets.

Instead, they put snipers and machine guns on the upper floors and the roofs - a safer deployment as the Soviet tanks could not elevate their guns that high. They also put men armed with panzerfausts in cellar windows to ambush tanks as they moved down the streets. These tactics were quickly adopted by the Hitler Youth and the First World War Volkssturm veterans.

To counter these tactics, Soviet sub-machine gunners rode the tanks and sprayed every doorway and window, but this meant the tank could not traverse its turret quickly. The other solution was to rely on heavy howitzers (152 mm and 203 mm) firing over open sights to blast defended buildings and to use anti-aircraft guns against defenders posted on the higher floors.

Soviet combat groups started to move from house to house instead of directly down the streets. They moved through the apartments and cellars blasting holes through the walls of adjacent buildings, while others fought across the roof tops and through the attics.

These tactics took the Germans lying in ambush for tanks in the flanks. Flamethrowers and grenades were very effective, but as the Berlin civilian population had not been evacuated these tactics inevitably killed many civilians.

First Chechen War

During the First Chechen War most of the Chechen fighters had been trained in the Soviet armed forces. They were divided into combat groups consisting of 15 to 20 personnel, subdivided into three or four-man fire teams.

A fire team consisted of an antitank gunner, usually armed with a Russian made RPG-7s or RPG-18s, a machine gunner and a sniper. The team would be supported by ammunition runners and assistant gunners.

To destroy Russian armoured vehicles in Grozny, five or six hunter-killer fire teams deployed at ground level, in second and third stories, and in basements. The snipers and machine gunners would pin down the supporting infantry while the antitank gunners would engage the armoured vehicle aiming at the top, rear and sides of vehicles.

Initially the Russians were taken by surprise. Their armoured columns that were supposed to take the city without difficulty as Soviet forces had taken Budapest in 1956 were decimated in fighting more reminiscent of the Battle of Budapest in late 1944.

As in the Soviet assault on Berlin, as a short term measure they deployed self-propelled anti-aircraft guns to engage the Chechen combat groups, as their tank's main gun did not have the elevation and depression to engage the fire teams and an armoured vehicle's machine gun could not suppress the fire of half a dozen different fire teams simultaneously.

In the long term the Russians brought in more infantry and began a systematic advance through the city, house by house and block by block, with dismounted Russian infantry moving in support of armour. In proactive moves the Russians started to set up ambush points of their own and then move armour towards them to lure the Chechen combat groups into ambushes.

As with the Soviets tank crews in Berlin in 1945, who attached industrially produced wire mesh screens to the outside of their turrets to reduce the damage done by German panzerfausts, some of the Russian armour was fitted quickly with a cage of wire mesh mounted some 25-30 centimetres away from the hull armor to defeat the shaped charges of the Chechen RPGs.

Close-Quarters Battle

The term close-quarter battle refers to fighting methods within buildings, streets, narrow alleys and other places where visibility and maneuverability are limited.

Both Close-Quarters-Battle (CQB) and Urban Operations (UO) are related to urban warfare, but while UO refers mainly to the macromanagement factor (i.e. sending troops, using of heavy armoured fighting vehicles, battle management), CQB refers to the micromanagement factor-namely: how a small squad of infantry troops should fight in urban environments and/or inside buildings in order to achieve its goals with minimal casualties.

As a doctrine, CQB concerns topics such as:

- Weapons and ammunition most suitable for the mission.
- Extra gear, such as bulletproof vests and night vision devices.
- Accurate explosives.
- Routines and drills for engaging the enemy, securing a perimeter, clearing a room, etc.
- Team maneuvers.
- Methods and tactics.

It should be noted that military CQB doctrine is different from police CQB doctrine, mainly because the military usually operates in hostile areas while the police operates within docile populations.

Armies that often engage in urban warfare operations may train most of their infantry in CQB doctrine.

6.8.7 Urban Field

The urban field is a form of urban habitat of relatively high density involving a good transportation system and a broad array of economic, social and recreational opportunities. Even though the sections of the urban field may still be in agricultural use, the area is nonetheless urbanized because anywhere within it a person is able to connect his/her home to telephones, radio and television facilities, electricity, gas, water supply systems and a network of freeway and primary roads.

An urban field is a field in the centre of an urban area, with a boundary wall and road frontage. Planning permission is not essential for a field to be classed as an 'urban field', although Outline Planning Permission is desirable. It is centered on and dominated to a certain extent by a metropolitan area of at least 200,000-300,000 people. It's outer limits can be defined by two things.

1. The maximum time or distance that most people are prepared to commute.
2. The time or distance that most people are prepared to spend traveling to or from weekly or weekend recreational activities. The daily commute perspective defines the "hardcore" of the urban field and results in regions of about a 40-50 mile radius from the central metropolitan area. The weekend recreational area perspective results in a much wider field of about 100 miles with far less determinate boundaries.

Extent of Urban Field

The actual extent of the urban field depends upon the freeway facilities available and the density of metropolitan development. In the South, Midwest and on the Prairies the fields may be quite extensive. In the East and the area bordering the lower Great Lakes where metropolitan densities are rather high, the urban field are much more restricted and may run into each other to form "galaxies". In these higher density areas the extent of the field, as defined as criterion 1, may be only 20-30 miles and the recreational field about 75 miles.

Implications of Urban Field

The planning implications of the urban field have become quite obvious. These large areas are inhabited and used by urbanites but they involve vast quantities of land and often very sensities environmental conditions. The management questions involve the preservation of valuable agricultural land and the protection of this land from abuse. A related issue that is becoming more important involves the pricing of farms and farmland out of the realm of wealthy individuals seeking weekend retreats or good speculative investments. The management of recreational resources and the provision of land for campsites and second homes has also become a question of concern, along with questions of conservation. Thus, the definition of an urban area according to the extent of its urban field has raised an enormous number of questions related to the management of agricultural, natural and recreational resources around a major metropolitan center.

CHAPTER

ECOLOGICAL URBANISM 7

7.1 INTRODUCTION

Ecological urbanism draws from ecology to inspire an urbanism that is more socially inclusive and sensitive to the environment. It is less ideologically driven, than green urbanism or sustainable urbanism. In many ways, ecological urbanism is an evolution of, and a critique of, Landscape Urbanism arguing for a more holistic approach to the design and management of cities. This type of urbanism has a central scope four main objectives, compactness, complexity, efficiency and stability. This model of Urbanism has as a goal to tackle the current challenges of society by intertwining sustainability and urban occupation models. The term appeared first in 1998 as "EcoUrbanism" in a book by Architect and Planner Miguel Ruano, who defined it as "the development of multi-dimensional sustainable human communities within harmonious and balanced built environments". The term was used later in April 2003 at a conference at the University of Oregon, and again in 2006 in a paper by Jeffrey Hou. The phrase was used by Mohsen Mostafavi in 2007 in Intervention Architecture and in a lecture at the Canadian Centre for Architecture, ecological urbanism as a project was largely started at Harvard University Graduate School of Design, with a conference, and exhibition, and book.

Arguing for a "new ethics and aesthetics of the urban," the 656-page Ecological Urbanism book, edited by Mohsen Mostafavi with Gareth Doherty, was published in May 2010 by Lars Müller Publishers. The book follows the conference, and exhibition, held at the GSD in 2009. The book has a long list of contributors, including Rem Koolhaas, Homi K. Bhabha, Mitchell Joachim, Andrea Branzi, and about 130 others. A blog during the conference is part of the book. According to Architecture Today, the book is "one of the few books that recognises and articulates how, if this systems-based approach is to be successful, it needs to design, integrate and express complex systems and social processes in ways that are fundamentally humane." The book has been reviewed and cited in many publications, including Metropolis Magazine, The Journal of Landscape Architecture, and Cities magazine. Events and discussions on the book have been held at the 2010 Venice Biennale, the Storefront for Art and Architecture in New York, and at the Van Alen Institute in New York.

In his introduction to the Ecological Urbanism, "Why Ecological Urbanism? Why Now?", extracted in Topos: The International Review of Landscape Architecture and Urban Design, Mostafavi asks: "Increased numbers of people and cities go hand in hand with a greater exploitation of the world's limited resources. Every year, more cities are feeling the devastating impacts of this situation. What are we to do? What means do we have as designers to address this challenging reality?

Jeb Brugmann in his book Welcome to the Urban Revolution: How Cities Are Changing the World says we need to become "masters of a stable, just, and ecological urbanism." For Brugmann, "The first step towards ecological urbanism is increasing the energy and nutrient productivity within the city, but the only way to move sufficiently from extractive mode to a sustainable productive mode is to think, design, and develop at the scale of the City."

There have been a number of recent conferences and lectures on the project of ecological urbanism, including: The New Aesthetics and Ecological Urbanism at Peking University in October 2010, New Zealand Institute of Landscape Architects Spring Lecture Series: Ecological Urbanism: A Prospectus for the Super City, in October 2010. The University of Washington's urban initiative included a seminar on Now Urbanism and talks "on Ecological Urbanism, Ecological Design for Healthy Cities, Networked Urbanism, and America's War on Immigrants". Eco-Urbanism: towards sustainable city living, was hosted by Nottingham University at the Shanghai Expo in August 2010.

In addition to courses at the GSD in 2008 and 2011, there have been courses on ecological urbanism at the Bergen School of Architecture, Oslo School of Architecture and Design, and Yale University. It is on numerous course syllabi, including Advanced Design Theories 2010 at Florida International University. Arizona State University expanded on the subject with a lecture by Charles Anderson, ASLA. Anderson stated that "ecology is concerned with the relationships between all organisms and the environment. Together and coupled with aesthetic and expressionist design principles, they form the foundation for urban design."

Mohsen Mostafavi, Gareth Doherty, Marina Correia, Ana Maria Duran Calisto, Giannina Braschi, and Luis Valenzuela produced a Latin American Ecological Urbanism project, resulting in the 2019 Spanish and Portuguese book "Urbanismo ecológico en América Latina".

Ecological urbanism has been criticized as an idea that is loosely defined from a set of flashy projects. These are expensive schemes with a commercial and esthetic purpose that satisfy a local or regional ambition to invest in ecology or sustainability without posing a more globally applicable approach. A true merger of landscape architecture with the field of Urban Ecology lacks. From this criticism Frederick Steiner introduced landscape ecological urbanism as an approach that can include the field of urban ecology and Wybe Kuitert has shown how such integrative planning and management of the city should rely on analysis. Discerning the potential quality of wild nature in the city is a first step to see how new urban ecology might be developed. Potential vegetation maps for a city are the tool to this end.

Ecological Urbanism is explained as a successor to Landscape Urbanism without the difference between the two approaches, and the terms used in the new approach, being defined. Jason King sees it as an inadequately explained addition to a list of 61 other 'Fill-in-the-blank Urbanisms'. Tom Turner welcomes landscape and ecological urbanism as 'the most significant contributions to landscape design theory since the landscape architecture profession was launched in the mid-nineteenth century' but is 'unpersuaded by the change of name'

7.2 URBAN VITALITY

Urban vitality is the quality of those spaces in cities that are capable of attracting heterogeneous people for different types of activities throughout varied time schedules. The areas of the city with high vitality are perceived as alive, lively or vibrant and they tend to attract people to carry out their activities, stroll or stay. However, the areas of low vitality repel people and can be perceived as unsafe.

The urban vitality index is a measure of this quality and in recent years it has become a fundamental tool for planning urban policies, especially for the intervention of spaces with low vitality. In addition, it is used for proper management of spaces with high vitality, as the success of certain areas can lead to processes of gentrification and touristification that, paradoxically, end up reducing the vitality that made them popular.

The concept of urban vitality is based on the contributions of Jane Jacobs, especially those of her most influential work, The Death and Life of Great American Cities. Jacobs criticized in the 1960s the modern and rationalist architecture defended by Robert Moses or Le Corbusier whose protagonist was the private car. She argued that these types of urban planning overlooked and oversimplified the complexity of human life in diverse communities. She opposed large-scale urban renewal programs that affected entire neighborhoods and built freeways through inner cities. Instead, she advocated for dense mixed-use development and walkable streets, with "eyes on the street" of passers-by helping to maintain public order.

Currently, the concept of urban vitality is revaluing Mediterranean urbanism and its history, in which public space, pedestrianity and squares are of great importance as centers of interaction and social cohesion, in opposition to the Anglo-Saxon urbanism of large urban infrastructures, long distances and car-centric.

7.3 GREEN URBANISM

Green urbanism has been defined as the practice of creating communities beneficial to humans and the environment. According to Timothy Beatley, it is an attempt to shape more sustainable places, communities and lifestyles, and consume less of the world's resources. Urban areas are able to lay the groundwork of how environmentally integrated and sustainable city planning can both provide and improve environmental benefits on the local, national, and international levels. Green urbanism is interdisciplinary, combining the collaboration of landscape architects, engineers, urban planners, ecologists, transport planners, physicists, psychologists, sociologists, economists and other specialists in addition to architects and urban designers.

7.3.1 Urbanization and Environmental Degradation

Urbanization and environmental consequences has always moved hand in hand. Odum in 1989 has called cities as 'parasites' on natural and domesticated environment, since it makes no food, cleans no air and cleans only a little amount of water for reuse and Mayur (1990) has argued that such disharmony may result in environmentally catastrophic events. Leitmann mentioned such critical urban environmental problems as the 'brown agenda' which deals with both environmental health and industrialization. He further pointed out that throughout the 19th century; developing countries were more concerned of the public health impacts of poor sanitation and pollution. Moreover, he figured out the links between cities and ecosystems into three phases. Early Urbanization phase, starting from 3000 BCE to 1800 CE, was of more productive agricultural techniques yielding a surplus that was able to support non-agricultural concentrations of people. In second phase, Urban Industrialization (1800 CE - 1950 CE), energy consumption, particularly fossil fuels, was increased rapidly with mechanization of production. Since the 1950s the city/environment relationship has entered into third phase, Global interdependence, with rapid population growth and globalization of economy. Cities became the nodal points for large and globally interconnected flows of resources, wastes, and labor. Also, environmental problems are local, regional and global in scale, with cities increasingly contributing to global environmental damage.

Since the 1960s, the number of people living within towns and cities has grown exponentially. According to the United Nations, 2009 is the year that the number of people living in urban areas surpassed those in rural areas. With the current urbanized growth rate, it is projected that by 2050, the global population living in urban areas will be at 68% or slightly over 6.5 billion, with a global population of 9.7 billion. With such a large population making the change to live in an urban area, it is vital to the health of cities to be able to provide enough resources and energy to the population by the means of environmentally sustainable resources. Meeting the growing production and consumption needs of urban populations causes an immense amount of strain on the surround suburban and rural communities. Nearby ecosystems can easily become compromised due to the physical expansion of urban areas.

Rydin (2010) accused the cities as both villains and victims of climate change pattern. Climate change affecting urban sustainability in regards to temperature increase which may exacerbate urban heat island (UHI) effect and rainfall patterns (Rydin, 2010). Some other cities may also go through environmental catastrophes, like cyclone and storm, coastal erosion, sea-level rise, ground instability and changes in biodiversity. The whole scenario called for an urgent need to focus on rebuilding the urban ecosystem with given emphasis on the human settlements.

7.3.2 History

A glimpse on the history of green urbanism of the U.S. as found in Karlenzig's, et al. 'How Green is Your City' book. The concept had a gradual start in the late 1800s, when some large cities of the United States (U.S.) started using advanced drinking water, sewage and sanitary systems. Consecutively, public parks and open spaces were implemented in New York City. At the end of the World War II, the US government offered its citizens affordable housing through easy loans to boost up the city population and also introduced a new federal Interstate System; combined with a rise of automobile ownership, this gave way to a novel way of life called 'Suburbia'. Meanwhile, in the 1950s, the inhabitants of other industrial cities, including Chicago, Detroit, St. Louis, Cleveland and Philadelphia, had already experienced greener suburban pastures. But all those green trees died because of old age or pollution, and were not replaced. The first book describing the comprehensive rebuilding of cities toward balance with nature is "Los Angeles: a History of the Future" (1982) by Paul Glover (activist). After a decade of the 'Urban Renaissance', the term used by Richard Rogers, came into light in 1990. Europe was never far behind to endorse urban sustainability. 'The Green Paper on the Urban Development' published in 1990 has been considered as a 'milestone' document in promoting sustainable city projects as a solution to global environmental role. Lehmann (2010) mentioned that since then, cities have engaged themselves in a global-scale competition with each other in three distinct areas. Firstly, to be regarded as an attractive, creative place and a cultural hub to attract highly skilled workers and Melbourne, Australia was strong competition with arts, museum and university; secondly, to get recognition as a place for secure investment, mention worthy, Dubai, Shanghai, and Singapore have topped in attracting and facilitating global investment capital; and thirdly, to become a leader of green vision by technological advancement and offering environmentally sound lifestyles and also providing green jobs; Hannover and Copenhagen have done well in this field.

After the Earth Summit in 1992, different terms, including, sustaining cities, sustainable cities, sustainable urbanism, green city, eco-towns, eco districts and eco-cities, have tried to reduce environmental impacts of cities and achieve sustainable development. Thus, to live more peacefully there. Both green and sustainable cities present fundamental opportunities to apply new technologies. For example, public transport, district heating, green buildings, and green design also bring major lifestyle changes such as, walking, bicycling, and reducing energy consumption. The major agenda of the above-mentioned cities are tackling global climate change, biodiversity loss, and also lifting themselves as 'hosts' of all environmental challenges.

It has been argued that the focus of these theories are mainly on adjusting the relationship between the city and nature and also creating new cities other than renovating existing cities. To address the gap, Timothy Beatley and Steffen Lehmann used the 'green urbanism' theory that aims to transform existing cities from fragmentation to compaction.

7.3.3 Principles of Green Urbanism

The three main contemporary planners who have contributed to Green Urbanism Thought are Steffen Lehmann, Timothy Beatley, and Peter Newman. Each has their own principles that define green urbanism, and what quality an urban space needs to attain to be green urbanist. All of these principles are based on the understanding that urbanization is a key driver of carbon emissions, resource depletion and environmental degradation. These are zero fossil-fuel energy use, zero waste and zero emissions especially aimed for low-to-no-carbon emissions.

Lehmann's Principles of Green Urbanism

Lehmann uses a strategic case study of the seaport city of Newcastle, New South Wales, Australia to build his definition of Green Urbanism. According to Lehmann, there are 15 such principles of green urbanism that are practical and holistic, including all the aspects needed to achieve sustainable development and encouraging best practice models. The principles are as follows:

1. **Climate and Context:** Based on climatic condition prior to selected city, every sustainable design project needs to maintain a complexity within biodiversity, eco-system or neighborhood layout. Enhance the opportunities offered by topographies and natural settings and use of the buildings' envelope to filter temperature, humidity, light, wind and noise.
2. **Renewable Energy for Zero CO_2 Emissions:** Transform city districts into local power stations of renewable energy sources including solar PV, solar thermal, wind on-and-off-shore, biomass, geothermal power, mini-hydro energy and other new technologies. Some most promising technologies are in building – integrated PV, urban wind turbines, micro CHP and solar cooling.
3. **Zero Waste City:** Waste prevention is better than the treatment or cleaning-up after waste is formed. So cities should adopt zero-waste urban planning in line with the manufacturing of metals, glass, plastics, paper into new products and better understanding of nutrient flows is needed to control global nitrogen cycle.
4. **Water:** Cities can be used as a water catchment area by educating the inhabitants in water efficiency, promoting rainwater collection and using waste water recycling

and storm water harvesting techniques. In terms of food yielding level, less water needed and drought resistant crops can be developed.

5. **Landscape, Gardens and Biodiversity:** Introduce inner-city gardens, urban farming/agriculture and green roofs to maximise the resilience of the eco-system through urban landscape thus to mitigate UHI effect. Plants can be used for air-purification and narrowing of roads for urban cooling. Moreover, preserving green space, gardens and farmland, maintaining a green belt around the city is necessity to absorb CO_2.
6. **Sustainable transport and good public space:** Compact and poly-centric cities: An integration of non-motorised transport, such as, cycling or walking and bi-cycle or pedestrian-friendly environment with safe bicycle ways, eco-mobility concepts and smart infrastructure that is electric vehicles, integrated transport system of bus transit, railway and bike stations, improved public space networks and connectivity and a focus on transport-oriented development (Green TODs).
7. **Local and sustainable materials with less embodied energy:** City construction by using regional, local materials with less embodied energy and applying pre-fabricated modular systems.
8. **Density and retrofitting of existing districts:** The city is with retrofitted districts, urban infill, and densification/intensification strategies for existing neighbourhoods.
9. **Green buildings and districts, using passive design principles:** The city, here, applies deep green building design strategies and offers solar access for all new buildings.
10. **Liveability, Healthy Communities and Mixed-Use Programmes:** The prime concern of the city is for affordable housing, mixed-use programmes and a healthy community.
11. **Local food and short supply chains:** High food security and urban agriculture by introducing 'eat local' and 'slow food' initiatives.
12. **Cultural heritage, identity and sense of place:** A sustainable city with high air quality, no pollution for good health, fosters resilient communities having public space networks and modern community facilities.
13. **Urban governance, leadership and best practices:** The city applies best practice for good urban governance through combined management and governance approaches and sustainable procurement methods, such as, environmental budgeting.
14. **Education, research and knowledge:** The city with education includes technical training and up-skilling, research, exchange of experiences and knowledge dissemination for all in sustainable urban development.

15. **Strategies for cities in developing countries:** Particular sustainability strategies are needed for cities in developing countries, such as, train local people to empower communities, creating new jobs and diversifying new job structures to harmonize the impacts of rapid urbanisation and globalisation.

7.3.4 Beatley's Vision of Green Urbanism

Beatley remarked that the vision of green urbanism includes programs, policies and creative design ideas for urban renewal and environment sustainability. Lehmann added the phrase also provides a proactive vision of what might be our zero-carbon, fossil fuel free future: overlapping mixed-use activities, living and working building typologies explored on the urban scale, infrastructures systems for renewable energies, public transport and individual energy-efficient building designs. According to Beatley, cities that exemplify green urbanism are:

- Cities must be conscious of their ecological effects on surrounding communities and their own natural resources. They must strive to reduce their cannon footprint and waste production.
- Cities should be designed to work with nature to help cleanse the city's air and water. Green urbanist cities should have plenty of sunlight, and greenspaces (e.g. rooftop gardens and parks).
- Cities should strive to create circler, not linear, waste flow. Breaking away from the current extraction and discard cycle.
- Cities must be self-sufficient by growing food, producing power, and cleaning drinking water.
- Cities should try to be more sustainable by prioritizing walking, biking, and public transit. All of which decreases the carbon emissions of the city.
- Cities should focus on its quality of life by creating livable conditions for all residents. Thus looking at the city from a neighborhood level to provide acceptable housing for all.

7.3.5 Newman's Seven Archetypal Cities

Peter Newman describes his principles of Green Urbanism through seven archetypal cities that each represent a different aspect of Green Urbanism. Newman believes that Green Urbanism can be used to create a more sustainable society that can fight against the ecological impacts of climate change. He discusses the importance of renewable energy, a circular economy, and sustainable infrastructure. These are Newman's seven Archetypal Cities:

1. **The Renewable City:** Is a city powered by renewable energy. Urban planning should be utilized to support wind power and solar energy, by creating rooftop solar paneling and off coast wind farms.

2. **The Carbon-Neutral City:** Is a city that strives to reduce carbon emissions, increase renewable energy, and offset remaining carbon emissions. Cities can decrease their carbon emissions by building energy efficient buildings, increasing the efficiency of public transportation, and lining roads with trees and greenery.
3. **The Distributed City:** Is a city that spreads energy and water production throughout different areas of the city. Thus decentralizing energy and water systems with smaller systems of production, like rooftop solar panels and water recycling systems.
4. **The Biophilic City:** Using the principles of Green Infrastructure to bring ecological features and green areas into the city in order to combat carbon emissions. Creating more green space and trees to trap air pollutants and urban agriculture to decrease a city's carbon footprint.
5. **Eco-Efficient City:** A city that strives to reduce their carbon footprint by reducing waste and decreasing its consumption. By using a circular economy to create a circular flow of waste, a city can reduce the amount of waste that eventually ends up in landfills or polluting surrounding nature and waterways.
6. **Placed-Based City:** It is a that prioritizes creating a local economy that is committed to the community, thus creating local jobs that then decrease the amount of carbon emissions used for commuting. The city then priorities localizing production of energy, food, and materials to create a strong place-based city.
7. **The Sustainable Transport City:** Is a city that creates a sustainable transportation system, by offering walkable transit-options that focus on using renewable energy. The city should strive to create a high-density urban form that makes walking, biking, and public transportation a more efficient option.

7.4 PLANS OF ACTION

7.4.1 Global Platform on Sustainable Cities

Cities play an important role in the concept of green urbanism. Many cities succeed in harming the environment versus preserving it. However, numerous cities are already taking action to enhance the sustainability of growing urban populations. Mayors and authorities of larger cities are pushing for greener development projects and governmental services to adhere to. Various organizations are calling for a plan of action to promote cities as natural places that have the capability of generating multiple benefits to the environment. A program had recently been launched that is known as the Global Platform on Sustainable Cities. The platform was created in March, 2016. Within four years, it has grown to comprise 28 cities throughout 11 countries. It is committed to finding various approaches to waste management, environmental conservation, transportation, and energy production and consumption. Urbanized areas have the largest energy consumption and pollute the environment the most.

7.4.2 'Die Energiewende': The Energy Turnaround in Germany

Some countries have quickly become global leaders in green energy during the previous decade. Germany held the title for the nation producing the most amount of solar energy in the world, up until China amped up their green energy production. Germany became one of the first industrialized nations to strive for an energy supply from renewable energy sources. The German society had been looking into green energy since the 1970s, but without much government support. After catastrophes such as Chernobyl and Fukushima, "Die Energiewende," otherwise known as the energy turnaround became a main political movement within German borders. During the 2010s, Germany saw tremendous leaps in green energy production. From 2012 – 2019, onshore and offshore wind capacity doubled from an installed capacity of 30,979MW to 61,357MW, and solar grew from 34,077MW install capacity to 49,061MW. Much of this has to do with laws and incentives that were passed during the leadership of Angela Merkel, who studied physics and earned her doctorate in quantum chemistry and has been nicknamed as "The Climate Chancellor". Part of the Energiewende goal is to reduce CO_2 emissions into the atmosphere by 90% by 2050. The advancement of renewable energy technologies and the reduction in costs has greatly helped Germany and other countries improve their renewable energy sectors. Germany has vowed to dismantle all 17 of their nuclear rectors. As of 2019, 6 nuclear reactors remain, with demolition dates set in 2021 and 2022 respectively.

7.4.3 Copenhagen: The World's First Carbon-Neutral Capital

Cities around the globe are diligently working to reduce the amount of carbon emissions being produced. Denmark has quickly become a buzzling metropolis of green energy and eco-friendly projects. Copenhagen has vowed to become the world's first carbon-neutral capital by 2025. It's a rather ambitious project, with most capitals planning to become carbon-neutral close to 2050, but Copenhagen is becoming a trailblazer and proving to others how easy it is to go green without going bankrupt. Officials are hoping to set an example to others and show that society also plays a large impact on achieving such goals. Copenhagen is constantly ranked as being the most bike-friendly city in the world. The city is working to offer all-electric public transportation that is constantly moving, encouraging more people to ride a bike or take public transportation instead of driving vehicles. To further shrink the carbon footprint of the city, Copenhagen has installed 62 wind turbines with a capacity to produce upwards of 158 megawatts. The city plans to have a wind turbine energy production capacity of 460 MW by 2025; more than doubling their capacity as of 2019.

7.4.4 Singapore: City in a Garden

Singapore has shown the world how green urbanism can be achieved in a densely populated city. Singapore is a city state that has an area of 728.6 km^2 with a population of 5.6 million making it the 2nd most densely populated country in the world. The city is renowned for its ability to bring nature and greenery to an urban environment. Singapore's current city motto is "Singapore – City in a Garden" which was changed from "Singapore – Garden City" highlighting that the city's first priority is its greenery. The city has an extensive 180 km park system that has paths and walkways interweaving throughout green spaces and gardens around the city. The city has bolstered its green urban structure through its Skyrise Greenery initiative which subsidises the building of roof and vertical gardens. Renewable energy has been growing throughout the city with 203 MWp in 2018 to 350 MWp in 2020, and the city wants to increase its capacity to over 1 GWp. The city also has an exemplary recycling system with almost all of construction waste being recycled, and extracting from the waste stream so very little eventually ends up in landfills. The Majority of rain and storm water is collected and used throughout the city. Singapore also has an esteemed public transit system with most public transit is faster than traffic down main corridors.

7.5.5 Practical Approaches

Many cities now have Sustainable Action Plans which is a roadmap towards sustainability. Green Urbanism has grown from textbook methodologies to living action plans that survive beyond the election cycles of city mayors and counsellors.

Green Urbanism poses the demand for an applicable method in planning and management of a city. Wybe Kuitert proposed analyzing the city as a landscape system to reach at a more comprehensive approach towards this end. The urban landscape connects the cultural components, like identity and history with the natural physics of a city, like its geography, water and natural ecology. As such it poses a vision that can be applied to any city, rich or poor. Discerning the potential quality of wild nature in the city is a first step to see what nature can give, if we only have an open eye for it. It is discovered through the potential vegetation map for the city. There is a lot of Conferences talking about this Field as 'Green Urbanism'.

7.5 SUSTAINABLE DEVELOPMENT

Sustainable development means attaining a balance between environmental protection and human economic development and between the present and future needs. It means equity in development and sectoral actions across space and time. It requires an integration of economic, social and environmental approaches towards development. Sustainable urban development refers to attaining social equity and environmental protection in urbanization while minimizing the costs of urbanization.

Sustainable Development (SD) is a pattern of growth in which resource use aims to meet human needs while preserving the environment so that these needs can be met not only in the present, but also for generations to come. The term was used by the Brundtland Commission which coined what has become the most often-quoted definition of sustainable development as development that "meets the needs of the present without compromising the ability of future generations to meet their own needs."

Sustainable development ties together concern for the carrying capacity of natural systems with the social challenges facing humanity. As early as the 1970s "sustainability" was employed to describe an economy "in equilibrium with basic ecological support systems." Ecologists have pointed to The Limits to Growth, and presented the alternative of a "steady state economy" in order to address environmental concerns.

The field of sustainable development can be conceptually broken into three constituent parts:

1. Environmental sustainability,
2. Economic sustainability and
3. Sociopolitical sustainability.

In 1987, the United Nations released the Brundtland Report, which included what is now one of the most widely recognised definitions:

"Sustainable development is development that meets the needs of the present without compromising the ability of future generations to meet their own needs." It contains within it two key concepts:

"The concept of 'needs', in particular the essential needs of the world's poor, to which overriding priority should be given; and

"The idea of limitations imposed by the state of technology and social organization on the environment's ability to meet present and future needs."

The United Nations 2005 World Summit Outcome Document refers to the "interdependent and mutually reinforcing pillars" of sustainable development as economic development, social development, and environmental protection. Based on the triple bottom line, numerous sustainability standards and certification systems have been established in recent years, in particular in the food industry. Well-known standards include Organic, Rainforest Alliance, Fairtrade, Utz Certified,Bird Friendly, and The Common Code for the Coffee Community.

Indigenous peoples have argued, through various international forums such as the United Nations Permanent Forum on Indigenous Issues and the Convention on Biological Diversity, that there are fourpillars of sustainable development, the fourth being cultural. The Universal Declaration on Cultural Diversity (UNESCO, 2001) further elaborates the concept by stating that "...cultural diversity is as necessary for humankind as biodiversity is for nature"; it becomes "one of the roots of development understood not simply in terms of economic growth, but also as a means to achieve a more satisfactory intellectual, emotional, moral and spiritual existence". In this vision, cultural diversity is the fourth policy area of sustainable development.

A useful articulation of the values and principles of sustainability can be found in the Earth Charter. It offers an integrated vision and definition of strong sustainability. The document, an ethical framework for a sustainable world, was developed over several years after the Rio Earth Summit in 1992 and launched officially in 2000. The Charter derives its legitimacy from the participatory process in which it was drafted, which included contributions from hundreds of organizations and thousands of individuals, and from its use since 2000 by thousands of organizations and individuals that have been using the Earth Charter as an educational instrument and a policy tool.

Economic Sustainability: Agenda 21 clearly identified information, integration, and participation as key building blocks to help countries achieve development that recognises these interdependent pillars. It emphasises that in sustainable development everyone is a user and provider of information. It stresses the need to change from old sector-centered ways of doing business to new approaches that involve cross-sectoral co-ordination and the integration of environmental and social concerns into all development processes. Furthermore, Agenda 21 emphasises that broad public participation in decision making is a fundamental prerequisite for achieving sustainable development.

According to Hasna Vancock, sustainability is a process which tells of a development of all aspects of human life affecting sustenance. It means resolving the conflict between the various competing goals, and involves the simultaneous pursuit of economic prosperity, environmental quality and social equity famously known as three dimensions with the resultant vector being technology, hence it is a continually evolving process; the 'journey' (the process of achieving sustainability) is of course vitally important, but only as a means of getting to the destination. However, the 'destination' of sustainability is not a fixed place in the normal sense that we understand destination. Instead, it is a set of wishful characteristics of a future system.

The concept has included notions of weak sustainability, strong sustainability and deep ecology.

Green development is generally differentiated from sustainable development in that Green development prioritizes what its proponents consider to be environmental sustainability over economic and cultural considerations. Proponents of Sustainable Development argue that it provides a context in which to improve overall sustainability where cutting edge Green development is unattainable. For example, a cutting edge treatment plant with extremely high maintenance costs may not be sustainable in regions of the world with fewer financial resources. An environmentally ideal plant that is shut down due to bankruptcy is obviously less sustainable than one that is maintainable by the community, even if it is somewhat less effective from an environmental standpoint.

Some research activities start from this definition to argue that the environment is a combination of nature and culture. The Network of Excellence "Sustainable Development in a Diverse World", sponsored by the European Union, integrates multidisciplinary capacities and interpretscultural diversity as a key element of a new strategy for sustainable development.

In fact, some researchers and institutions have even pointed out that a fourth dimension should be added to the three dimensions of sustainable development, since these three dimensions do not seem to be enough to reflect the complexity of contemporary society. In this context, the Agenda 21 for culture and the United Cities and Local Governments (UCLG) Executive Bureau lead the preparation of the policy statement "Culture: Fourth Pillar of Sustainable Development", passed on 17 November 2010, in the framework of the World Summit of Local and Regional Leaders - 3rd World Congress of UCLG, held in Mexico City. This document inaugurates a new perspective and points to the relation between culture and sustainable development through a dual approach: developing a solid cultural policy and advocating a cultural dimension in all public policies.

Still other researchers view environmental and social challenges as opportunities for development action. This is particularly true in the concept of sustainable enterprise that frames these global needs as opportunities for private enterprise to provide innovative and entrepreneurial solutions. This view is now being taught at many business schools including the Center for Sustainable Global Enterprise atCornell University and the Erb Institute for Global Sustainable Enterprise at the University of Michigan.

The United Nations Division for Sustainable Development lists the following areas as coming within the scope of sustainable development:

Sustainable development is an eclectic concept, as a wide array of views fall under its umbrella. The concept has included notions of weak sustainability, strong sustainability and deep ecology. Different conceptions also reveal a strong tension between ecocentrism and anthropocentrism. Many definitions and images of sustainable development coexist. Broadly defined, the sustainable development mantra enjoins current generations to take a systems approach to growth and development and to manage natural, produced, and social capital for the welfare of their own and future generations.

During the last ten years, different organizations have tried to measure and monitor the proximity to what they consider sustainability by implementing what has been called sustainability metrics and indices.

Sustainable development is said to set limits on the developing world. While current first world countries polluted significantly during their development, the same countries encourage third world countries to reduce pollution, which sometimes impedes growth. Some consider that the implementation of sustainable development would mean a reversion to pre-modern lifestyles.

Others have criticized the overuse of the term:

"The word sustainable has been used in too many situations today, and ecological sustainability is one of those terms that confuse a lot of people. You hear about sustainable development, sustainable growth, sustainable economies, sustainable societies, sustainable agriculture. Everything is sustainable."

7.5.1 Economic Sustainability

The Venn diagram of sustainable development shown above has many versions, but was first used by economist Edward Barbier (1987). However, Pearce, Barbier and Markandya (1989) criticized the Venn approach due to the intractability of operationalizing separate indices of economic, environmental, and social sustainability and somehow combining them. They also noted that the Venn approach was inconsistent with the Brundtland Commission Report, which emphasized the interlinkages between economic development, environmental degradation, and population pressure instead of three objectives. Economists have since focused on viewing the economy and the environment as a single interlinked system with a unified valuation methodology. Intergenerational equity can be incorporated into this approach, as has become common in economic valuations of climate change economics. Ruling out discrimination against future generations and allowing for the possibility of renewable alternatives to petro-chemicals and other non-renewable resources, efficient policies are compatible with increasing human welfare, eventually reaching a golden-rule steady state. Thus the three pillars of sustainable development are interlinkages, intergenerational equity, and dynamic efficiency.

Arrow et al. (2004) and other economists have advocated a form of the weak criterion for sustainable development - the requirement than the wealth of a society, including human-capital, knowledge-capital and natural-capital (as well as produced capital) not decline over time. Others, including Barbier 2007, continue to contend that strong sustainability - non-depletion of essential forms of natural capital - may be appropriate.

7.5.2 Three Types of Capital in Sustainable Development

The sustainable development debate is based on the assumption that societies need to manage three types of capital (economic, social, and natural), which may be non-substitutable and whose consumption might be irreversible. Daly (1991), for example, points to the fact that natural capital can not necessarily be substituted by economic capital. While it is possible that we can find ways to replace some natural resources, it is much more unlikely that they will ever be able to replace eco-system services, such as the protection provided by the ozone layer, or the climate stabilizing function of the Amazonian forest. In fact natural capital, social capital and economic capital are often complementarities. A further obstacle to substitutability lies also in the multi-functionality of many natural resources. Forests, for example, not only provide the raw material for paper (which can be substituted quite easily), but they also maintain biodiversity, regulate water flow, and absorb CO_2.

Another problem of natural and social capital deterioration lies in their partial irreversibility. The loss in biodiversity, for example, is often definite. The same can be true for cultural diversity. For example with globalisation advancing quickly the number of indigenous languages is dropping at alarming rates. Moreover, the depletion of natural and social capital may have non-linear consequences. Consumption of natural and social capital may have no observable impact until a certain threshold is reached. A lake can, for example, absorb nutrients for a long time while actually increasing its productivity. However, once a certain level of algae is reached lack of oxygen causes the lake's ecosystem to break down suddenly.

Market Failure

If the degradation of natural and social capital has such important consequence the question arises why action is not taken more systematically to alleviate it. Cohen and Winn (2007) point to four types of market failure as possible explanations: First, while the benefits of natural or social capital depletion can usually be privatized the costs are often externalized (i.e. they are borne not by the party responsible but by society in general). Second, natural capital is often undervalued by society since we are not fully aware of the real cost of the depletion of natural capital. Information asymmetry is a third reason-often the link between cause and effect is obscured, making it difficult for actors to make informed choices. Cohen and Winn close with the realization that contrary to economic theory many firms are not perfect optimizers. They postulate that firms often do not optimize resource allocation because they are caught in a "business as usual" mentality.

7.5.3 The Business Case for Sustainable Development

The most broadly accepted criterion for corporate sustainability constitutes a firm's efficient use of natural capital. This eco-efficiency is usually calculated as the economic value added by a firm in relation to its aggregated ecological impact. This idea has been popularised by the World Business Council for Sustainable Development (WBCSD) under the following definition: "Eco-efficiency is achieved by the delivery of competitively priced goods and services that satisfy human needs and bring quality of life, while progressively reducing ecological impacts and resource intensity throughout the life-cycle to a level at least in line with the earth's carrying capacity."

Similar to the eco-efficiency concept but so far less explored is the second criterion for corporate sustainability. Socio-efficiency describes the relation between a firm's value added and its social impact. Whereas, it can be assumed that most corporate impacts on the environment are negative this is not true for social impacts. These can be either positive or negative. Depending on the type of impact socio-efficiency thus either tries to minimize negative social impacts or maximise positive social impacts in relation to the value added.

Both eco-efficiency and socio-efficiency are concerned primarily with increasing economic sustainability. In this process they instrumentalize both natural and social capital aiming to benefit from win-win situations. However, as Dyllick and Hockerts point out the business case alone will not be sufficient to realise sustainable development. They point towards eco-effectiveness, socio-effectiveness, sufficiency, and eco-equity as four criteria that need to be met if sustainable development is to be reached..

7.6 SUSTAINABLE AGRICULTURE

Sustainable Agriculture can be simply defined as environmentally friendly methods of farming that allow the production of crops or livestock without damage to the farm as an ecosystem. Apart from this, it also prevents the adverse effects on soil, water supplies, biodiversity, or other surrounding natural resources. The concept of sustainable agriculture is an intergenerational one in which we pass on a conserved or improved natural resource base instead of one which has been depleted or polluted.

7.6.1 The Elements of Sustainable Agriculture

Agroforestry : According to the World Agroforestry Centre, agroforestry is a collective name for land use systems and practices in which woody perennials are deliberately integrated with crops and/or animals on the same land management unit. The integration can be either in a spatial mixture or in a temporal sequence. There are normally both ecological and economic interactions between woody and non-woody components in agroforestry.

Mixed Farming : Many farmers in tropical & temperate countries survive by managing a mix of different crops or animals. The best known form of mixing occurs probably where crop residues are used to feed the animals and the excreta from animals are used as nutrients for the crop. Other forms of mixing takes place where graving under fruit trees keeps the grass short or where manure from pigs is used to feed the fish. Mixed farming exists in many forms depending on external and internal factors. External factors are: Weather Patterns, Market Prices, Political Stability and Technological Development. Internal factors relate to Local Soil Characteristics, Composition of family and Farmer's Ingenuity. Mixed Farming provides farmers with (a) an opportunity to diversify risk from single-crop production; (b) to use labour more efficiently; (c) to have a source of cash for purchasing farm inputs; (d) to add value to crop or crop by-product; (e) combining crops and livestocks.

Multiple Cropping : The process of growing two or more crops in the same piece of land, during the same season is called Multiple Cropping. It can be rightly called a form of polyculture. It can be - (a) Double Cropping (the practice where the second crop is planted after the first has been harvested); (b) Relay Cropping (the practice where a second crop is started along with the first one, before it is harvested).

Crop Rotation : The process of growing two or more dissimilar or unrelated crops in the same piece of land in different seasons is known as Crop Rotation. This process could be adopted as it comes with a series of benefits like - (a) avoid the build up of pests that often occurs when one species is continuously cropped; (b) the traditional element of crop rotation is the replenishment of nitrogen through the use of green manure in sequence with cereals and other crops; (c) Crop rotation can also improve soil structure and fertility by alternating deep-rooted and shallow-rooted plants; (d) it is a component of polyculture.

7.6.2 Criticisms of Sustainable Development

The concept of "Sustainable Development" raises several critiques at different levels.

Consequences

John Baden views the notion of sustainable development as dangerous because the consequences have unknown effects. He writes: "In economy like in ecology, the interdependence rule applies. Isolated actions are impossible. A policy which is not carefully enough thought will carry along various perverse and adverse effects for the ecology as much as for the economy.

Many suggestions to save our environment and to promote a model of 'sustainable development' risk indeed leading to reverse effects." Moreover, he evokes the bounds of public action which are underlined by the public choice theory: the quest by politicians of their own interests, lobby pressure, partial disclosure etc.

He develops his critique by noting the vagueness of the expression, which can cover anything. It is a gateway to interventionist proceedings which can be against the principle of freedom and without proven efficacy. Against this notion, he is a proponent of private property to impel the producers and the consumers to save the natural resources. According to Baden, "the improvement of environment quality depends on themarket economy and the existence of legitimate and protected property rights." They enable the effective practice of personal responsibility and the development of mechanisms to protect the environment. The State can in this context "create conditions which encourage the people to save the environment."

Vagueness of the Term

Some criticize the term "sustainable development", stating that the term is too vague. For example, bothJean-Marc Jancovici and the philosopher Luc Ferry express this view. The latter writes about sustainable development: "I know that this term is obligatory, but I find it also absurd, or rather so vague that it says nothing." Luc Ferry adds that the term is trivial by a proof of contradiction: "who would like to be a proponent of an "untenable development! Of course no one! The term is more charming than meaningful. Everything must be done so that it does not turn into Russian-type administrative planning with ill effects." sustainable development has become obscured by conflicting world views, the expansionist and the ecological, and risks being co-opted by individuals and institutions that perpetuate many aspects of the expansionist model.

Basis

Sylvie Brunel, French geographer and specialist of the Third World, develops in A qui profite le developement durable a critique of the basis of sustainable development, with its binary vision of the world, can be compared to the Christian vision of Good and Evil, an idealized nature where the human being is an animal like the others or even an alien. Nature - as Rousseau thought - is better than the human being. It is a parasite, harmful for the nature. But the human is the one who protects the biodiversity, where normally only the strong survive.

Moreover, she thinks that the core ideas of sustainable development are a hidden form of protectionism by developed countries impeding the development of the other countries. For Sylvie Brunel, sustainable development serves as a pretext for protectionism and "I have the feeling that sustainable development is perfectly helping out capitalism".

De-growth

The proponents of the de-growth reckon that the term of sustainable development is an oxymoron. According to them, on a planet where 20% of the population consumes 80% of the natural resources, a sustainable development cannot be possible for this 20%: "According to the origin of the concept of sustainable development, a development which meets the needs of the present without compromising the ability of future generations to meet their own needs, the right term for the developed countries should be a sustainable de-growth".

Measurability

In 2007 a report for the U.S. Environmental Protection Agency stated: "While much discussion and effort has gone into sustainability indicators, none of the resulting systems clearly tells us whether our society is sustainable. At best, they can tell us that we are heading in the wrong direction, or that our current activities are not sustainable. More often, they simply draw our attention to the existence of problems, doing little to tell us the origin of those problems and nothing to tell us how to solve them." Nevertheless a majority of authors assume that a set of well defined and harmonised indicators is the only way to make sustainability tangible. Those indicators are expected to be identified and adjusted through empirical observations (trial and error) see also Ecological footprint.

The most common critiques are related to issues like data quality, comparability, objective function and the necessary resources. However a more general criticism is coming from the project management community: How can a sustainable development be achieved at global level if we cannot monitor it in any single project?

The Cuban-born researcher and entrepreneur Sonia Bueno suggests an alternative approach that is based upon the integral, long-term cost-benefit relationship as a measure and monitoring tool for the sustainability of every project, activity or enterprise.

Furthermore this concept aims to be a practical guideline towards sustainable development following the principle of conservation and increment of value rather than restricting the consumption of resources.

7.7 SUSTAINABILITY

Sustainability is the capacity to endure. For humans, sustainability is the long-term maintenance of well being, which has environmental, economic, and social dimensions, and encompasses the concept of stewardship, the responsible management of resource use. In ecology, sustainability describes how biological systems remaindiverse and productive over time, a necessary precondition for human well-being. Long-lived and healthy wetlands and forests are examples of sustainable biological systems.

Healthy ecosystems and environments provide vital goods and services to humans and other organisms. There are two major ways of managing human impact on ecosystem services. One approach is environmental management; this approach is based largely on information gained from earth science, environmental science, and conservation biology. Another approach is management of consumption of resources, which is based largely on information gained from economics.

Human sustainability interfaces with economics through the social and ecological consequences of economic activity. Moving towards sustainability is also a social challenge that entails, among other factors, international and national law, urban planning and transport, local and individual lifestyles andethical consumerism. Ways of living more sustainably can take many forms from reorganising living conditions, to reappraising work practices, or developing new technologies that reduce the consumption of resources.

The word sustainability is derived from the Latin sustinere. Dictionaries provide more than ten meanings for sustain, the main ones being to "maintain", "support", or "endure". However, since the 1980s sustainability has been used more in the sense of human sustainability on planet Earth and this has resulted in the most widely quoted definition of sustainability and sustainable development, that of the Brundtland Commission of the United Nations on March 20, 1987: "sustainable development is development that meets the needs of the present without compromising the ability of future generations to meet their own needs."

At the 2005 World Summit it was noted that this requires the reconciliation of environmental, social and economic demands - the "three pillars" of sustainability. This view has been expressed as an illustration using three overlapping ellipses indicating that the three pillars of sustainability are not mutually exclusive and can be mutually reinforcing. The three pillars - or the "triple bottom line" - have served as a common ground for numerous sustainability standards and certification systems in recent years, in particular in the food industry. Standards which today explicitly refer to the triple bottom line include Rainforest Alliance,Fairtrade, UTZ Certified, and The Common Code for the

Coffee Community. The triple bottom line is also recognized by the ISEAL Alliance - the global association for social and environmental standards.

The triple bottom line as defined by the UN is not universally accepted and has undergone various interpretations. What sustainability is, what its goals should be, and how these goals are to be achieved are all open to interpretation. For many environmentalists the idea of sustainable development is an oxymoron as development seems to entail environmental degradation. Ecological economist Herman Daly has asked, "what use is a sawmill without a forest?" From this perspective, the economy is a subsystem of human society, which is itself a subsystem of the biosphere, and a gain in one sector is a loss from another. This can be illustrated as three concentric circles.

A universally accepted definition of sustainability remains elusive because it is expected to achieve many things. On the one hand it needs to be factual and scientific, a clear statement of a specific "destination". The simple definition "sustainability is improving the quality of human life while living within the carrying capacity of supporting eco-systems", though vague, conveys the idea of sustainability having quantifiable limits. But sustainability is also a call to action, a task in progress or "journey" and therefore a political process, so some definitions set out common goals and values.The Earth Charter speaks of "a sustainable global society founded on respect for nature, universal human rights, economic justice, and a culture of peace."

To add complication, the word sustainability is applied not only to human sustainability on Earth, but to many situations and contexts over many scales of space and time, from small local ones to the global balance of production and consumption. It can also refer to a future intention: "sustainable agriculture" is not necessarily a current situation but a goal for the future, a prediction. For all these reasons sustainability is perceived, at one extreme, as nothing more than a feel-good buzzword with little meaning or substance but, at the other, as an important but unfocused concept like "liberty" or "justice". It has also been described as a "dialogue of values that defies consensual definition".

Some researchers and institutions have pointed out that these three dimensions are not enough to reflect the complexity of contemporary society and suggest that culture could be included in this development model.

7.7.1 History of Sustainability

The history of sustainability traces human-dominated ecological systems from the earliest civilizations to the present. This history is characterized by the increased regional success of a particular society, followed by crises that were either resolved, producing sustainability, or not, leading to decline.

In early human history, the use of fire and desire for specific foods may have altered the natural composition of plant and animal communities. Between 8,000 and 10,000 years ago, Agrarian communities emerged which depended largely on their environment and the creation of a "structure of permanence."

The Western industrial revolution of the 17th to 19th centuries tapped into the vast growth potential of the energy in fossil fuels. Coal was used to power ever more efficient engines and later to generate electricity. Modern sanitation systems and advances in medicine protected large populations from disease. In the mid-20th century, a gathering environmental movement pointed out that there were environmental costs associated with the many material benefits that were now being enjoyed. In the late 20th century, environmental problems became global in scale. The 1973 and 1979 energy crises demonstrated the extent to which the global community had become dependent on non-renewable energy resources.

In the 21st century, there is increasing global awareness of the threat posed by the human-induced enhanced greenhouse effect, produced largely by forest clearing and the burning of fossil fuels.

7.7.2 Principles and Concepts

The philosophical and analytic framework of sustainability draws on and connects with many different disciplines and fields; in recent years an area that has come to be called sustainability science has emerged. Sustainability science is not yet an autonomous field or discipline of its own, and has tended to be problem-driven and oriented towards guiding decision-making.

Scale and Context

Sustainability is studied and managed over many scales (levels or frames of reference) of time and space and in many contexts of environmental, social and economic organization. The focus ranges from the total carrying capacity (sustainability) of planet Earth to the sustainability of economic sectors, ecosystems, countries, municipalities, neighbourhoods, home gardens, individual lives, individual goods and services, occupations, lifestyles, behaviour patterns and so on. In short, it can entail the full compass of biological and human activity or any part of it. As Daniel Botkin, author and environmentalist, has stated: “We see a landscape that is always in flux, changing over many scales of time and space.”

7.7.3 Consumption-Population, Technology, Resources

A major driver of human impact on Earth systems is the destruction of biophysical resources, and especially, the Earth’s ecosystems. The total environmental impact of a community or of humankind as a whole depends both on population and impact per person, which in turn depends in complex ways on what resources are being used, whether or not those resources are renewable, and the scale of the human activity relative to the carrying capacity of the ecosystems involved. Careful resource management can be applied at many scales, from economic sectors like agriculture, manufacturing and industry, to work organizations, the consumption patterns of households and individuals and to the resource demands of individual goods and services.

One of the initial attempts to express human impact mathematically was developed in the 1970s and is called the I PAT formula. This formulation attempts to explain human consumption in terms of three components: population numbers, levels of consumption (which it terms "affluence", although the usage is different), and impact per unit of resource use (which is termed "technology", because this impact depends on the technology used). The equation is expressed:

$I = P \times A \times T$

Where: I = Environmental impact, P = Population, A = Affluence, T = Technology

7.8 SUSTAINABILITY MEASUREMENT

Sustainability measurement is a term that denotes the measurements used as the quantitative basis for the informed management of sustainability. The metrics used for the measurement of sustainability (involving the sustainability of environmental, social and economic domains, both individually and in various combinations) are evolving: they include indicators, benchmarks, audits, sustainability standards and certification systems like Fairtrade and Organic, indexes and accounting, as well as assessment, appraisal and other reporting systems. They are applied over a wide range of spatial and temporal scales.

Some of the best known and most widely used sustainability measures include corporate sustainability reporting, Triple Bottom Line accounting, World Sustainability Society and estimates of the quality of sustainability governance for individual countries using the Environmental Sustainability Index and Environmental Performance Index.

7.8.1 Population

According to the 2008 Revision of the official United Nations population estimates and projections, the world population is projected to reach 7 billion early in 2012, up from the current 6.9 billion (May 2009), to exceed 9 billion people by 2050. Most of the increase will be in developing countries whose population is projected to rise from 5.6 billion in 2009 to 7.9 billion in 2050. This increase will be distributed among the population aged 15-59 (1.2 billion) and 60 or over (1.1 billion) because the number of children under age 15 in developing countries is predicted to decrease. In contrast, the population of the more developed regions is expected to undergo only slight increase from 1.23 billion to 1.28 billion, and this would have declined to 1.15 billion but for a projected net migration from developing to developed countries, which is expected to average 2.4 million persons annually from 2009 to 2050. Long-term estimates of global population suggest a peak at around 2070 of nine to ten billion people, and then a slow decrease to 8.4 billion by 2100.

Emerging economies like those of China and India aspire to the living standards of the Western world as does the non-industrialized world in general. It is the combination of population increase in the developing world and unsustainable consumption levels in the developed world that poses a stark challenge to sustainability.

7.8.2 Carrying Capacity

At the global scale scientific data now indicates that humans are living beyond the carrying capacity of planet Earth and that this cannot continue indefinitely. This scientific evidence comes from many sources but is presented in detail in the Millennium Ecosystem Assessment and the planetary boundaries framework. An early detailed examination of global limits was published in the 1972 book Limits to Growth, which has prompted follow-up commentary and analysis. The Ecological footprint measures human consumption in terms of the biologically productive land needed to provide the resources, and absorb the wastes of the average global citizen. In 2008 it required 2.7global hectares per person, 30% more than the natural biological capacity of 2.1 global hectares. The resulting ecological deficit must be met from unsustainable extra sources and these are obtained in three ways: embedded in the goods and services of world trade; taken from the past; or borrowed from the future as unsustainable resource usage.

7.8.3 Global Human Impact on Biodiversity

At a fundamental level energy flow and biogeochemical cycling set an upper limit on the number and mass of organisms in any ecosystem. Human impacts on the Earth are demonstrated in a general way through detrimental changes in the global biogeochemical cycles of chemicals that are critical to life, most notably those of water, oxygen, carbon, nitrogen and phosphorus.

The Millennium Ecosystem Assessment is an international synthesis by over 1000 of the world's leading biological scientists that analyses the state of the Earth's ecosystems and provides summaries and guidelines for decision-makers. It concludes that human activity is having a significant and escalating impact on the biodiversity of world ecosystems, reducing both their resilience and biocapacity. The report refers to natural systems as humanity's "life-support system", providing essential "ecosystem services". The assessment measures 24 ecosystem services concluding that only four have shown improvement over the last 50 years, 15 are in serious decline, and five are in a precarious condition.

7.9 SUSTAINABLE CITY

A sustainable city, or eco-city is a city designed with consideration of environmental impact, inhabited by people dedicated to minimization of required inputs of energy, water and food, and waste output of heat, air pollution - CO_2, methane, and water pollution. Richard Register first coined the term "ecocity" in his 1987 book, Ecocity Berkeley: Building Cities for a Healthy Future. Other leading figures who envisioned the sustainable city are architect Paul F Downton, who later founded the company Ecopolis Pty Ltd, and authors Timothy Beatley and Steffen Lehmann, who have written extensively on the subject. The field of industrial ecology is sometimes used in planning these cities.

There remains no completely agreed upon definition for what a sustainable city should be or completely agreed upon paradigm for what components should be included. Generally, developmental experts agree that a sustainable city should meet the needs of the present without sacrificing the ability of future generations to meet their own needs. The ambiguity within this idea leads to a great deal of variation in terms of how cities carry out their attempts to become sustainable. However, a sustainable city should be able to feed itself with minimal reliance on the surrounding countryside, and power itself with renewable sources of energy. The crux of this is to create the smallest possibleecological footprint, and to produce the lowest quantity of pollution possible, to efficiently use land; compost used materials, recycle it or convert waste-to-energy, and thus the city's overall contribution to climate change will be minimal, if such practices are adhered to.

It is estimated that over 50% of the world's population now lives in cities and urban areas. These large communities provide both challenges and opportunities for environmentally conscious developers, and there are distinct advantages to further defining and working towards the goals of sustainable cities. Humans are social creatures and thrive in urban spaces that foster social connections. Because of this, a shift to more dense, urban living would provide an outlet for social interaction and conditions under which humans can prosper. Contrary to common belief, urban systems can be more environmentally sustainable than rural or suburban living. With people and resource located so close to one another it is possible to save energy and resources things such as food transportation and mass transit systems. Finally, cities benefit the economy by locating human capital in one relatively small geographic area where ideas can be generated.

These ecological cities are achieved through various means, such as:

- Different agricultural systems such as agricultural plots within the city. This reduces the distance food has to travel from field to fork. Practical work out of this may be done by either small scale/private farming plots or through larger scale agriculture.
- Renewable energy sources, such as wind turbines, solar panels, or bio-gas created from sewage. Cities provide economies of scale that make such energy sources viable.
- Various methods to reduce the need for air conditioning, such as planting trees and lightening surface colors, natural ventilation systems, an increase in water features, and green spaces equaling at least 20% of the city's surface. These measures counter the "heat island effect" caused by an abundance of tarmac and asphalt, which can make urban areas several degrees warmer than surrounding rural areas-as much as six degrees Celsius during the evening.
- Improved public transport and an increase in pedestrianization to reduce car emissions. This requires a radically different approach to city planning, with integrated business, industrial, and residential zones. Roads may be designed to make driving difficult.

- Optimal building density to make public transport viable but avoid the creation of urban heat islands.
- Solutions to decrease urban sprawl, by seeking new ways of allowing people to live closer to the workspace. Since the workplace tends to be in the city, downtown, or urban center, they are seeking a way to increase density by changing the antiquated attitudes many suburbanites have towards inner-city areas. One of the new ways to achieve this is by solutions worked out by the Smart Growth Movement.
- Green roofs
- Zero-emission transport
- Zero-energy building
- Sustainable urban drainage systems or SUDS
- Energy conservation systems/devices
- Xeriscaping - garden and landscape design for water conservation
- Key Performance Indicators - development and operational management tool providing guidance and M&V for city administrators.

7.10 INFRASTRUCTURE

Buildings provide the infrastructure for a functioning city and allow for many opportunities to demonstrate a commitment to sustainability. A commitment to sustainable architecture encompasses all phases of building including the planning, building, and restructuring.

7.10.1 Eco-Industrial Park

The purpose of an eco-industrial park is to connect a number of firms and organizations to work together to decrease their environmental impact while simultaneously improving their economic performance. The community of businesses accomplishes this goal through collaboration in managing environmental and resource issues, such as energy, water, and materials. The components for building an eco-industrial park include natural systems, more efficient use of energy, and more efficient material and water flows Industrial parks should be built to fit into their natural settings in order to reduce environmental impacts, which can be accomplished through plant design, landscaping, and choice of materials. For instance, there is an industrial park in Michigan built by Phoenix Designs that is made almost entirely from recycled materials. The landscaping of the building will include native trees, grasses, and flowers, and the landscaping design will also act as climate shelter for the facility. In choosing the materials for building an eco-industrial park, designers must consider the life-cycle analysis of each medium that goes into the building to assess their true impact on the environment and to ensure that they are using the material with the least impact. Some examples of efficient energy strategies in industrial parks are flows of steam or heated water from one plant to another, steam connections from firms to provide heating for homes in the area, and using renewable energy such as wind and solar power. In terms of material flows, the companies in an eco-industrial park may have common waste treatment facilities, a means for transporting by-products from one plant to another,or anchoring the park around resource recovery companies that are recruited to the location or started from scratch. To create more efficient water flows in industrial parks, the processed water from one plant can be reused by another plant and the parks infrstructure can include a way to collect and reuse storm water runoff.

7.10.2 Urban Farming

Urban farming is the process of growing and distributing food, as well as raising animals, in and around a city. According to the RUAF Foundation, urban farming is different from rural agriculture because "it is integrated into the urban economic and ecological system: urban agriculture is embedded in -and interacting with- the urban ecosystem. Such linkages include the use of urban residents as labourers, use of typical urban resources (like organic waste as compost and urban wastewater for irrigation), direct links with urban consumers, direct impacts on urban ecology (positive and negative), being part of the urban food system, competing for land with other urban functions, being influenced by urban policies and plans, etc". There are many motivations behind urban agriculture, but in the context of creating a sustainable city, this method of food cultivation saves energy in food transportation and saves costs. In order for urban farming to be a successful method of sustainable food growth, cities most allot a common area for community gardens or farms, as well as a common area for a farmers market in which the foodstuffs grown within the city can be sold to the residents of the urban system

7.10.3 Urban Infill

Many cities are currently in a shift from the suburban sprawl model of development to a return to urban dense living. This shift in geographic distribution of population leads to a denser core of city residents. These residents provide a growing demand in many sectors that is reflected in the architectural fabric of the city. This new demand can be supplied by new construction or historic rehabilitation. Sustainable cities will opt for historical rehabilitation wherever possible. Having people live in higher densities not only gives economies of scale but also allows for infrastructure to be more efficient.

7.10.4 Walkable Urbanism

Walkable urbanism is a development strategy in opposition to suburban sprawl. It advocates housing for a diverse population, a full mix of uses, walkable streets, positive public space, integrated civic and commercial centers, transit orientation and accessible open space. The most clearly defined form of walkable urbanism is known as the Charter of New Urbanism.

7.10.5 Individual Buildings (LEED)

LEED, or Leadership in Energy and Environmental Design, is an internationally-recognized green building certification system. LEED recognizes whole building sustainable design by identifying key areas of excellence including: Sustainable Sites, Water Efficiency, Energy and Atmosphere, Materials and Resources, Indoor Environmental Quality, Locations & Linkages, Awareness and Education, Innovation in Design, Regional Priority. In order for a building to become LEED certified sustainability needs to be prioritized in design, construction, and use. One example of sustainable design would be including a certified wood like bamboo. Bamboo is fast growing and has an incredibly replacement rate after being harvested. BY the far the most credits are rewarded for optimizing energy performance. This promotes innovative thinking about alternative forms of energy and encourages increased efficiency.

7.10.6 Urban Strategic Planning

Although there is not an international policy regarding sustainable cities and there are not established international standards, there is an organization, the United Cities and Local Governments(UCLG) that is working to establish universal urban strategic guidelines. The UCLC a democratic and decentralized structure that operates in Africa, Asia, Eurasia, Europe, Latin America, North America, Middle East, West Asian and a Metropolitian section work to promote a more sustainable society. The 60 members of the UCLG committee evaluate urban development strategies and debate theses experiences to make the best recommendations. Additionally, the UCLG accounts for differences in regional and national context.

7.10.7 Development

Recently, local and national governments and regional bodies such as the European Union have recognized the need for a holistic understanding of urban planning. This is instrumental to establishing an international policy that focuses on cities challenges and the role of the local authorities responses. Generally, in terms of urban planning, the responsibility of local governments are limited to land use and infrastructure provision excluding inclusive urban development strategies. The advantages of urban strategic planning include an increase in governance and cooperation that aides local governments in establishing performance based-management, clearly identify the challenges facing local community and more effectively responding on a local level rather than national government and finally it improves institutional responses and local decision making. Additionally, it increase dialogue between stakeholders and develops consensus-based solutions, establishing continuity between sustainability plans and the change in local government, it places environmental issues as the priority for the sustainable development of the city and serves as a platform to develop concepts and new models of housing, energy and mobility.

7.10.8 Obstacles

The City Development Strategies (CDS) has evolved to address new challenges and to provide space for innovative polices that involves all stakeholders. The inequality in spatial development and socio-economic classes paired with recent concerns of poverty reduction and climate change are new factors in achieving global sustainable cities. According to the UCLG there are differences between regional and national conditions, framework and practice that are overcome in the international commitment to communication and negotiation with other governments, communities and the private sector to continual to develop through innovative and participatory approaches in strategic decisions, building consensus and monitoring performance management and raising investment.

7.10.9 Social Factors of Sustainable Cities

According to the UN Habitat, around half of the world's population is concentrated in cities that is set to rise to 60% which a couple decades. The UCLG has specifically identified 13 global challenges to establishing sustainable cities: demographic change and migration, globalisation of the job market, poverty and unmet Millennium Development Goals, segregation, spatial patterns and urban growth, metropolisation and the rise of urban regions, more political power for local authories, new actors for developing a city and providing services, decline in public funding for development, the environment and climate change, new and accessible building technologies, preparing for uncertainty and limits of growth and global communications and partnerships.

7.11 SUSTAINABLE DESIGN

Sustainable design is the philosophy of designing physical objects, the built environment, and services to comply with the principles of economic, social, and ecological sustainability.

The intention of sustainable design is to "eliminate negative environmental impact completely through skillful, sensitive design". Manifestations of sustainable design require no non-renewable resources, impact the environment minimally, and relate people with the natural environment.

Beyond the "elimination of negative environmental impact", sustainable design must create projects that are meaningful innovations that can shift behaviour. A dynamic balance between economy and society, to generate long-term relationships between user and object or service and finally respectful and be mindful of the environmental and social differences.

7.11.1 Examples of Sustainable Design

Sustainable Planning

Urban planners that are interested in achieving sustainable development or sustainable cities use various design principles and techniques when designing cities and their infrastructure. These include Smart Growth theory, Transit-oriented development, sustainable urban infrastructure and New Urbanism. Smart Growth is an urban planning and transportation theory that concentrates growth in infill sites within the existing infrastructure of a city or town to avoid urban sprawl; and advocates compact, transit-oriented development, walkable,bicycle-friendly land use, including mixed-use development with a range of housing choices. Transit-oriented development attempts to maximise access to public transport and thereby reduce the need for private vehicles. Public transport is considered a form of Sustainable urban infrastructure, which is a design approach which promotes protected areas, energy-efficient buildings, wildlife corridors and distributed, rather than centralized, power generation and waste water treatment. New Urbanism is more of a social and aesthetic urban design movement than a green one, but it does emphasize diversity of land use and population, as well as walkable communities which inherently reduce the need for automotive travel.

Both urban and rural planning can benefit from including sustainability as a central criterion when laying out roads, streets, buildings and other components of the built environment. Conventional planning practice often ignores or discounts the natural configuration of the land during the planning stages, potentially causing ecological damage such as the stagnation of streams, mudslides, soil erosion, flooding and pollution. Applying methods such as scientific modelling to planned building projects can draw attention to problems before construction begins, helping to minimise damage to thenatural environment.

Cohousing is an approach to planning based on the idea of intentional communities. Such projects often prioritize common space over private space resulting in grouped structures that preserve more of the surrounding environment.

Watershed assessment of carrying capacity; estuary, riparian zone restoration and groundwater recharge for hydrologic cycle viability; and other opportunities and issues about Water and the environment show that the foundation of smart growth lies in the protection and preservation of water resources. The total amount of precipitation landing on the surface of a community becomes the supply for the inhabitants. This supply amount then dictates the carrying capacity - the potential population - as supported by the "water crop."

Sustainable Architecture

Sustainable architecture is the design of sustainable buildings. Sustainable architecture attempts to reduce the collective environmental impacts during the production of building components, during the construction process, as well as during the lifecycle of the building. This design practice emphasizes efficiency of heating and cooling systems; alternative energy sources such as solar hot water, appropriate building siting, reused or recycled building materials; on-site power generation - solar technology, ground source heat pumps, wind power; rainwater harvesting for gardening, washing and aquifer recharge; and on-site waste management such as green roofs that filter and control stormwater runoff. This requires close cooperation of the design team, the architects, the engineers, and the client at all project stages, from site selection, scheme formation, material selection and procurement, to project implementation.

Sustainable architects design with sustainable living in mind. Sustainable vs green design is the challenge that designs not only reflect healthy processes and uses but are powered by renewable energies and site specific resources. A test for sustainable design is - can the design function for its intended use without fossil fuel - unplugged. This challenge suggests architects and planners design solutions that can function without pollution rather than just reducing pollution. As technology progresses in architecture and design theories and as examples are built and tested, architects will soon be able to create not only passive, null-emission buildings, but rather be able to integrate the entire power system into the building design. In 2004 the 59 home housing community, the Solar Settlement, and a 60,000 sq ft (5,600 m^2) integrated retail, commercial and residential building, the Sun Ship, were completed by architect Rolf Disch in Freiburg, Germany. The Solar Settlement is the first housing community world wide in which every home, all 59, produce a positive energy balance.

An essential element of Sustainable Building Design is indoor environmental quality including air quality, illumination, thermal conditions, and acoustics. The integrated design of the indoor environment is essential and must be part of the integrated design of the entire structure. ASHRAE Guideline 10-2011 addresses the interactions among indoor environmental factors and goes beyond traditional standards.

Sustainable Landscape and Garden Design

Sustainable landscape architecture is a category of sustainable design and energy-efficient landscaping concerned with the planning and design of outdoor space. Design techniques include planting trees to shade buildings from the sun or protect them from wind, using local materials, on-site composting and chipping to reduce green waste hauling, and also may involve using drought-resistant plantings in arid areas and buying stock from local growers to avoid energy use in transportation. Areas of the garden and landscape can also be allowed to grow wild to encourage bio-diversity.

Sustainable Graphic Design

Sustainable graphic design considers the environmental impacts of graphic design products throughout a life cycle that includes: raw material; transformation; manufacturing; transportation; use; and disposal. Techniques for sustainable graphic design include: reducing the amount of materials required for production; using paper and materials made with recycled, post-consumer waste; printing with low-VOC inks; and using production and distribution methods that require the least amount of transport.

Sustainable Agriculture

Sustainable agriculture adheres to three main goals:

- Environmental health,
- Economic profitability,
- Social and economic equity.

A variety of philosophies, policies and practices have contributed to these goals. People in many different capacities, from farmers to consumers, have shared this vision and contributed to it. Despite the diversity of people and perspectives, the following themes commonly weave through definitions of sustainable agriculture.

There are strenuous discussions - among others by the agricultural sector and authorities - if existing pesticide protocols and methods of soil conservation adequately protect topsoil and wildlife. Doubt has risen if these are sustainable, and if agrarian reforms would permit an efficient agriculture with fewer pesticides, therefore reducing the damage to the ecosystem.

Energy Sector

Sustainable technology in the energy sector is based on utilizing renewable sources of energy such as solar, wind, hydro, bioenergy, geothermal, and hydrogen. Wind energy is the world's fastest growing energy source; it has been in use for centuries in Europe and more recently in the United States and other nations. Wind energy is captured through the use of wind turbines that generate and transfer electricity for utilities, homeowners and remote villages. Solar power can be harnessed through photovoltaics, concentrating solar, or solar hot water and is also a rapidly growing energy source.

The availability, potential, and feasibility of primary renewable energy resources must be analyzed early in the planning process as part of a comprehensive energy plan. The plan must justify energy demand and supply and assess the actual costs and benefits to the local, regional, and global environments. Responsible energy use is fundamental to sustainable development and a sustainable future. Energy management must balance justifiable energy demand with appropriate energy supply. The process couples energy awareness, energy conservation, and energy efficiency with the use of primary renewable energy resources.

Water Sector

Sustainable water technologies have become an important industry segment with several companies now providing important and scalable solutions to supply water in a sustainable manner.

Beyond the use of certain technologies, Sustainable Design in Water Management also consists very importantly in correct implementation of concepts. Among one of these principal concepts is the fact normally in developed countries 100% of water destined for consumption, that is not necessarily for drinking purposes, is of potable water quality. This concept of differentiating qualities of water for different purposes has been called "fit-for-purpose". This more rational use of water achieves several economies, that are not only related to water itself, but also the consumption of energy, as to achieve water of drinking quality can be extremely energy intensive for several reasons.

Sustainable Technologies

Sustainable technologies use less energy, fewer limited resources, do not deplete natural resources, do not directly or indirectly pollute the environment, and can be reused or recycled at the end of their useful life. There is a significant overlap with appropriate technology, which emphasizes the suitability of technology to the context, in particular considering the needs of people in developing countries. However, the most appropriate technology may not be the most sustainable one; and a sustainable technology may have high cost or maintenance requirements that make it unsuitable as an "appropriate technology," as that term is commonly used.

7.12 SUSTAINABLE DRAINAGE SYSTEMS

Sustainable Drainage Systems, sometimes known as Sustainable Urban Drainage Systems (SUDS), are designed to reduce the potential impact of new and existing developments with respect to surface water drainage discharges.

7.12.1 Background of SUDS

The idea behind SUDS is to try to replicate natural systems that use cost effective solutions with low environmental impact to drain away dirty and surface water run-off through collection, storage, and cleaning before allowing it to be released slowly back into the environment, such as into water courses. This is to counter the effects of conventional drainage systems that often allow for flooding, pollution of the environment - with the resultant harm to wildlife - and contamination of groundwater sources used to provide drinking water. The paradigm of SUDS solutions should be that of a system that is easy to manage, requiring little or no energy input (except from environmental sources such as sunlight, etc.), resilient to use, and being environmentally as well as aesthetically attractive. Examples of this type of system are reed beds and other wetland habitats that collect, store, and filter dirty water along with providing a habitat for wildlife.

Originally the term SUDS described the UK approach to sustainable urban drainage systems. These developments may not necessarily be in "urban" areas, and thus the "urban" part of SUDS is now usually dropped to reduce confusion. Other countries have similar approaches in place using a different terminology such as Best Management Practice (BMP) and low-impact development in the United States, and water-sensitive urban design in Australia.

SUDS use the following techniques:

- Source control
- Permeable paving such as pervious concrete
- Storm water detention
- Storm water infiltration
- Evapo-transpiration (e.g. from a green roof)

A common misconception of SUDS systems is that they reduce flooding on the development site. In fact the SUDS system is designed to reduce the impact that the surface water drainage system of one site has on other sites. For instance, sewer flooding is a problem in many places. This happens when flows entering a sewer exceed its capacity and it overflows. The SUDS system aims to minimise or eliminate discharges from the site, thus reducing the impact, the idea being that if all development sites incorporated SUDS then urban sewer flooding would be less of a problem. Unlike traditional urban stormwater drainage systems, SUDS can also help to protect and enhance ground water quality.

7.13 URBAN RUNOFF

Urban runoff is surface runoff of rainwater created by urbanization. This runoff is a major source of water pollution in urban communities worldwide. Impervious surfaces (roads, parking lots and sidewalks) are constructed during land development. During rain storms and otherprecipitation events, these surfaces (built from materials such as asphalt andconcrete), along with rooftops, carry pollutedstormwater to storm drains, instead of allowing the water to percolate through soil. This causes lowering of the water table(because groundwater recharge is lessened) and flooding since the amount of water that remains on the surface is greater. Most municipal storm sewer systems discharge stormwater, untreated, to streams, rivers and bays.

7.13.1 Pollutants in Water Runoff

Water running off these impervious surfaces tends to pick up gasoline, motor oil, heavy metals, trash and other pollutants from roadways and parking lots, as well as fertilizers and pesticides from lawns. Roads and parking lots are major sources of polycyclic aromatic hydrocarbons(PAHs), which are created as combustion by products of gasoline and other fossil fuels, as well as of the heavy metals nickel, copper, zinc, cadmium, and lead. Roof runoff contributes high levels of synthetic organic compounds and zinc. Fertilizer use on residential lawns, parks and golf courses is a significant source ofnitrates and phosphorus in urban runoff.

As stormwater is channeled into storm drains and surface waters, the natural sediment load discharged to receiving waters decreases, but the water flow and velocity increases. In fact, the impervious cover in a typical city creates five times the runoff of a typical woodland of the same size.

7.14 SUSTAINABLE IMPLANT

Sustainable implant is an urban typology that acts as a decentralized infrastructure provision hub on the neighborhood or district scale. Sustainable implants provide integrated infrastructure services that maintain cycles of energy, water and material, as well as provides social and economic returns. The concept originates from Arjan van Timmeren's research, Autonomy & Heteronomy (2006), as an answer to the problem of scale versus innovation in infrastructure; wherein infrastructure benefits from increasing returns to scale but suffer from extremely slow rate of change and turnover. To answer this problem, the sustainable implant is an instrument for mid-scale facilitation of alternative system innovation. The sustainable implant is a synthesis of techniques for sustainable processing of urban flows within an ecological processing device. The objective of a sustainable implant is to generate qualitative and quantitative improvements for utility service provision.

A sustainable implant cannot be regarded as a fixed design that can be repeated. The device comprises a guiding principle for a sustainable solution to the mainly non-sustainable flows in new or existing neighborhoods making use of E. F. Schumacher's appropriate technology, matching technology to need. The sustainable implant facilitates the joint performance of the community as a cognitive agent within transition processes into sustainable development. On a neighborhood or district level, the Sustainable Implant entails the design of a more sustainable main structure for the transportation of water, wastewater, nutrients, energy, materials and/or waste.

The sustainable implant is matched to the social culture, the micro-climate, and the client's ambitions. The technical morphology the Sustainable Implant is formulated by comparing relevant alternatives against the conventional approach. The maintenance of clean urban cycles is analyzed to prove the fulfillment of the sustainability ambition. Beyond technocracy, the Sustainable Implant is to be socially calibrated and economically profitable.

The sustainable implant originated from the desire to find harmony between autarky and returns to scale in the field of infrastructure by means of decentralization of utility provision on an appropriate scale such that a more sustainable urban development can be achieved. In contrast to sustainable architecture, the sustainable implant is related to the flow of materials and energy through the built environment. A sustainable implant will often use decentralized sanitation and reuse for enhanced reuse of water, energy and nutrients, and waste.

In his research, Arjan van Timmeren formulated the following five questions:

- To what extent are the current technical (infra)structures decisive for the possibilities and impossibilities of sustainable development?

- Can the central or decentralized solution of the essential flows generate further processes of preservation at a higher scale level?
- Is there an optimal scale for self-sufficiency per flow, and, if there is, what is this optimal scale?
- To what extent can user participation and involvement increase by solving sustainability issues?
- Should the various techniques for the optimization of the flows be combined in one device and can this be done, or should they be integrated separately into existing (infra)structures or buildings?

It is of importance to recognize the need for connection or interconnection with networks at higher scale levels. In fact, the pursuit of full closure or "autarky" even counteracts the principle of the "external economies of scale". By contrast, the arrangement of the cluster within the larger network can respond to a wish for self-sufficiency, creating some sort of a "quasi-autarky", with as a direct advantage the use as a fall-back scenario of the increasing dependence on other areas, countries or even continents.

Such an alternative network principle can also help breaking through the historical relations between the internal organization within the organizations themselves, and the connections to each other and to the more general social structures in the specific localities.

CHAPTER

URBAN MORPHOLOGY 8

8.1 INTRODUCTION

Urban morphology is the study of the formation of human settlements and the process of their formation and transformation. The study seeks to understand the spatial structure and character of a metropolitan area, city, town or village by examining the patterns of its component parts and the ownership or control and occupation. Typically, analysis of physical form focuses on street pattern, lot pattern and building pattern, sometimes referred to collectively as urban grain. Analysis of specific settlements is usually undertaken using cartographic sources and the process of development is deduced from comparison of historic maps.

Special attention is given to how the physical form of a city changes over time and to how different cities compare to each other. Another significant part of this subfield deals with the study of the social forms which are expressed in the physical layout of a city, and, conversely, how physical form produces or reproduces various social forms.

The essence of the idea of morphology was initially expressed in the writings of the great poet and philosopher Goethe (1790). However, the term as such was first used in bioscience. Recently it is being increasingly used in geography, geology, philology and other subject areas. In geography, urban morphology as a particular field of study owes its origins to Lewis Mumford, James Vance and Sam Bass Warner. Peter Hall and Michael Batty of the UK and Serge Salat, France, are also central figures.

Urban morphology is considered as the study of urban tissue, or fabric, as a means of discerning the environmental level normally associated with urban design. Tissue comprises coherent neighborhood morphology and functions. Neighborhoods exhibit recognizable patterns in the ordering of buildings, spaces and functions, variations within which nevertheless conform to an organizing set of principles. This approach challenges the common perception of unplanned environments as chaotic or vaguely organic through understanding the structures and processes embedded in urbanisation. Complexity science has provided further explanations showing how urban structures emerge from the uncoordinated action of multiple individuals in highly regular ways. Amongst other things this is associated with permanent energy and material flows to maintain these structures.

8.2 SOME LOCKED CONCEPTS

Urban morphology approaches human settlements as generally unconscious products that emerge over long periods, through the accrual of successive generations of building activity. This leaves traces that serve to structure subsequent building activity and provide opportunities and constraints for city-building processes, such as land subdivision, infrastructure development, or building construction. Articulating and analysing the logic of these traces is the central question of urban morphology.

Urban morphology is not generally object-centred, in that it emphasises the relationships between components of the city. To make a parallel with linguistics, the focus is placed on an active vocabulary and its syntax. There is thus a tendency to use morphological techniques to examine the ordinary, non-monumental areas of the city and to stress the process and its structures over any given state or object, therefore going beyond architecture and looking at the entire built landscape and its internal logic.

Roger Trancik discusses three major theories of urban spatial design and urban mythology which can guide analysis:

1. Figure and Ground theory
2. Linkage theory
3. Place Theory

Figure and Ground theory is founded on the study of the relationship of land coverage of buildings as solid mass to open voids Each urban environment has an existing pattern of solid and voids, and figure and ground approach to spatial design is an attempt to manipulate these relationships by adding to, subtracting from, or changing the physical geometry of the pattern. The objective of these manipulations is to clarify the structure of urban space in a city or district by establishing a hierarchy of spaces of different sizes that are individually enclosed but ordered directionally in relation to each other. Linkage theory focuses on lines formed by streets, pedestrian ways, linear open spaces or other linking elements that physically connect the parts of the city. Place theory operates upon structured systems of human needs and usage.

8.3 URBAN POPULATION

Urban population refers to the population inhabiting areas that have a greater population density than rural areas and are overall more compact than rural areas. In simple terms, it's the people living in cities.

As well as being more densely populated than rural areas, in wealthier countries there is generally a higher percentage of people living in urban centers than on rural land. Thus, the percentage of people living in urban centers compared to the rural population can be an indication of the overall development and wealth of a country. Going forward, as world population increases, organizations like the United Nations predict that the percentage of people living in urban centers will steadily increase as people migrate from rural areas into cities.

8.3.1 Effects of Increased Urban Population

The continuing growth of urban populations has both positive and negative side effects on people's lives and the environment. Urbanization can, at times, improve both the economy and education as people move to the cities in search of higher-income jobs and better schooling. A prime example of this would be in Beijing, where many young people from rural areas move in search of jobs. As well, urban centers such as this one often contain more high-ranked schools and universities, leading to a higher level of education among citizens, and access to health care and social services is often better in cities.

However, there are numerous environmental side effects of urbanization that are detrimental. Urbanization will lead to a dramatic increase in metal use, specifically with metals such as steel for construction. The increased urban population will also increase the demand for energy, with more access to electricity and other amenities, especially due to rural-urban migration. This will have a large impact on energy demand worldwide. More specific issues will also arise, such as pollution, increased cost and difficulty of providing public transportation, and destruction of wildlife habitat.

8.3.2 Demographic Trends and Urbanization

Global demographic trends are at a turning point. As population growth rates continue to decline in every region of the world, and in the face of an unprecedented aging of global population, one question is top of mind for everyone – what do these trends mean for the world's cities?

This report – Demographic Trends and Urbanization – aims to frame that conversation by tracing the relationship between urbanization and relevant demographic indicators and how demography shapes a country's desired development outcomes. The report explores how population growth, aging and migration affect the growth trajectories of cities, and how and why city-level demographic trends diverge from national and regional trends.

The analysis of demographic trends is particularly relevant as larger shares of people move to or live in urban areas. The pattern, pace and causes of rural-urban migration continue to be diverse, and significantly influenced by the search for opportunity in cities, the flight from conflict, poverty and climate events, amongst other issues.

To date, the use of city-level demographic trends to project cities' needs in terms of infrastructure and development remains rare. Most demographic studies tend to focus on the analysis of fiscal trends and implications related to labor force, health and care services and retirement spending.

This report presents an opportunity to start thinking about what the city of tomorrow will look like, and what it will need in terms of infrastructure and services. Not all cities will grow endlessly, so today's investments must account for of the demographic reality of tomorrow.

Cities' demographic trends also vary vastly by regions: Some cities, particularly in some large middle-income and high-income countries, are contending with an unprecedented aging of their population, based on low fertility trends and an increase in life expectance. On the other hand, many low-income countries especially in Sub-Sahara Africa - which are also among the least urbanized – have the highest urban growth rates in the world.

8.3.3 Regional Highlights

In Europe and Central Asia, urban population declined in over half of low- and middle-income countries between 1991 and 2011. The rise in average age in Eastern Europe and Central Asia (ECA) is largely attributable to a decrease in fertility rather than to an increase in longevity, which makes aging trends in ECA unique. The emigration of young people is reinforcing these trends. The region and its cities will continue to see diverse and contrasting patterns of growth and decline as population aging is expected to accelerate across major cities.

The Latin America and the Caribbean region is among the most urbanized regions in the world, having reached an urban majority by 1960 - a milestone achieved globally only in 2008. Average growth rates in cities today are declining, and some city centers are losing population to the urban periphery. Urbanization is shifting towards small and intermediate cities. The region is starting to age rapidly, a trend that is especially pronounced in urban areas.

The Middle East and North Africa (MENA), is already 64% urbanized. Much of the region's future urban growth—in absolute terms—is projected to occur in the region's primary cities although faster growth—in relative terms—is projected to occur in the region's secondary cities. Conflict and climate-driven migration as well as cross-border movement driven by search for economic opportunity are unique considerations in the region and have been overriding some secular urbanization trends in a subset of countries. Aging is not yet an issue in MENA but is set to rapidly increase over the coming decades. The region's youth bulge, if managed well, could present a significant demographic dividend with potential higher growth rates, especially in urban areas.

Africa is the fastest growing and youngest region in the world. There is great diversity of demographic trends among African countries and regions. The region has the highest urban growth rates in the world, but a relatively low share of population living in urban areas. As life expectancy increases, between 2015 and 2050, Africa will have the fastest increase in the world in the number of elderly persons. Small and intermediate cities, not the largest ones, will absorb a bulk of the region's urban population growth in the future.

In South Asia, a combination of increasing life expectancy, falling fertility rates, and low old-age dependency ratios has created the conditions for a "youth bulge" generation. The region has disproportionately fewer women, especially at birth, compared with the global average, which could have long-term implications when it comes to labor market participation, for instance. Despite the common perception, urban population growth is mainly driven by natural growth and in-situ reclassification rather than rural-to-urban migration. Congestion and weak urban management may also undermine South Asian cities' pull factors.

The East Asia and Pacific (EAP) region is experiencing rapid urban growth, second only to Sub-Saharan Africa. EAP became 50% urban in 2018, with most urban inhabitants residing in secondary cities. The region also has the largest regional population of elderly but there is an enormous diversity of aging-levels within the region, which are correlated with national income levels. Migration likely has a significant role in driving urbanization levels in EAP.

Three key demographic forces have reshaped the overall U.S. population in recent years: growing racial and ethnic diversity, increasing immigration and rising numbers of older adults. But these trends are playing out differently in the nation's rural, urban and suburban communities, touching some more than others.

Likewise, recent U.S. population growth also has been uneven. Urban counties have grown at roughly the overall national rate of 13% since 2000. Suburban and small metropolitan areas have grown more briskly. Rural counties have lagged, and half of them have fewer residents now than they did in 2000.

According to a Pew Research Center analysis of census data, since 2000, U.S. urban and suburban populations have grown at least as much as they did over the prior decade. But the total rural population has grown less than it did in the 1990s, when rising numbers fed hope of a modest "rural rebound." As a result, a somewhat smaller share of Americans now live in rural counties.

More recently, the Census Bureau's population estimates for 2017 show a one-year uptick in the nation's rural population, though not enough to make up for previous declines. Analysis by demographer Kenneth M. Johnson attributed the increase to gains in rural communities on the edge of metropolitan areas, while more remote counties continued to lose population.

8.4 WHAT IS AN URBAN, SUBURBAN OR RURAL COUNTY?

The flow of people in and out of different types of U.S. counties is affecting their size and composition. Since 2000, more people left rural counties for urban, suburban or small metro counties than moved in from those areas. Because there were not enough new immigrants to offset those departures, rural counties as a group grew only because they had more births than deaths.

At the national level, non-Hispanic whites make up the majority of the population, but a key demographic shift is underway: Whites are a shrinking share of the population and expected to be less than half by midcentury as other groups grow more rapidly. Whites have become a minority of the population in most urban counties since 2000, while remaining the majority in 90% of suburban and small metro counties and 89% of rural ones.

Another key demographic trend, the rise in immigration in recent decades, has raised the foreign-born share of the U.S. population overall and has increased the share in each type of county, although to varying degrees. Immigrants, along with their children and grandchildren, have accounted for the majority of U.S. population growth since 1965. But immigrants are more concentrated in cities and suburbs than in rural areas. On the flip side, the majority of rural counties now have fewer U.S.-born residents than in 2000, a key factor in their dwindling populations.

A third major population driver – the aging of the giant Baby Boom generation – also has varying impacts on different county types. Rural areas have a higher share of adults who are ages 65 and older than urban or suburban counties. But suburban counties have experienced the sharpest increases in the number of older adults since 2000.

The analysis in this chapter relies mainly on Census Bureau data. Current numbers for county characteristics come from the American Community Survey (ACS) combined data for 2012-2016, the latest available. Current numbers for natural increase/decrease and migration flows come from population estimates for 2014, the most comparable year to the ACS data because it is the midpoint of the combined ACS data used in this chapter. See Methodology for more detail.

8.4.1 Suburbs Growing more Rapidly than Rural or Urban Areas

About 46 million Americans live in the nation's rural counties, 175 million in its suburbs and small metros and about 98 million in its urban core counties.

As a group, the population in rural counties grew 3% since 2000, less than their 8% growth in the 1990s. Urban county population rose 13% since 2000 and the population in suburban and small metro counties went up 16%, growth rates somewhat higher than in the 1990s. The share of U.S. residents who live in rural counties declined in the 1990s and since 2000, but rose in suburban counties during both periods and held steady in urban counties.

Although the rural population as a whole has grown since 2000, the majority of populations in individual rural counties have not. Since the turn of the century, the population declined in 52% of rural counties – 1,024 of 1,969. Among the hardest hit counties were those where the economy is based on farming, about a fifth of rural counties.

Population growth in rural counties since 2000 has trailed other areas

Population (in millions)

	2000	2012-16	Change (%)
Urban	87	98	13
Suburban	150	175	16
Rural	45	46	3

Note: County categories based on the National Center for Health Statistics Urban-Rural Classification Scheme for Counties.
Source: Pew Research Center analysis of 2000 decennial census SF3 data and 2012-2016 American Community Survey data.
"What Unites and Divides Urban, Suburban and Rural Communities"

PEW RESEARCH CENTER

Fig. 1. Suburbs Growing more rapidly.

8.4.2 Growth Factors Vary for Cities, Suburbs and Rural Areas

There are four main drivers of population gain or loss at the county level: births, deaths, new immigrants coming from abroad or leaving, and people moving to or from other U.S. counties (including immigrants already living in the U.S.). The census numbers show that these factors are affecting cities, suburbs and rural communities differently.

Urban areas gained 1.6 million net new migrants since 2000, with a surplus of immigrants more than offsetting a loss of people who moved out to suburbs or rural areas. As a group, urban counties had 9.8 million more births than deaths, further bolstering their populations.

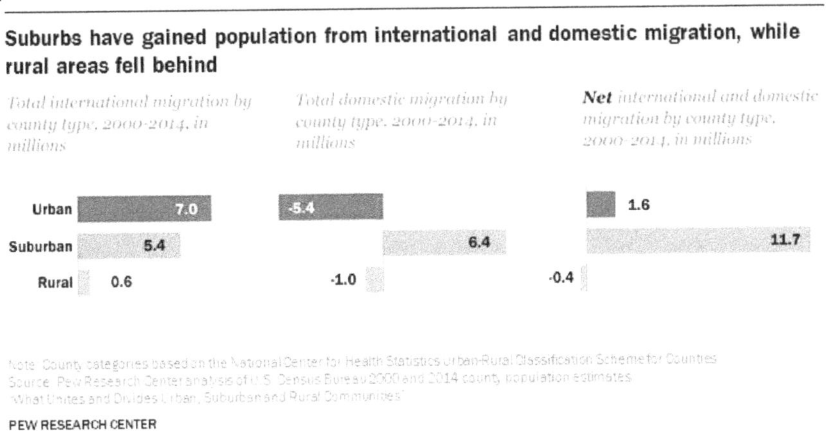

Fig. 2. Growth Factors.

Suburban and small metro counties have grown since 2000 because of gains in all the drivers of population change. They gained 11.7 million new residents by drawing former residents of U.S. urban and rural areas, as well as immigrants from abroad. On top of that, they had 12.1 million more births than deaths.

It was a different picture for rural counties, however, where move-outs since 2000 exceeded move-ins. As a group, they had a net loss of 380,000 people who moved out. The loss would have been larger – more than 950,000 people – had it not been partly offset by about 600,000 new immigrants. The total population of rural counties grew only through natural increase – that is, they had 1.2 million more births than deaths.

8.4.3 Rural Population Loss Largest in Midwest

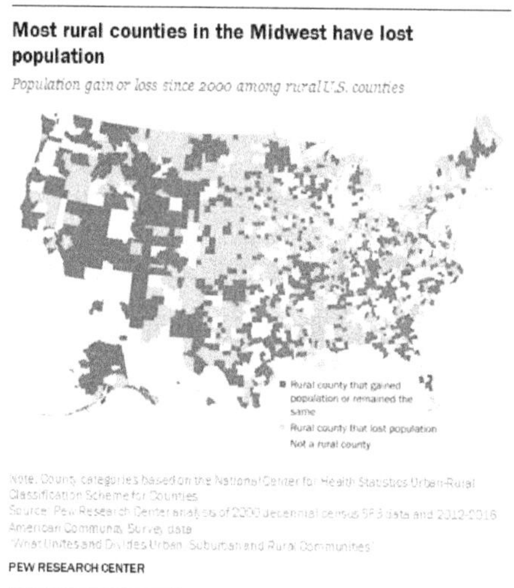

Fig. 3. Rural population.

Patterns of births, deaths, migration and immigration vary greatly among regions, and generally illustrate the long-term trend of Americans favoring the Sunbelt states of the South and West over Northeastern or Midwestern states. These regional differences persist within each county type.

Among rural counties, a majority in the Northeast and Midwest lost population since 2000, while a majority in the South and especially the West gained population. One factor behind the regional difference is that rural counties in the Northeast and Midwest were more likely than other rural areas to have more deaths than births. These counties also were more likely to have experienced a net loss of migrants – more people moving out than moving in.

The population trends of rural counties are linked to their economic profiles. As a group, the nation's 391 rural farming counties – heavily concentrated in the Great Plains – have lost total population since 2000, while rural counties with other types of economies gained population.

A majority of rural counties – especially in the Northeast and Midwest – have more people moving out than moving in

% of counties ...

	... losing population since 2000	... with more deaths than births since 2000	... with more people moving out than moving in since 2000
Urban	**15**	**3**	**50**
Northeast	13	7	93
Midwest	42	0	92
South	11	4	26
West	0	0	14
Suburban	**18**	**13**	**34**
Northeast	23	20	40
Midwest	24	9	48
South	16	16	29
West	5	7	19
Rural	**52**	**41**	**68**
Northeast	54	53	63
Midwest	68	48	85
South	44	37	57
West	34	31	58

Note: County categories based on the National Center for Health Statistics Urban-Rural Classification Scheme for Counties.
Source: Pew Research Center analysis of U.S. Census Bureau 2000 and 2014 county population estimates.
"What Unites and Divides Urban, Suburban and Rural Communities"

PEW RESEARCH CENTER

Fig. 4. A majority of rural counties.

The total population of rural counties with recreation-based and government-based economies grew more since 2000 than the populations of other rural county types. One reason recreation-based counties grew was that they had a net gain of new residents who moved from other U.S counties, the only rural county type to have a gain in net domestic migration. An analysis by the Population Reference Bureau found that rural recreation-based counties were especially likely to have growing numbers of residents 65 and older, while rural farming-based counties were losing residents in that age group.

Among urban areas, the Midwest had the largest share of population-losing counties since 2000: 42% of urban counties in this region, including the ones that encompass Chicago, Detroit, and Cleveland, lost population.

Among suburban and small metro counties, about a quarter of the ones in the Northeast and Midwest lost population since 2000, a higher share than in other regions. A majority of Northeast and Midwest suburban counties had a net gain of migrants, but that was mainly due to immigration. A majority had a net loss of residents to urban or rural U.S. counties during this period.

Older adults are a higher share of the population in rural areas than in urban and suburban counties

A key demographic trend shaping the makeup of local populations, as well as the nation as a whole, is the rising number of older Americans. The Baby Boom generation, born between 1946 and 1964, began turning 65 in 2011, and all will have reached that age by 2030.

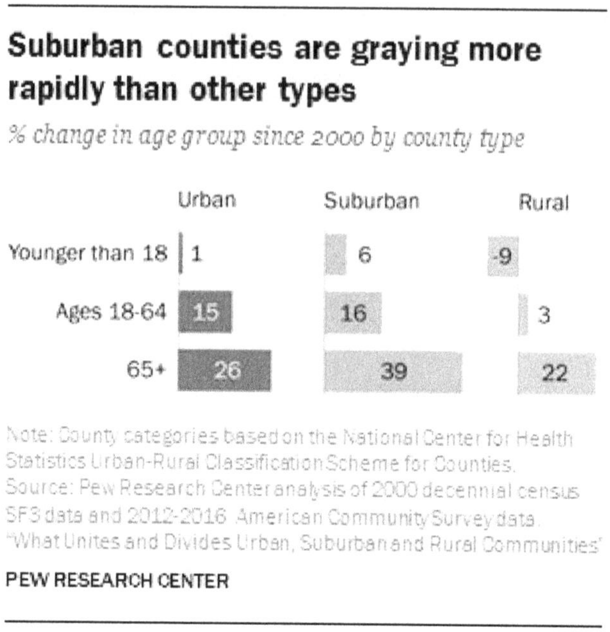

Fig. 5. Percent change in age group.

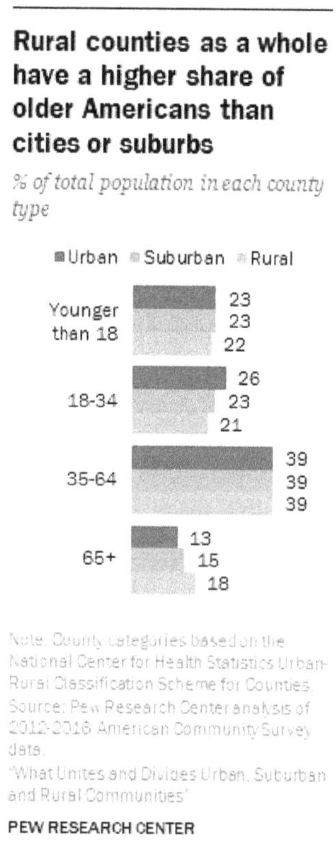

Fig. 6. Percent of total population.

While the population is aging in all three county types, this is happening more rapidly in U.S. suburban and small metro counties. The 65-and-older population grew 39% in the suburbs since 2000, compared with 26% in urban and 22% in rural counties.

Nationally and in each county type, the older adult population grew more sharply since 2000 than any other age group – young children, school-age children, young adults or middle-aged adults. In rural areas, the population younger than 18 declined during this period. As a result, in each county type, adults ages 65 and older now make up a larger share of the total population than in 2000.

As a group, rural counties skew older than suburban and urban counties: 18% of rural residents are 65 or older vs. 15% in suburban and small metro counties and 13% in cities.

Rural counties also have a smaller share of young adults than urban or suburban populations.

8.4.4 Urban and Suburban Counties are Becoming More Racially and Ethnically Diverse at a Much Faster Pace Than Rural Counties

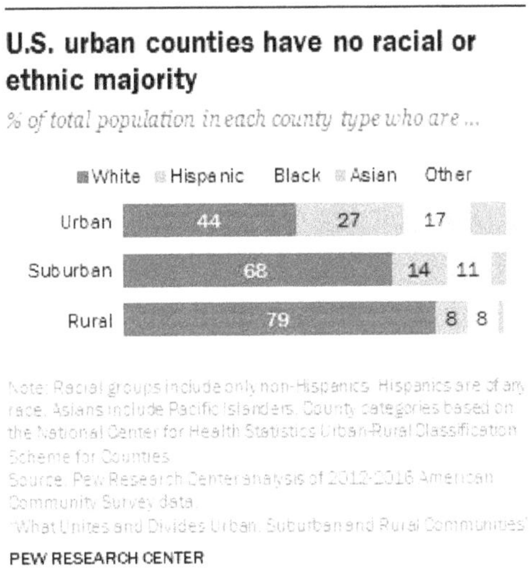

Fig. 7. Urban and suburban Counties are more racially.

The nation is becoming more racially and ethnically diverse, but these changes have been more muted in rural counties compared with urban and suburban ones. The white share of the population fell 8 percentage points since 2000 in the suburbs, 7 points in the urban core and only 3 points in rural counties.

Overall, the U.S. population remains majority white, but not so in urban areas as a group. Among urban residents, 44% are white, compared with 68% in suburban and small metro counties and 79% in rural counties. In fact, whites have become the minority in most urban counties (53% of them are majority nonwhite) since 2000; only about one-in-ten suburban (10%) and rural (11%) counties are majority nonwhite.

While the share of the non-Hispanic whites has declined, the number of whites grew in the nation overall – and in suburban counties as a whole – since 2000. But the white population did not grow as sharply as other groups did, leading to a decline in the white share of the total U.S. and suburban populations.

In urban counties, the decline in the share of the white population was due both to a decrease in the number of whites and an increase in the size of other populations, chiefly Hispanics.

In rural counties, the white population also decreased and other groups also increased in size, but the impact was more modest on the white share of the population because whites are so large a proportion of rural residents.

8.4.5 Immigrants are a Rising Share of the Population in Each County Type

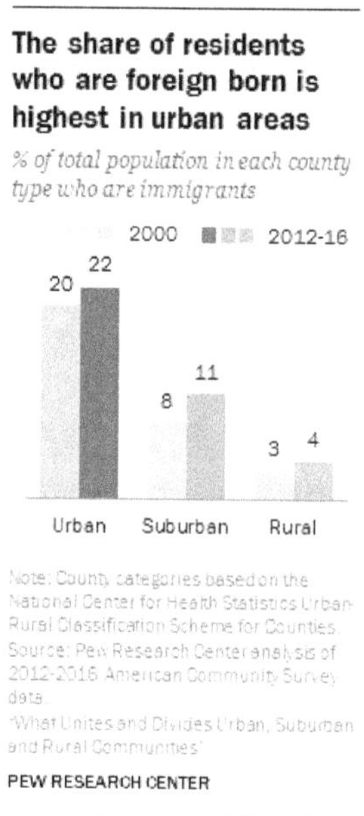

Fig. 8. Rising share of the population.

The foreign-born population is not evenly distributed across county types; immigrants tend to be concentrated in big metropolitan areas. In fact, about half live in urban counties, where they make up a higher share of the total than in suburban or rural counties.

The immigrant share of the population grew since 2000 in the nation as a whole and in each county type. Immigrants were responsible for a larger share of overall growth in rural (37%) and urban (38%) counties than in suburban (26%) ones.

Although rural counties as a group have more U.S.-born residents than in 2000, the majority of individual rural counties have fewer U.S.-born residents than in 2000. There is a large overlap between rural counties that have fewer U.S.-born residents than in

2000 and those that have fewer total residents than in 2000. In the vast majority of rural counties that lost population – 1,011 out of 1,025 – the number of U.S.-born residents declined, and there were not enough new immigrants to offset the loss.

Gaps in Poverty, Education and Employment Across County Types

In addition to the three major demographic shifts that are reshaping urban, suburban and rural counties in the U.S. – the aging of the population, changing racial and ethnic makeup and the influx of new immigrants – there are significant differences in other important metrics across community types. These relate to the economic well-being of their residents.

8.4.6 Poverty has Increased more Sharply in Suburbs Than in Urban or Rural Counties

Overall number of poor residents has increased most in suburban areas

Population in poverty (in millions)

	2000	2012-16	Change (%)
Urban	12	16	31
Suburban	15	23	51
Rural	6	8	23

Note: County categories based on the National Center for Health Statistics Urban-Rural Classification Scheme for Counties.
Source: Pew Research Center analysis of 2000 decennial census SF3 data and 2012-2016 American Community Survey data.
"What Unites and Divides Urban, Suburban and Rural Communities"

PEW RESEARCH CENTER

Fig. 9. Povertiy has increased more Sharply.

Overall, the poverty rate is somewhat higher in rural (18%) and urban (17%) counties than in suburban ones (14%). Poverty rates have risen in all three county types since 2000.

The number of people living in poverty also has risen across community types, but the size of the poor population rose more sharply in suburban counties than in urban or rural ones.

About half the U.S. poor population (49%) lives in suburban and small metro counties, while 34% live in cities and 17% in rural areas.

But looking at the share of counties where at least a fifth of the population is poor – a measure known as concentrated poverty – rural areas are at the top. About three-in-ten rural counties (31%) have concentrated poverty, compared with 19% of cities and 15% of suburbs. The number of counties with concentrated poverty grew for all three county types since 2000.

8.4.7 Growing Share of Residents of All County Types have College Degrees

The share of adults who are college educated has increased in each community type

% of population 25 and older by education

	Year	<HS grad	HS grad	Some college	Bachelor's+
Urban					
	2012-16	15	23	27	35
	2000	22	24	27	28
Suburban					
	2012-16	11	28	30	31
	2000	17	29	28	25
Rural					
	2012-16	15	36	30	19
	2000	24	36	25	15

Note: "Some college" includes those with an associate degree and those who attended college but did not obtain a degree. County categories based on the National Center for Health Statistics Urban-Rural Classification Scheme for Counties.
Source: Pew Research Center analysis of 2000 decennial census SF3 data and 2012-2016 American Community Survey data.
"What Unites and Divides Urban, Suburban and Rural Communities"
PEW RESEARCH CENTER

Fig. 10. Growing share of residents.

Growing shares of residents ages 25 and older have graduated from college in all types of U.S. communities since 2000, though growth since 2000 was not as sharp as during the 1990s. Rural communities lag in the share of the population with a college degree.

Today, 35% of urban residents and 31% in the suburbs have a bachelor's degree or more education, compared with 19% in rural counties. Rural areas also trail urban and suburban areas in their share of residents with postgraduate degrees.

In urban and suburban counties overall, college graduates outnumber residents with a high school diploma and no further education, but in the total rural population there are more high school graduates than college graduates. The share of residents who did not graduate from high school has dwindled in all three county types.

8.4.8 Rural Counties Lost Prime-Age Workers, While Urban and Suburban Areas Gained Them

Prime-age employment has declined in rural areas overall

Employed 25- to 54-year-olds (in millions)

	2000	2012-16	Change (%)
Urban	29	32	12
Suburban	51	53	4
Rural	14	12	-11

Note: County categories based on the National Center for Health Statistics Urban-Rural Classification Scheme for Counties. Employment refers to civilian employment.
Source: Pew Research Center analysis of 2000 decennial census SF3 data and 2012-2016 American Community Survey data.
"What Unites and Divides Urban, Suburban and Rural Communities"

PEW RESEARCH CENTER

Fig. 11. Rural counties lost prime age workers.

Rural counties also trail other types of communities, especially urban counties, on key measures of employment of prime-age workers – those 25 to 54 years old. For example, 71% of rural residents of prime working age are employed, compared with 77% in both urban and suburban counties.

The number of employed adults in this age group rose in urban areas as well as in suburban and small metro areas since 2000, but declined in rural counties overall. Rural counties now are home to a smaller share of the nation's prime-age workers than in 2000.

The growth in the prime-age working population was particularly sharp in urban areas. As a result, urban counties now are home to a larger share of the nation's prime-age workers than in 2000.

In suburban areas, which also saw an increase in the number of prime-age workers since 2000, the picture looks somewhat less rosy when analyzed through another lens – the experience of individual counties. Although the number of these workers grew for suburban counties as a group, a majority of suburban counties (59%) have fewer of them than in 2000. Among rural counties, 88% lost prime-age workers since 2000. Only 29% of urban counties did.

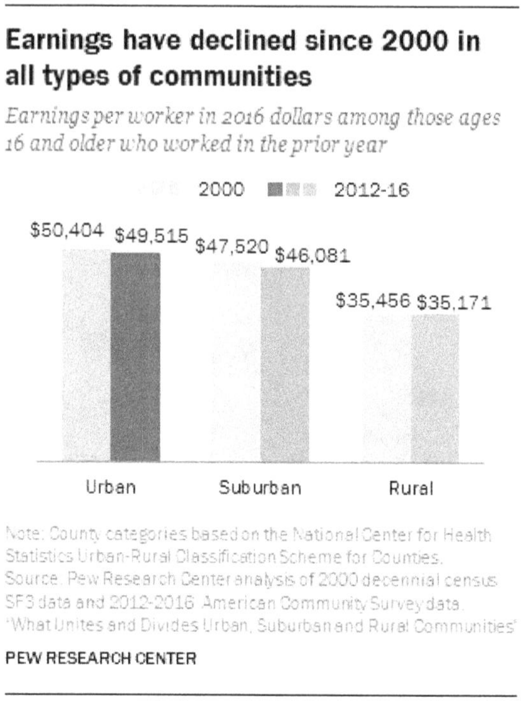

Fig. 12. Earning have declined since 2000.

Another measure of economic health – average earnings per worker – is highest in urban counties and lowest in rural ones. These average earnings are lower now than in 2000 for all county types, reflecting lingering effects of the 2007-2009 recession, though average earnings in rural areas declined the least.

CHAPTER

URBANIZATION AND MIGRATION 9

9.1 INTRODUCTION

Human migration is the movement of people from one place to another with intentions of settling, permanently or temporarily, at a new location. The movement often occurs over long distances and from one country to another, but internal migration is also possible; indeed, this is the dominant form of human migration globally. Migration is often associated with better human capital at both individual and household level, and with better access to migration networks, facilitating a possible second move. Age is also important for both work and non-work migration. People may migrate as individuals, in family units or in large groups. There are four major forms of migration: Invasion, Conquest, Colonization and Emigration/Immigration.

Persons moving from their home due to forced displacement (such as a natural disaster or civil disturbance) may be described as displaced persons or, if remaining in the home country, internally-displaced persons. A person who seeks refuge in another country can, if the reason for leaving the home country is political, religious, or another form of persecution, make a formal application to that country where refuge is sought and is then usually described as an asylum seeker. If this application is successful, this person's legal status becomes refugee.

In contemporary times, migration governance has become closely associated with state sovereignty. States retain the power of deciding on the entry and stay of non-nationals because migration directly affects some of the defining elements of a State.

Migration, whether internal or international, has always been one of the forces driving the growth of urbanization and bringing opportunities and challenges to cities, migrants and governments. Increasingly, municipal authorities are becoming recognized as key actors in managing migration and have started including migration in their urban planning and implementation.

For cities to better manage migration, data on migration and urbanization are essential. However, these data are not always available or – if available – not used or accessible at the urban level, nor disaggregated, comprehensive or comparable, particularly in low-income countries.

Data could improve urban planning and delivery of public services, as well as help measure progress toward Sustainable Development Goals (SDGs) related to cities and migration, implement the Global Compacts on Migration and Refugees, which emphasize the role of cities as stakeholders in migration, and fulfill migration-related commitments in the UN Habitat's New Urban Agenda.

Depending on the goal and reason for relocation, people who migrate can be divided into three categories: migrants, refugees, and asylum seekers. Each category is defined broadly as the mixed circumstances that motivate a person to change their location.

As such, migrants are traditionally described as persons who change the country of their residence for general reasons and purposes. These purposes may include the search for better job opportunities or healthcare needs. This term is the most generally defined one as anyone changing their geographic location permanently can be considered a migrant.

Contrastly, refugees is not defined and described as persons who do not willingly relocate. The reasons for the refugees' migration usually involve war actions within the country or other forms of oppression, coming either from the government or non-governmental sources. Refugees are usually associated with people who must unwillingly relocate as fast as possible; hence, such migrants will likely relocate undocumented.

Asylum seekers are associated with persons who also leave their country unwillingly, yet, who also do not do so under oppressing circumstances such as war or death threats. The motivation to leave the country for asylum seekers might involve an unstable economic or political situation or high rates of crime. Thus, asylum seekers relocate predominantly to escape the degradation of the quality of their lives.

Nomadic movements usually are not regarded as migrations, as the movement is generally seasonal, there is no intention to settle in the new place, and only a few people have retained this form of lifestyle in modern times. Temporary movement for travel, tourism, pilgrimages, or the commute is also not regarded as migration, in the absence of an intention to live and settle in the visited places.

9.2 DEFINITION

These key concepts and definitions are important to understanding urbanization and migration.

Urban is problematic to define and no single globally accepted definition of what constitutes an urban settlement exists. What national statistical offices define as "urban" varies from country to country and often varies over time within countries. Some countries define urban based on a minimum population threshold and population density, while other countries use an administrative definition of what constitutes an urban area. Yet others, include more criteria such as proportion of workforce employed in non-agricultural sectors and availability of infrastructure or of education, health and other services. Most urban population thresholds fall between one and five thousand inhabitants.

Urbanization or "urban transition" refers to "a shift in a population from one that is dispersed across small rural settlements, in which agriculture is the dominant economic activity, towards one that is concentrated in larger and denser urban settlements characterized by a dominance of industrial and service activities".

Urbanization generally occurs as a result of one or more of the following processes:

- Natural population growth;
- When more people move from rural to urban areas;
- When the boundaries of what is considered urban are extended: and/or
- From the creation of new urban centres.

"Very often, urbanization is primarily the result of migration".

Urban population growth is often confused with urbanization but is a distinct concept. Urban growth can take place without any urbanization if urban and rural areas are both growing at the same rate. Urban growth is the increase in the absolute number of people living within defined urban areas. It is defined as "the increase in the proportion of the urban population over time as part of the whole population. Urban growth comes from demographic growth and international and internal migration.

Urban-to-urban migration, rural-to-rural migration, rural-to-urban migration, and urban-to-rural migration: These types of migration refer to the movement of people from one urban or rural area to a different urban or rural area. These types of migration may happen within a national border or involve crossing an international boundary.

City is also difficult to define because what constitutes an urban settlement varies and there are no standardized international criteria for determining the city. Multiple boundary definitions can exist for any given city. In general, there are three concepts:

- City proper describes a city according to an administrative boundary.
- Urban agglomeration considers the extent of the contiguous urban area, or built-up area, to delineate the city's boundaries and is therefore bigger than the city proper.
- Metropolitan area defines its boundaries according to the degree of economic and social interconnectedness of nearby areas, identified by interlinked commerce or

commuting patterns, for example. Metropolitan areas are bigger than the city proper, and can be as big as urban agglomeration, but distinct. While both urban agglomerations and metropolitans can be defined by size, density and contiguity, the definition of metropolitan areas includes other criteria such as administrative functions, and industries and services available, to name a few.

Cities can also be categorized by population thresholds (IOM, 2015, based on UN DESA):

- Small cities have up to 1 million inhabitants
- Medium-sized cities are cities with 1 to 5 million inhabitants
- Large cities are cities with 5 to 10 million inhabitants
- Megacities are cities with 10 million or more inhabitants

9.3 RECENT TRENDS

A majority and growing proportion of the world's population is living in urban areas. The share of the world's population living in urban areas is expected to increase from 55 percent in 2018 to 60 percent in 2030. In 1950, 30 percent of the global population was living in urban areas.

In 2018, Northern America was the most urbanized region in the world, with 82 percent of its population living in urban areas. This was followed by Latin America and the Caribbean (81%), Europe (74%) and Oceania (68%). The lowest levels of urbanization are found in Asia (50%) and Africa (43%), however with large differences among some countries.

Most of the world's fastest growing cities are in Asia and Africa. Between 2018 and 2050, the urban population of Africa is projected to triple and that of Asia to increase by 61 percent, so that by 2050 most of the world's urban population will be concentrated in Asia (52%) and Africa (21%).

Urban population growth has been propelled by the growth of cities of all sizes. In 2018, 33 megacities hosted 13 percent of the global urban population (UN, 2018). By 2030, the number of megacities is projected to increase to 41, with 14 percent of urban dwellers worldwide residing in megacities.

9.3.1 Urbanization and Migration

Approximately one in five international migrants are estimated to live in just 20 cities -Beijing, Berlin, Brussels, Buenos Aires, Chicago, Hong Kong SAR, China, London, Los Angeles, Madrid, Moscow, New York, Paris, Seoul, Shanghai, Singapore, Sydney, Tokyo, Toronto, Vienna and Washington DC.

For 18 of these cities, international migrants represented around 20 percent of the total population.

The share of foreign-born persons in the total population in some cities exceeds the global average (around 3.5%) by a large margin. Dubai has an foreign born population of close to 83 percent, while in Brussels it is 62 percent, in Toronto 46 percent, New York 37 percent, and Melbourne 35 percent, to name a few examples.

Different types of migration play a role in urban growth and diversity, but to different extents. In the developed countries, one of the main sources of population diversity is international migration, while in the developing countries it is most likely internal migration (IOM, 2015), in addition to demographic growth through births outnumbering deaths.

In some countries, rural-to-urban migration and reclassification of what is considered urban together accounted for more than half of the urban growth, such as in China and Thailand (80%), Rwanda (79%), Indonesia (68%) and Namibia (59%). Circular and temporary migration is found in many urban parts of fast-urbanizing Asian and African countries, especially China and India as well as Ghana and Kenya.

9.4 CHALLENGES OF MIGRANTS IN THE CITIES

The reasons for migration can be many and they are often driven by multiple motives, mainly for employment among males and females, marriage especially among females. Eisenstadt, SN 1953, defined migration as the physical transition of an individual or a group from one society to another. This transition normally involves abandoning one social setting and entering another, a different one.

There is a significant contribution to the growth of Indian cities from internal migration. The Indian Constitution guarantees freedom of movement and freedom to settle within the territory of India as a fundamental right of all citizens (Article 19). But this article only remains in the constitution and is not followed by the citizens of the country. Though the freedom to settle is followed by the migrants, few self-proclaimed communities hinder their freedom of movement. These discriminations are rampant especially in the metropolitan cities as there is a huge amount of population migrating to these cities for employment and a better standard of living. Migrants are more vulnerable to discrimination and exploitation as many of them are poor, illiterate and live in slums and hazardous locations prone to disaster and natural calamities. There is a lack of urban policies and programmes catering to the needs and settlements of migrants.

The growth of the cities eventually attracted more migrants and housing societies have often denied homes on the basis of religion, caste, food preferences, and even professions.

A recent incident that happened in Nagpur in the colony where I reside, a Marathi speaking flat owner entered into a verbal fight with a tenant and his family. The tenant's previous residence was in Hyderabad, his native town and he had migrated for work along with his family. The issue was about parking cars in the allotted slot and the tenant had encroached his allotted parking space for which he proudly mentioned that he has paid an amount of one lakh to the developer. It all got zeroed down to one aspect that migrants from South do not have the ability to adjust and stay in harmony with other residents of the apartment. This derogatory statement shows the ideas that people have against migrants from the southern part of the country and how the discrimination against them occur. The psychology of people has divided India into four parts of the country namely- South, North, North-East, and West. These cases happen with migrants who relocate to a place with different culture and local language. These are the invisible and unidentified problems that are unaccounted for and migrants eventually isolate themselves from other residents of the city.

Mumbai, the capital of Maharashtra is influenced by strong anti-migrant political parties who believe that migrants are stealing job opportunities of locals and Marathi culture is deteriorating due to them. This blame game has given locals a chance to discriminate people who have migrated from cities especially from North Indian states like Uttar Pradesh and Bihar.

The recent tragedy caused by the stampede that occurred due to the collapse of Elphinstone Road, Railway Station Bridge was also dumped on the excessive migrants in the city. The lack of attention towards the stability of the structure is given a blind eye and proper infrastructure was not considered as the need of the hour for the political party. Another incident that came into the limelight was the denial of tenancy to a Muslim working woman in Mumbai. Housing and rental issues related to religion show that people are not ready to welcome migrants in the city and migrants face difficulty for residence. Brihanmumbai Municipal Corporation (BMC) had passed a resolution seeking to deny key clearances to builders and housing societies that discriminate against buyers on the basis of caste, religion or food habits. But there was no further hearing about the resolution. This is how the critical issues get dissolved in the system and no further action is taken in these matters.

As per the records of 2011 Census, the largest migrant-sending states are Uttar Pradesh and Bihar amongst the north Indian states. These states also have the highest percentages of people residing in villages with 18.6 percent and 11.1 percent respectively. Other major migrant-sending states include Rajasthan, Andhra Pradesh, Madhya Pradesh, Jharkhand, Chhattisgarh, and Orissa. Various states across central and northern India such as Maharashtra, West Bengal, Delhi, Gujarat, and Haryana receive a substantial flow of labour migrants from Uttar Pradesh. Migrants from Bihar can be seen in the majority in the states of Delhi and West Bengal. It is because the major migrant-sending states possess very low indices of economic and social development and the major migrant-receiving states act as the growing economic magnets in India. The metropolitan cities of India – Mumbai, Delhi, and Kolkata are the largest destinations in India for internal migrants. Many of the migrants to these cities are intrastate migrants, relocating from rural areas of Maharashtra and West Bengal.

There are strong reasons for internal migration. Migrant labourers come to Kerala from the north Indian states in search of job for better wages and are often blamed for the crimes that take place in the city. Poverty does lead to crime but it is not necessarily the migrant population participating in it. Migrants largely contribute to the economy of the state which is overlooked and less known to the citizens. Migration helps a city prosper but the city lacks provision of infrastructure and security to the migrants. If we take migrants out of the cities, it might lead to a sudden drop in the social and economic development which will in turn affect the locals. There are no government policies that talk about the facilities to be given to migrants or for their protection from exploitation. Thus migrants become a vulnerable cluster in the society exposed to insecurities and discrimination.

The women and children who migrate along with men face hardships in their own way. Social exclusion from festivals and language barriers are a hindrance to the lives of migrants. 'Country-one family' is a far-fetched dream because 'City-one family' in the first place seems hard to accomplish as there is no right to control the migration of people and the discrimination on the basis of caste, sex, religion, and the region will take several generations to curb it.

With the issues still prevailing in cities related to internal migration, it leaves me with a thought to ponder on – 'whose city is it, anyway'.

9.4.1 Facets of Migration

Never in India's post-Independence history till 2004-05 did the absolute number of workers in agriculture fall; the Lewisian turning point took over a half-century to arrive. During 2004-05 and 2011-12, non-agri job growth was as high as 7.5 million per annum. At the same time, the number of young entrants into the labour force was only about two million per annum. The remaining five million plus workers were migrants from agriculture, and were mostly absorbed in construction activity.

Net migrant flows at the all-India level have averaged close to nine million annually (between 2011-12 and 2016-17), peaking around 2013-14, considerably above levels suggested by the Census (of 6.9 million in 2011). The largest recipient was the Delhi region, which accounted for more than half of migration in 2015-16,. This is consistent with our finding that of the five million leaving agriculture per annum there were 3.5 million from Uttar Pradesh and Bihar alone.

The Report by the Working Group on Migration has identified 54 districts with a high level of inter-State out-migration intensity. These districts account for half the male inter-State out-migration in the country. Of these, 36 districts are concentrated in Eastern Uttar Pradesh and Bihar, with certain districts in other States — like Nadia and Midnapore (West Bengal), Ganjam (Odisha), Gulbarga (Karnataka), Jalgaon (Maharashtra), Pali (Rajasthan), and a few in western Uttar Pradesh.

Internal migration has been rising over time, nearly doubling in the 2000s relative to the 1990s. This suggests the rewards have become greater than the costs and risks that migration entails.

9.4.2 Structure of Jobs

The jobs challenge created by migration will be monumental in scope. In the manufacturing sector in urban areas, 38 percent of the male workforce is composed of migrant workers, with a similar share in modern services. In contrast to the stereotype of migrants being largely in low-income occupations like street vending, etc., they are employed across all sectors and are essential for manufacturing growth. It also contributes to the very high share of informality in India's workforce. There is high likelihood that due to the sharp increase in construction work, especially in urban areas, work-related migration is turning more short-term. Though the proportion of short-term migrants is much lower than long-term migrants, they are definitely drawn from the lower consumption quintiles.

There is a certain concentration of migrants in specific sectors. Moreover, in construction, the concentration of ST and SC categories is rather high. The SCs tend to be more landless, poorest, and with the least education. Hence it is not surprising that they are found to be in manual work in construction. But NSSO (2018) data shows that construction generated only 4.4 million new jobs over 2012-18, compared with 25 million between 2004-2011.

9.5 FASTER URBANISATION

This migration has been and will, to a greater extent than before, be accompanied by faster urbanisation in India. Asia is going through a historic demographic transformation from a rural society to an urban society that is far larger than any transformation seen in the past, in any other part of the globe. By 2025, the majority of Asia's population will be urban. India's urban population is expected to grow from 410 million in 2014 to 814 million by 2050.

Rapidly growing cities, increasing slum populations, and corrupt officials could combine with high open unemployment among educated youth and under-employment among less educated adults and can cause violent social conflict. Latin America's wave of urbanisation is roughly 65 years ahead of Asia's.

Argentina, Brazil, Mexico and Venezuela were unable to manage rapid growth of illegal, unserviced settlements, and failed to provide adequate services. In many cases urban gangs filled a gap left by weak local governments. The difference in the Indian case will be that the total population involved will be much large. This combination, together with delayed action, will be explosive.

Most unfortunately, contrary to successful cities in the world, India has been locking itself into more dispersed patterns of urbanisation. Car ownership is increasing so fast that carbon emissions could increase by 2.5 times over current levels in PRC and by four times in India by 2035. Lower densities are leading to sprawl, which is leading to higher rates of motorisation, leading to more sprawl in a vicious cycle.

Higher density cities are less expensive on a per capita basis than are low-density cities. Clearly, the country needs a far more sophisticated planning framework, to imagine and implement a new urban vision, to ensure that India prepares to receive its migrants in its burgeoning cities.

What governments, both Union and State, will have to focus on is infrastructure for middle-tier cities, since it will be much too expensive to invest in the 193 tiny towns with populations below half a million; but at present, these tiny towns are home to half of India's urban residents (as against a quarter of China's urban residents). In other words, the missing middle in respect of urban India will need to be filled, to attract the migrant population to middle-tier cities.

9.5.1 The Education Challenge

Major rural to urban migration went hand in hand with the economic growth of the 19th to mid-20th century in today's high-income countries. As high-income countries transition to urbanised, ageing societies, these movements have subsided. Today, the largest internal population movements occur in low- and middle-income countries, particularly China and India.

When male migrants move, they leave behind children going to school. If families migrate all together, children must be found schools in urban locales. In India, 10.7 million children aged six to 14 lived in rural households with a seasonal migrant in 2013. About 28 percent of youth aged 15 to 19 in these households were illiterate or had not completed primary school, compared with 18 percent of the cohort overall. Out-of-school children will need to be diverted to vocational schooling/training.

A difficulty is that historically, barely 2.4 percent of India's workforce has acquired vocational skills formally. Together with low levels of general academic education, the minuscule workforce with any formal vocational skills is a mega challenge if the manufacturing share of GDP is to rise above 17 percent, where it is stuck for the last quarter century (since 1991).

The three mega-challenges will be hard enough to handle even with synergy between actions. Without urban planning to ensure synergy between actions, the demographic dividend will peter out.

9.6 THE NEW MIGRATION

Is there an evolving pattern of migratory behavior and composition that warrants a reconsideration of our prevailing models? The first round of migration models presumed that movement -- permanent movement -- was induced by prevailing wage differentials and economic opportunities. Thus one observed transatlantic migration from Old to New World as part of a permanent population redistribution within and across generations. To be sure the emptying of the countryside in the 19th century made London, Paris and New York into huge urban agglomerations.

The conventional push-pull model was eventually placed by one a bit more subtle, one that took into account the inefficiencies of markets in many developing settings. Rather than producing adjustment and equilibrium, migration spurred the growth of the informal sector as rural origin persons settled, at least temporarily, for wages and employment chances below the hoped-for formal sector offering. Thinking about migration in high income settings also shifted. New thinking began to take into account more of the non-pecuniary determinants of migration: climate; social composition of origin and destination; access to amenities. Such thinking made it possible to understand certain paradoxes such as the movement away from some of the economic regions with the highest hourly wages and toward other, lower wage but more amenity-rich locales.

During much of this time, policymaking often held a series of assumption about the process of migration and urban growth. A common observation was that urbanization was rapid and driven by rural-urban migration. Consistent with the dual economy (Harris-Todaro) model, the perception was that migrants had a difficult time adjusting in the urban area and were often unemployed or underemployed. Percolating through the time periods as well, was an old demographic adage about the composition of the migrant stream: that it was predominantly, unmarried young males. Concurrently, rapid urban growth fueled concern about denigration of the surrounding physical environment.

These models, trends, and to some degree social realities were consequential for policy. Rural-urban migration was seen as a problem. Urbanization was seen as too rapid, and efforts were considered to slow it or shift the balance of growth to other areas. Notable for instance, was China's policy of encouraging development in mid-size cities during the 1980s. Growth pole or satellite town developments were other offered responses. Of course, an alternative line of thinking evolved, arguing the public sector was the cause of some of the urban ills, with its disproportionate investment in cities and resultant urban bias.

Is our stock of knowledge, our set of models, adequate to understand population movement and redistribution today, especially in contemporary developing economies? Some rethinking is in order. Two major changes have come to many systems of population distribution in the last decade or two: (1) the revolution in technology of transportation and communication; and (2) the restructuring of national economies to allow more market activity. Both of these changes have received considerable attention elsewhere. They have been discussed in the case of migration, but it is worth tracing through how they reshape the nature and composition of migratory flows.

The changes in the technology of communication and transportation have made it easier for migrants to stay in touch with their origin communities. This is more than maintaining simple social ties. The tightness and stability of these connections can reinforce the implicit contracts that generate sharing of resources across locations. Most notably these are remittances. Despite the high level of interest in remittances, it remains to be clearly documented that these technological developments help maintain a continuous flow. There are related influences. The technologies of communication help impart knowledge of job market opportunities within and across national borders.

The "New Migration" includes circulation. Migrants oscillate between origin and destination. From various parts of the wold circulatory migration patterns have been identified. Often these circulatory movements are timed with the agricultural season. In West Africa, Thailand, and Mexico, for instance, individuals remain in the origin region from planting through tending to harvest, and then depart for the cities during the off season. Limited evidence suggests that the movements are repeated, but not necessarily every season. Landing a "good" job in one year may lead to a longer stay in the destination. This is just another way of managing the informal-formal sector issue.

It may be the case also that the ability to store earnings (in banks) and move funds geographically is contributing to the new migration. One marker of economic development is the improvement in financial infrastructures. If one can move money across international borders, then the ease with which one can be a new migrant and remain a connected member of the origin community and household rises. While there have been efforts to account for the value of international flows of remittances, there is little understanding of whether improvements in financial infrastructure help generate and support migration in the first place.

Data coming from various field sites suggest that the conventional demographic profile of the rural-urban migrant may be shifting as well. Migrants are still young. While many are male and single, there seems to be an increasing fraction of migrants who are female and a larger pool of family migrants. The work on Mexican migration to the United States finds substantial fractions of females in the migratory flow, fractions that increase with time. Our own work in Ghana challenges the notion that migrants are detached from families in the destination. What we need to know more about is the timing of the movements of family members. It is probably the case that frontier or first-wave migration is predominantly young single males, but how exactly the stream is altered after that is not well known.

The second major impetus for the New Migration is economic restructuring. Many countries have reoriented their economies in the direction of more free market activity. To argue that this trend is universal or that the movement is to an unfettered marketplace would be silly. Nevertheless, in several important ways the shift is on, and population distribution is a manifest component of this shift. The most notable case is China. Where once all residence was controlled by registration permit, the years since market reform have enabled individuals to relocate to areas of economic opportunity. This has created a huge pool --a "floating population" in the tens of millions of persons -- living apart from their place of formal registration. While often referred to as "temporary" migrants, the length of residence in destination can now approach several years. Considerable controversy swirled about the motivations of these temporary migrants (including the claim that women were moving to avoid the structures of the one-child family planning policy), but the migration seems to be economically driven.

This kind movement, a migration problem, in the wake of the relaxation of strictures on economic activity and hocusing has been seen in other settings as well. Viet Nam is now going through a process similar to that experienced by China. While residence registration was never as strictly controlled as in China, the economic restructuring (Doi Moi) has generated internal migration. In Ethiopia the fall of the Derg and its more authoritarian and socialist ways ushered in a period of economic relaxation. This loosening not only lets people move to new locations, but also generates differential economic growth by region, producing labor force opportunities to which workers respond.

Even in economies without government restrictions on residence and movement, there have been patterns of population movement that are similar in many respects. The undocumented movement of the Mexican origin population to the United States and parallel movement of former colonial populations to high-income economies of Europe have created similar floating populations, each with its own stamp for the particular migratory flow and condition of reception in the host society. Again, "temporary" migration is sometime sustained by circulation, at some risk of being caught, and rarely is temporary or guest-worker migration, in fact, temporary. Circulation may be substituting for return.

The evidence from Ghana indicates that the country's pursuit of structural adjustment has resulted in substantial shifts in regional activity, even as the overall growth of economic activity outpaces other sub-Saharan African nations. In the age-old way this induced movement, directed differentially to some urban areas.

9.6.1 Migration's Consequences

Migration from rural to urban areas generates a series of concerns, including worries about environmental stress and social adaptation of the migrants themselves. Since migration feeds urbanization, and since urban growth is associated with industrial development and land consumption, migration is often held culpable in environmental degradation. Although the link is there, it is not clear how strong that link is.

Direct public policies regarding environmental conditions, the underlying infrastructure for transportation, and the national level of income may have much more to say about the amount of insult visited upon the environment than the amount of rural-urban migration per se. As income rises, so does consumption of consumer goods, transportation, and land. These all can lead to more pollution and sprawl in any country. But as the level of income rises so does the demand for a cleaner local environment, and so there is an element of feedback in all of this.

There is another demographic component of the comparison. It is useful to remember that a large fraction, maybe nearly half, of urban growth is generated just by natural increase of the urban population. Thus, stemming urbanward migration will not stem urban growth. This reminds us that in the absence of migration but in the presence of positive population growth rates, there is more "population pressure" in both urban and rural areas. Migration may be more implicated than its true demographic contribution would warrant. Not that theincreasingly intensive use of rural and quasi-rural areas can lead to soil erosion, deforestation, and the like. This might lead one to call for stronger emphasis on fertility reduction measures, but the demographic community seems somewhat agnostic about the empirical connection between population growth and environmental conditions.

The other major area of concern in urbanward migration is that of the absorption of migrants into the host community. Migrants have always generated apprehension about their ability to mix into the receiving society. Migrants are seen as adapting slowly or not at all. Empirical evidence runs counter to this.

In many studies of immigrant adaptation in the United States, the first generation exhibits substantial differences from the native population along socioeconomic lines: income, education, language ability, etc. By the second generation, however, differences are narrowed considerably. Even if one does not adjust for background characteristics, the second generation gap is modest compared to the first generation gap. But when one does adjust for the level of resources for members of the second generation, the gap narrows even more. Work with nationally representative longitudinal data on immigrant achievement in school illustrates the point. Once we control for family characteristics (parental presence and supervision, socioeconomic status) and starting point in school, the trajectory of subsequent school achievement does not differ for immigrant children.

Interestingly we find a closing of the gap among migrants to Kumasi, Ghana. The first generation has lower status than members of the indigenous group in the same residential area, but members of the second generation show very little shortfall at all. In a parallel vein we find a process of occupational adaptation that differentiates the first generation more than the second. Market adjustment in China generated a huge flow of persons, both officially recognized and documented and "temporary" or undocumented. Despite the observation of migrants on streetcorners and under bridges, the evidence again is that migrants are driven by economic considerations. The claim was made in China that female migrants to the city were motivated by a desire to avoid the family planning policies enforced in their origin community. In fact such rural-urban migrants were found to have birth rates no higher than urban natives of the same educational level and age. Here again, migrants were not out of line with expectation; they had come for jobs and merged into the urban fabric.

9.6.2 Concluding Policy Considerations

There are several policy implications of this redistribution, but I will concentrate on just three points.

Migration is Only Part of the Picture

As we have discussed migration accounts for roughly half of urban growth in most developing country cities. Most demographic analyses confirm that overall urban growth rates are closely ties to national population growth rates. Declining fertility rates overall will help reduce urban population pressure. High income economies experience the mirror image effect. Most of these nations have very low rates of natural increase. Consequently international migration contributes a substantial fraction of total year-to-year national population growth. The lesson is that it is important to have the comparative framework clear when making assertions about the impact of migration.

Informal Migration is a Broad-Based Phenomenon

Undocumented, circulate, and temporary migration is now a world-wide phenomenon. It exists as internal migration in some restructuring economies (such as in China) but also appears as international migrant flows in other settings (such as Mexican origin migration to the United States). It is useful for policy-makers to think about the parallels in these movements, rather than concentrate only on the differences.

Recognize Migratory Realities

It is time for policy-makers to respond to migration rather than try to control migration. This is the case for internal migration and displacements. The opening of markets within and between countries will produce further internal and international redistribution. For voluntary movement, this recommendation means that efforts should go into adjustment efforts for migrants. Efforts might be made to improve access to information about health care, get migrant children into health and educational facilities, and provide language assistance as appropriate. For forced migrants the story is somewhat different. Here the migrants are outside their original location, often in a bounded settlement or camp. The policy need here is to realize that these refugee movements are often not temporary. This means a whole set of long-term services (health, family planning, productive work) must beprovided. For people displaced within national borders, sometimes without the assistance of international agencies, the relocation can be permanent.

CHAPTER

URBAN SOCIETY 10

10.1 INTRODUCTION

Urban society is characterized by a sharp economic disparity between a small number of wealthy families and a large number of petty traders, artisans and the poor. Urban planning is, therefore, necessary to reduce this disparity. Urban, city or town planning deals with the design of the built environment from the municipal or the metropolitan perspective. It is also concerned with the smaller scale of development in the city or town including architecture and urban design.

The Greek Hippodamus is often considered the Father of City Planning for his design of the city called Melitus, though examples of cities permeate antiquity. In India, people built planned cities right from the ages of the Harappan culture. But all these cities of antiquity were based on agricultural income and were built mostly because of the rise in population. Gradually, the nature of settlements or urban centres underwent a qualitative change, a transition from a locally focused, isolated, economically and socially self-contained, and basically agricultural community to global, economically and socially dynamic and basically industrial or service-oriented community.

Today's urban centres are the basis of economic activities. People mostly migrate to these centres to earn their livelihood. Among the migrants, some settle down and some others move on to larger cities or towns in search of better opportunities. To improve the economic and physical infrastructure and also to provide essential services and facilities to the existing and forthcoming population, planning is essential.

Planning is also required to condense the load of larger cities or towns including metropolitan areas as it is observed that people usually migrate from rural areas to small urban centres, from small and medium towns to larger cities or metropolitan centres in search of livelihood. This can be done by providing similar employment opportunities and standard of living in the smaller towns and cities.

Urban planning is mostly concerned with the physical development process; spatial planning plays a significant role in the identification and development of new areas. It helps in creating a new spatio-economic order leading to the rational use of land and other resources. From the late 19th century, modern urban planning started placing stress on the socioeconomic aspects of the urban centres.

Therefore, nowadays, planning includes both the social aspects such as segregation, culture and social change, and also the physical aspects of the urban land. In addition, todays urban plan is also forced to comply with rapid economic changes that may come about due to liberalization and globalization processes. Thus, urban planning for all sizes of the towns and cities has become compulsory in today's urban centres.

To understand what city planning is, it is necessary to understand first what planning is. Planning is future-oriented; it is doing things now consciously, and deliberately to shape the future welfare, environment and condition of humans and/or non-humans. Myerson and Banfield describes a plan as 'a course of action which can be carried into effect, which can be expected to lead to the attainment of the ends sought, and which someone (or organization) intends to carry into effect.'

Urban or city planning is a process for rational, controlled physical and social change in relation to socially and corporately determined goals and objectives rooted in societal or community values. Ideally, an urban plan is based on the best social, economic and physical data available and the best professional, technical and intellectual ideas available. Realistically, it is mitigated by political and cultural considerations.

There are two aspects in urban planning—social planning and physical planning. Social planning involves the changes in and control of social institutions and human communities, while physical planning involves the changes in and control of land uses and building design and determining the shape and pattern of urban spaces, with the use of zoning laws and building and subdivision regulations. Planning also sometimes includes social policies announced by the central and state governments such as housing, education, mental health, criminal justice, community organization and community development.

Urban sociology is an important. Discipline of Sociology, it is a branch of Sociology. In urban Sociology, as in general sociology, we study social relation though to a limited extent & in a shorter context. The kind of social relation studied in Urban Sociology, as is evident from the name, are the relations which occur in an Urban Society. Thus Urban sociology is a study of urban society. The urban society is studied & investigated according to sociological methods & techniques, which are of course scientific. The study of urban society can be general as well as specific. In every city we find certain features which are to be found in every urban society and are therefore, to be regarded general. These are distinct from the features of a rural society. for example, in cities, the relations among individuals are importantersonal. Besides the general features which are common to all metropolises, there are certain specific features peculiar to one or more towns but inapplicable to others. Urban sociology studies general & specific features.

An urban community may be defined as a group of people having a certain specialized economic recognizable characteristics. In the above definition of an urban community the social characteristics is the most distinguishing & vital. The following are the special or distinctive features of urban community;

1. Social heterogeneity
2. Moderate social contacts
3. Social tolerance
4. Moderate control
5. Social dynamism
6. Voluntary association
7. Importantersonality
8. Lack of communal feeling
9. Lease morals
10. Unbalanced individuals
11. Prevalence of crimes
12. Dynamic life
13. Rapid change and
14. Artificiality in life

Urban Sociology is a study of the communities having above mentioned charges.

A major shift in population from rural to urban areas is predicted for the coming decades, and the United Nations' projections show hundreds of millions more urban dwellers, especially in Asia and Africa.

Accommodating this many more urban dwellers efficiently and equitably will be difficult – but thinking that urban growth rates are unprecedented and rising is wildly wrong and gives the wrong policy messages.

Rather than curbing the rate of urbanisation, we need to look at the real challenges and opportunities that exist. We welcome the growing number of positive accounts of the contribution that urbanisation can make to economic productivity and environmental sustainability – although official attitudes to urbanisation in the urbanising countries remain broadly negative.

Urbanisation is not just about what happens in the cities but is part of broader transformations under way. Although people and their activities tend to be classed as either 'rural' or 'urban', in fact the links between rural and urban locations, people and activities are key.

Urbanisation needs to be seen as part of wider transformations that include the change in the global economy and that include both urban and rural areas. We need to look too at the migration of people from one area to another, and at the growth rates of urban population. Also key is that urbanisation benefits wider regions, and not just the urban settlements themselves.

10.2 URBAN VERSUS RURAL

One way to understand city life is to think about it in contrast to rural life. Let's start with some basics. First, what is a city? The U.S. Census defines a city as a place that has a minimum population of 2,500. But most likely, when you think of a city you probably think of a place like New York City or Paris.

Cities are much more populous than rural places. Cities also have more density, meaning that more people are concentrated in a particular place. Density is a measure of how many people are living, working, or passing through a particular area. Dhaka, Bangladesh is the densest urban area in the world, with about 118,000 people per square mile. As you might imagine, this is much denser than a rural area! Just for some context, Alaska is the least dense state in the United States, with about 1.3 people living per square mile.

Let's turn to some of the other ways that cities are distinct places.

10.2.1 Culture, Work, Politics

Cities aren't simply just places where more people live. There are a number of other important ways that cities are different from rural areas. For one thing, cities are more heterogenous than rural areas, meaning that they are more diverse in terms of the different racial and ethnic groups who live there, the cultures you might encounter, and the variety in things like food and style of dress. For example, if you've ever walked through the streets of New York City you've probably seen aspects of different cultures from all over the world. You could probably get almost any kind of food you wanted at any time of day or night, and there is a concentration of important cultural institutions like museums and theatres.

Urban areas are also more heterogenous in terms of the type of employment that is found within them. Cities have a higher concentration of many different kinds of industry than rural areas. Specifically, cities have a higher concentration of high-skilled and knowledge based work. There are more opportunities for high-skilled work in cities and often more opportunities for education. This also leads to greater social and economic mobility within cities compared to rural areas.

Another way that urban areas differ from rural areas is in voting patterns and politics. In general, urban areas in the United States tend to favor more progressive policies and politicians. Voters in cities generally vote for more left-leaning candidates than voters in rural areas. Even in generally conservative states, like Texas, the urban areas tend to vote more progressively than the rural areas. This has a lot to do with education levels and density. Research has shown that as density increases and levels of education increase, places tend to vote more progressively. Cities are both more dense and more highly educated than rural areas.

10.3 URBAN COMMUNITY

The first difficulty with which we meet in the study of urban community is regarding the definition of the word 'urban'. The difficulty lies in the fact that the term 'community' denotes two conditions: first, the physical condition, second, the social condition. The physical condition may not necessarily give rise to social condition. Generally, by an urban area we mean an area with a high density of population.

But a village with an average number of persons per room as great as the city may not be called urban since it has too small an absolute population and too small an inhabited area. Then a village with too large an area may not be called urban, the population being small. Therefore, a city cannot be defined in terms of density of population. Account must also be taken of absolute population and absolute area.

But there is no uniformity of standards in different countries, in the context of 'India, the census reports to date have defined (i) those places as urban areas which have a local authority like municipality, cantonment board, notified area committees, (ii) all other places which satisfy the requirements of a minimum population of 5,000 at least 75 percent of the male working population engaged in non-agricultural pursuits and a density of population of at least 900 persons per square Km.

Taking up the social aspect of urban community, the city is a way of life. The word 'urbane' suggests this way of life; it indicates fashionable living, wide acquaintance with things and people and political manner of speech. But is urban way of life limited only to the urban population? As we know, the rural people also have come under the influence of urban ways of life.

The rural areas can become urbanized. While there may be causal connection between the demographic features and social features of an urban community, a country can be demographically urban but socially more rural than the one which is demographically rural and socially more urban. For example, Chile has a greater population living in cities than does Canada, but its people are more rural than Canadians.

Thus, it is difficult to frame a definition of the term 'urban'. While the distinction between "rural" and "urban" is a familiar one and is commonly recognized in everyday language, the criteria employed are not very exact and scientifically precise. Sorokin and Zimmermann used occupational criteria to distinguish the 'rural' from 'urban' who said. "The principal criterion of the rural society or population is occupational, collection and cultivation of plants and animals. Through it, rural society differs from other, particularly, urban, populations engaged in different occupational pursuits."

These writers mentioned other characteristics also in which rural and urban communities differ: size, density, heterogeneity, social differentiation and stratification, mobility, environment and systems of interaction, but according to them, these characteristics are causally connected.

The first 'variable', the agricultural occupation carries the other variables with it. These authors also recognize that the above traits differ in degree and not in kind. They observe: "In reality the transition from a purely rural community to an urban one is not abrupt but gradual.

There is no absolute boundary line which would show a clear cut cleavage between the rural and the urban community. Many differential characteristics of the rural and urban community would consist not so much in the presence of certain traits in rural, and their absence in urban communities, as much as in a quantitative increase of these characteristics."

The above conception has been given the name of "rural-urban continuum" by recent writers suggesting that the differences between the two types of community are gradual and continuous, and not qualitative differences per se, resulting in a simple dichotomy.

Thus, Stuart A. Queen and David B. Carpenter claim that "there is a continuous gradation from rural to urban rather than a simple rural-urban dichotomy." According to Gist and Halbert," The familiar dichotomy between 'rural' and 'urban' is more of a theoretical concept than a division based upon the facts of community life." Every village possesses some elements of the city while every city carries some features of the villages. The slum areas of a city are no better than a village. As remarked by MacIver, "But between the two there is no sharp demarcation to tell where the city ends and country begins."

Generally speaking, the urban communities are identified as large, dense and heterogeneous and the rural communities as small, less dense and homogeneous, yet no hard and fast line can be drawn. While the extremes can be easily identified, problems arise in drawing a hard and fast line.

Similarly, occupational criteria would also present the problem of demarcation. At what point should the proportion engaged in agriculture be set in order to determine whether a community is rural or urban? Should a simple majority be regarded as sufficient to classify a community one way or the other?

Thus, the characteristics of size, density, heterogeneity and occupation exhibit differences in degree from place to place and Lime to Lime. Sharp and absolute divisions between rural and urban communities do not exist and, therefore, we must recognize the fact that rural and urban communities are polar types and that they may be found at many points to the so-called "continuum" that extends from one extreme pole Lo the other.

10.3.1 Important Features of Urban Community in India

By virtue of its size and population, the city cannot be a primary group. The inhabitants of a city do not come into primary contact with each other. Some of them interact with each other without even knowing each others name. Superficial manners of politeness and mutual convenience may evolve in the city but they are mostly mechanical. The city dweller usually treats the other person as a stranger.

A person may live in the city for a long time but still may not being one-third of his/her neighbourhood. Thus, anonymity is a common feature of a city life: is the urban world putting a premium on varied recognition and urban contacts that are segmental in nature. Therefore, Lee remark best describes urban anonymity; Anonymity is a loss of identity in a city teeming with millions. Many urbanites live in a social void or vacuum in which institutional norms are not effective in controlling or regulating their social behavior. Although they are aware of the existence of many institutional organizations and many people around them, they do not feel a sense of belongingness to any one group or community. Socially, they are in the midst of plenty, but alone.'

Some of the important features of urban community are: (i) Namelessness (ii) Homelessness (iii) Class Extremes (iv) Social Heterogeneity (v) Social Distance (vi) Energy and Speed.

Namelessness

The urban groups have, as Bogardus observes, a reputation for namelessness. By virtue of its size and population, the city cannot be a primary group. The inhabitants of a city do not come into primary contact with each other.

They meet and speak without knowing each other's name. Although superficial manners of politeness and mutual convenience evolve in the city, they are mechanical. The city dweller treats the strangers he meets as animated machines rather than as human beings.

A citizen may live for several years in a city and may not know the names of one-third of the people who live in the same city area. The urban world puts a premium on varied recognition. "In short, urban contacts are segmental. It is part of persons, not whole persons. Lee remarks." Anonymity is a loss of identity in a city teeming with millions.

Many urbanites live in a social void or vacuum in which institutional norms are not effective in controlling or regulating their social behaviour. Although they are aware of the existence of many institutional organizations and many people around them, they do not feel a sense of belongingness to any one group or community. Socially, they are poor in the midst of plenty.

Homelessness

Homelessness is another disturbing feature of city community. The house problem in a big city is very acute. Many low class people pass their nights on pavement. The middle class people have built insufficient accommodation, a room or two and that also on the sixth or seventh floor. The child does not get any play space. The city environment puts a premium on childlessness.

Class Extremes

Class extremes characterize urban community. In a city are found the richest as well as the poorest people, the people rolling in luxury and living in grand mansions as well as the people living on pavements and hardly getting two meals a day. The best forms of ethical behaviour and the worst racketeering are both to be found in cities. Superior creativeness and chronic unemployment are alike urban features. The city is the home of opposites.

Social Heterogeneity

The city is more heterogeneous than the village. It has "been the melting pot of races, peoples, and cultures, and is a most favourable breeding ground of new biological and cultural hybrids. It has not only tolerated but rewarded individual differences. It has brought together people from the ends of the earth because they are different and thus useful to one another rather than because they are homogeneous and likeminded." The personal traits, the occupations, the cultural life and the ideas of the members of the urban community vary widely than those of the rural inhabitants.

Social Distance

Social distance is a product of anonymity and heterogeneity. The city dweller feels lonely. There is masking of one's true feelings. Most routine social contacts are impersonal, and segmented. Formal politeness takes the place of genuine friendliness. Urbanites become night dwellers, not neighbours.

Energy and Speed

Energy and speed are the final traits of a city. People with ambition work at a tremendous speed, day and night, which stimulate others also to work similarly. Stimulation and inter-stimulation are endless. People indulge in too many activities and inconceivable efforts which ultimately eat their nerves and kill their energies.

Urban life produces greater emotional tension and insecurity than does rural life. Cities may be called consumers of population in the sense that due to congestion, pollution, insanitariness and unhealthiness they adversely affect health of the inhabitants.

It may be of interest to note that the death rate in rural areas is lower despite the fact that rural areas spend very little money on public health while cities spend much and in cities facilities for preserving health, such as hospitals and medicine specialists, are many and excellent. Sickness rates are higher in cities. There are more cases of insanity and heart failures in urban community than in the rural one.

Some other features of urban community are discussed as follows:

i. Social heterogeneity: As the urban society is diverse and complex, it is considered as a heterogeneous society. In urban communities, there is a confluence of many cultures.

ii. Secondary means of social control: Formal means of social control such as police, law and courts are more powerful and are essential in regulating the behaviour of the people.

iii. Social mobility: The urban community is more open. The chances for social mobility are higher. The achieved status of an individual is given more promi-nence than the ascribed one. The status of an individual is determined by his own works and economic status and not by his birth into a particular caste unlike the rural community. The barriers based on the caste are breaking down to an extent.

iv. Lack of community feeling: In urban societies, there is a lack of community feeling. People are so busy with their own personal works that they do not have time to think about others.

v. Division of labour and specialization are important features of urban community.

vi. Nuclear families exist more in number than the joint families. Also, there is a lack of unity and harmony among the members of the families.

vii. Most of people in the cities are dissatisfied and discontented, which gradually leads to the social disorganization. Due to the heterogeneous nature of the cities, the chances of conflicts are more. They may arise in the form of class conflicts, communalism, factionalism, etc.

viii. People in the urban community are more progressive and broad-minded. They accept changes and are more exposed to the developments in science and technology.

10.3.2 Urbanism as a Way of Life

Early sociologists in Europe and the United States of America extensively studied the rise and growth of cities. They presented their views on urbanism as a way of life. A brief description of Ferdinand Tonnies, Emile Durkheim, Georg Simmel and the Chicago School of thoughts and ideas is given below.

Ferdinand Tonnies: Gemeinschaft and Gesellschaft

German Sociologist, Ferdinand Tonnies, in the late 19th century, studied how life in the cities differs from the life in the rural areas. He developed two concepts based on this contrast—one he called Gemeinschaft (meaning Community') and another he called Gesellschaft (meaning association).

Gemeinschaft: Tonnies used the German word to refer to a type of social organization by which people are bound closely together by kinship and tradition. The Gemein- schalt of the people in the rural village joins people into a single primary group.

Gesellschaft: Tonnies says that Gemeinschaft is absent in the modern city: on the contrary, urbanization fosters Gesellschaft. What he means is that, in a city, a type of social organization is formed by which people come together only on the basis of individual self-interest. In this ways of life, individuals are motivated by their own needs rather than a drive to enhance the well-being of everyone. City dwellers display little sense of community or common identity and look to others mostly as a means of advancing their individual goals. Thus, Tonnies saw in urbanization the erosion of close, enduring social relations in favour of the fleeting and temporary ties typical of business.

Emile Durkheim: Mechanical and Organic Solidarity

The French Sociologist Emile Durkheim agreed with Tonnies' thought about the cities but he did not think that urban people lacked social bonds. According to him, the urban people simply organize themselves into social life differently from their rural counterpart. He described traditional, rural life as mechanical solidarity', i.e., social bonds are based on the common sentiments and shared moral values.

He argued that traditional societies were 'mechanical' and held together by the fact that everyone was more or less the same, and hence had things in common. In traditional societies, argues Durkheim, the collective—consciousness entirely subsumes individual—consciousness—norms are strong and behaviour is well regulated. With its emphasis on tradition, Durkheim's concept of mechanical solidarity bears a striking similarity to Tonnies' Gemeinschaft.

According to him, urbanization on the other hand erodes this mechanical solidarity. Instead, it generates a new type of bonding, which he termed as 'organic solidarity.' Here, the social bonds are based on specialization and interdependence. This concept parallels Tonniess Gesellschaft, but there is a difference. Like Tonnies, Durheim also thought that industrial cities undermined tradition, but he took a more positive view of this change and said that urban centres create a new kind of solidarity.

In modern societies, he argued the highly complex division of labour resulted in 'organic' solidarity. Different specializations in employment and social roles created dependencies that tied people to one another, since people no longer could count on filling all of their needs by themselves. In 'mechanical' societies, for example, subsistence fanners live in communities, which are self-sufficient and knit together by a common heritage and common job. In modern 'organic' societies, workers earn money, and must rely on other people who specialize in certain products (groceries and clothing) to meet their needs.

The result of increasing division of labour, according to Durkheim, is that individual consciousness emerges distinct from collective consciousness—often finding itself in conflict with collective consciousness. This is especially true of societies which have been built on likeness, but their social organization is based on difference. Hence, according to

Durkheim, urban society in contrast to rural society offers more individual choice, moral tolerance and personal privacy.

Georg Simmel: The Blase Urbanite: German sociologist Georg Simmel offered a micro-analysis of cities, studying how urban life shapes peoples attitudes and behaviour. According to Simmel, individuals experience the city as an intense crush of people, objects and events. To prevent being overwhelmed by all this stimulation of urban living, people develop a 'blase attitude', tuning out much of what goes on around them. Such detachment does not mean that city dwellers lack compassion for others, although they may sometimes seem cold and heartless.' Rather, as Simmel saw it, a blase attitude is simply a strategy for social survival by which people devote their time and energy to what really matters to them.

The Chicago School: Robert Park and Louis Wirth: Sociologists Robert Park and Louis Wink in the United States, belonging to the Chicago School, sought to give urban studies a street-level perspective by getting out (or walking on the streets) and studying real cities. They said that the city is more sophisticated but also more superficial, more challenging but less protective; more exciting but also more impersonal.

A bert park concluded from the study that the city is an organized mosaic of distinctive ethnic communities, commercial centres and industrial districts. According to him, over time these 'natural areas' develop and change in relation to each other. To him, a city is a living organism—a human kaleidoscope.

This interpretation of city life was summed up by Louis Wirth by defining the city as a setting with a large, dense and socially diverse population.

According to Wirth, the main effects of urbanism are as follows:

i. The creation of impersonal, transitory and superficial social contacts—in other words, secondary rather than the primary relations.

ii. The emergence of an 'urban personality'—a flexible and multiple self.

iii. The rise of formal social controls and special-purpose associations.

iv. The formation of segregated groups with divergent lifestyles.

He believed that urbanism weakens social bonds and this effect accounts for the social disorganization that occurs in cities. Living among millions of others, urbanites come into contact with many more people than rural residents do. Thus, city people notice others not in terms of who they are but in terms of what they do—like bus driver and grocery store owner. Specialized urban relations are sometimes quite pleasant for all concerned.

But one should remember that self-interest rather than friendship is the main reason for the interaction. Finally, limited social involvement coupled with great social diversity makes city dwellers more tolerant than rural villages. In rural communities, often jealousy may enforce narrow traditions but the heterogeneous population of a city rarely shares any single code of moral conduct.

10.4 URBAN SOCIAL ACTIONS

Urban social actions guided by urban social relationships are secondary and specific in native. They are segmentary in character and mainly aim at the fulfillment of a single function of an urbanite's life. Thus while in rural areas imitation of one's fathers work is work and education combined in one, in urban centers specialized and formal education paves the way for one's career in a technical job.

10.4.1 Urban Social Relations

Urban social relations are many and diverse and formal in character. Urban social institutions leased on these are again specialized agencies which serve particular functions, say education through schools, law enforcement by the city govt. etc. Elaborate procedure is laid down to guide the activities of these institutions and membership is through necessity and availability of the service. Thus site density heterogeneity and specialization are the typically urban characteristics. Urban sociology studies human social life in relation to these factors.

Urban Problems

On the other hand we have novel and peculiar urban problems which a necessary concomitant of urbanization such as housing, sanitation, slums, preusions and recreation, pollution of all types, fluidity of the law and order situation, increase in rise, vice and deviance, various physical and mental disorders, increasing suicides and divorces, traffic problems, zoning anomalies etc. The study of all these forms is an important corollary of the study of urban socio. Thus urban socio is a study of all aspects related to man's modification of environment to subserve his varied purposes with all its implications.

Urban sociology studies all aspects of a city life such as its size, density of population social organization, problems etc. Urban studies is a specialized discipline which been recently developing in urban socio which exclusively deals with an analysis of urban problems. Urban problems occupy the major attention of urban sociologists today because of its complexity and its craving for an immediate solution. They arise at an instant much as unanticipated increase in traffic at a place in a single day becomes a problem.

Thus we see that the scope of Urban Sociology is very vast because although urban socio is a recent science, urbanism forms a major part of contemporary life patterns which is slowly encroaching upon and engulfing the existing forms of rural life. In short, urban socio-studies the nature and characteristics of a city, its size, density and characters of its people, its spatial pattern and changes thereon; the peculiar type of relations and social interactions formed in the urban milieu, the nature of primary social institutions such as family in an urban setting, the novel instructions of commercial reaction, education etc. It moreover studies all the implications and complexities and consequences of urban social problems, the various urban social groupings including deviant groups and the various aspects of urban social attitudes. On short, urban sociology is a developing science which studies all aspects of urban life.

Features of Urban Society

i. The urban society is heterogeneous known for its diversity and complexity.

ii. It is dominated by secondary relations.

iii. Formal means of social control such as law, legislation, police, and court are needed in addition to the informal means for regulating the behavior of the people.

iv. The urban society is mobile and open. It provides more chances for social mobility. The status is achieved than ascribed.

v. Occupations are more specialized. There is widespread division of labor and specialization opportunities for pursuing occupations are numerous.

vi. Family is said to be unstable. More than the family individual is given importance. Joint families are comparatively less in number.

vii. People are more class -conscious and progressive. They welcome changes. They are exposed to the modern developments in the fields of science and technology.

viii. Urban community is a complex multigroup society.

ix. The urban community replaced consensus by dissensus.The social organization is atomistic and illdefined. It is characterized by disorganization, mental illness and anomie.

x. Mass education is widespread in the city increasing democratization of the organizations and institutions demand formal education.

10.5 IMPORTANT CHARACTERISTICS OF URBAN COMMUNITY

A sociological analysis of urban community contains several salient features. They are as follows:

1. Size: As a rule, in the same country and at the same period, the size of an urban community is much larger than that of a rural community. In other words, urbanity and size of a community are positively correlated.

2. Density of Population: Density of population in urban areas is greater than in rural communities. Urbanity and density are positively correlated.

3. Family: So far as urban community is concerned, greater importance is attached to the individual than to the family. Nuclear families are more popular in urban areas.

4. Marriage: In case of urban community there is a preponderance of love marriages and inter-caste marriages. One also comes across a greater number of divorces. Sons and daughters enjoy considerable freedom in choosing their life partners.

5. Occupation: In the urban areas, the major occupations are industrial, administrative and professional in nature. Divisions of labour and occupational specialization are very much common in towns/cities/metropolises.

6. Class Extremes: In the words of Bogardus, "Class extremes characterize the city." A town and a city house the richest as well as the poorest of people. In a city, the slums of the poor exist alongside the palatial bungalows of the rich, amidst the apartments of the middle class members. The most civilized modes of behaviour as well as the worst racketeering are found in the cities.

7. Social Heterogeneity: If villages are the symbol of cultural homogeneity, the cities symbolize cultural heterogeneity. The cities are characterized by diverse peoples, races and cultures. There is great variety in regard to the food habits, dress habits, living conditions, religious beliefs, cultural outlook, customs and traditions of the urbanites.

8. Social Distance: Social distance is the result of anonymity and heterogeneity. Most of one's routine social contacts in a town or city are impersonal and segmentary in character. In the urban community social responses are incomplete and halfhearted. There is utter lack of personal involvement in the affairs of others.

9. System of Interaction: Georg Simmel held that the social structure of urban communities is based on interest groups. The circles of social contact are wider in the city than in the country. There is a wider area of interaction system per man and per aggregate. This makes city life more complex and varied. The city life is characterized by the predominance of secondary contacts, impersonal, casual and short-lived relations. Man, at any rate, the man in the street, virtually loses his identity being treated as a "number" having a certain "address".

10. Mobility: The most important feature of urban community is its social mobility. In urban areas the social status of an individual is determined not by heredity or birth but by his merit, intelligence and perseverance. Urbanity and mobility are positively correlated.

11. Materialism: In the urban community the social existence of man revolves round wealth and material possessions. The worth of an urbanite today is being judged not by what he is but by what he has. Status symbols in the form of financial assets, salaries, costly home appliances count a lot for the urbanites.

12. Individualism: The urbanites attach supreme importance to their own welfare and happiness. They hesitate to think or act for the good of others.

13. Rationality: In urban community there is emphasis on rationality. People are inclined to reason and argue. Their relationship with others is governed, for the most part, by the consideration of gain or loss. Relationship takes place on a contractual basis. Once the contract is over, human relationship automatically comes to a close.

14. Anonymity: As Bogardus observes, the "Urban groups have a reputation for namelessness." By virtue of its size and population, the urban community cannot be a primary group. Here nobody knows anybody and nobody cares for anybody. The urbanites do not care for their neighbours and have nothing to do with their miseries or pleasures.

15. Norm and Social Role Conflict: The urban community is characterized by norm and social role conflict. Factors such as the size, density and heterogeneity of the population, extreme occupational specialisation and the class structure prevalent in the urban context lead to such a state of affairs.

In the absence of uniform and fixed social norms, individuals or groups often seek divergent ends. This has a considerable share in causing social disorganization.

16. Rapid Social and Cultural Change: Rapid social and cultural change characterize urban life. The importance attached to traditional or sacred elements has been relegated to the background. The benefits of urban life have effected changes in respect of norms, ideologies and behaviour patterns.

17. Voluntary Associations: The urban community is characterized by impersonal, mechanical and formal social contacts occurring among the people. Naturally they have a strong desire for developing genuine social relationships to satisfy their hunger for emotional warmth and sense of security. They form associations, clubs, societies and other secondary groups.

18. Formal Social Control: Social control in urban community is essentially formal in nature. Individual's behaviour is regulated by such agencies as police, jails, law courts etc.

19. Secularization of Outlook: In cities ritual and kinship obligations are diluted. Caste and community considerations yield to economic logic. This results in secularization of outlook.

20. Urban Areas Provide: impulses for modernization in society as a whole.

10.6 MAJOR PROBLEMS OF URBANISATION IN INDIA

Although India is one of the less urbanized countries of the world with only 27.78 percent of her population living in urban agglomerations/towns, this country is facing a serious crisis of urban growth at the present time. Whereas urbanisation has been an instrument of economic, social and political progress, it has led to serious socio-economic problems.

The sheer magnitude of the urban population, haphazard and unplanned growth of urban areas, and a desperate lack of infrastructure are the main causes of such a situation. The rapid growth of urban population both natural and through migration, has put heavy pressure on public utilities like housing, sanitation, transport, water, electricity, health, education and so on.

Poverty, unemployment and under employment among the rural immigrants, beggary, thefts, dacoities, burglaries and other social evils are on rampage. Urban sprawl is rapidly encroaching the precious agricultural land. The urban population of India had already crossed the 285 million mark by 2001. By 2030, more than 50 percent of India's population is expected to live in urban areas.

10.6.1 Urban Sprawl

Urban sprawl or real expansion of the cities, both in population and geographical area, of rapidly growing cities is the root cause of urban problems. In most cities the economic base is incapable of dealing with the problems created by their excessive size. Massive immigration from rural areas as well as from small towns into big cities has taken place almost consistently; thereby adding to the size of cities.

The first large flow of migration from rural to urban areas was during the "depression" of late 1930s when people migrated in search of jobs. Later, during the decade 1941-51, another a million persons moved to urban places in response to wartime industrialisation and partition of the country in 1947.

During 1991-2001, well over 20 million people migrated to cities. The greatest pressure of the immigrating population has been felt in the central districts of the city where the immigrants flock to their relatives and friends before they search for housing. Population densities beyond the "old city" decline sharply.

Brush (1968) has referred to this situation in the central parts of the cities as "urban impulsion" which results from concentration of people in the centre of the city close to their work and shopping. Incidentally many of the fastest growing urban centres are large cities.

This is due to the fact that such large cities act as magnets and attract large number of immigrants by dint of their employment opportunities and modern way of life. Such hyperurbanisation leads to projected cities sizes of which defy imagination. Delhi, Mumbai, Kolkata, Chennai, Bangalore, etc. are examples of urban sprawl due to large scale migration of people from the surrounding areas.

In several big cities wealthy people are constantly moving from the crowded centres of the cities to the more pleasant suburbs where they can build larger houses and enjoy the space and privacy of a garden around the house. In some cities, the outskirts are also added to by squatters who build makeshift shacks of unused land although they have no legal right to the land. The difficulty of restricting town growth in either case is immense and most towns and cities are surrounded by wide rings of suburbs.

Historically suburbs have grown first along the major roads leading into the town. This type of growth is known as ribbon settlement. Such sites are first to be developed because of their location near the road gives them greater accessibility. But soon the demand for suburban homes causes the land between ribbon settlements to be built and made accessible by constructing new roads.

This type of development is known as 'infil'. Simultaneously small towns and villages within the commuting distance of major cities are also developed for residential purposes. In this way towns are continuously growing and in some areas the suburbs of a number of neighbouring towns may be so close together as to form an almost continuous urban belt which is called conurbation. Urban sprawl is taking place at the cost of valuable agricultural land.

10.6.2 Overcrowding

Overcrowding is a situation in which too many people live in too little space. Overcrowding is a logical consequence of over-population in urban areas. It is naturally expected that cities having a large size of population squeezed in a small space must suffer from overcrowding. This is well exhibited by almost all the big cities of India.

Fig. 1. Overcrowding.

For example, Mumbai has one-sixth of an acre open space per thousand populations though four acre is suggested standard by the Master Plan of Greater Mumbai. Metropolitan cities of India are overcrowded both in 'absolute' and 'relative' terms. Absolute in the sense that these cities have a real high density of population; relative in the sense that even if the densities are not very high the problem of providing services and other facilities to the city dwellers makes it so.

Delhi has a population density of 9,340 persons per sq km which is the highest in India. This is the overall population density for the Union territory of Delhi. Population density in central part of Delhi could be much higher. This leads to tremendous pressure on infrastructural facilities like housing, electricity, water, transport, employment, etc. Efforts to decongest Delhi by developing ring towns have not met with the required success.

10.6.3 Housing

Overcrowding leads to a chronic problem of shortage of houses in urban areas. This problem is specifically more acute in those urban areas where there is large influx of unemployed or underemployed immigrants who have no place to live in when they enter cities/towns from the surrounding areas.

An Indian Sample Survey in 1959 indicated that 44 percent of urban households (as compared to 34 percent of rural families) occupied one room or less. In larger cities the proportion of families occupying one room or less was as high as 67 percent.

Moreover, the current rate of housing construction is very slow which makes the problem further complicated. Indian cities require annually about 2.5 million new devellings but less than 15 percent of the requirement is being constructed.

The Census of India 2001 concluded the first ever and the largest survey of household amenities and assets which points a never-before profile of problem relating to housing in India. The outcome is both instructive and amusing. Taking India as whole, there are 179 million residential houses, i.e., about six people to each house.

Thirty-nine percent of all married couples in India (about 86 million) do not have an independent room to themselves. As many as 35 percent (18.9 million) urban families live in one-room houses.

For about a third of urban Indian families, a house does not include a kitchen, a bathroom, a toilet—and in many cases there is no power and water supply. Only 79 percent (42.6 million) urban household live in permanent (pucca) houses. 67 percent (36 million) of the urban houses are owned by the households while 29 percent (15 million) are rented.

Several factors are responsible for the above mentioned sad state of affairs with respect to housing problems faced by the urban people. The major factors are shortage of building materials and financial resources, inadequate expansion of public utilities into sub-urban areas, poverty and unemployment of urban immigrants, strong caste and family ties and lack of adequate transportation to sub-urban areas where most of the vacant land for new construction is located.

10.6.4 Unemployment

The problem of unemployment is no less serious than the problem of housing mentioned above. Urban unemployment in India is estimated at 15 to 25 percent of the labour force. This percentage is even higher among the educated people.

It is estimated that about half of all educated urban unemployed are concentrated in four metropolitan cities (Delhi, Mumbai, Kolkata, and Chennai). Furthermore, although urban incomes are higher than the rural incomes, they are appallingly low in view of high cost of living in urban areas.

One of the major causes of urban unemployment is the large scale migration of people from rural to urban areas. Rural-urban migration has been continuing for a pretty long time but it has not always been as great a problem as it is today. The general poverty among the rural people pushes them out to urban areas to migrate in search of livelihood and in the hope of a better living.

But the growth of economic opportunities fails to keep pace with the quantum of immigration. The limited capacity of urban areas could not create enough employment opportunities and absorb the rapid growth of the urban labour force. Efforts made by the central and the state governments to create employment opportunities in rural areas and to check the large scale rural-urban migration have not met with much success.

10.6.5 Slums and Squatter Settlements

The natural sequel of unchecked, unplanned and haphazard growth of urban areas is the growth and spread of slums and squatter settlements which present a striking feature in the ecological structure of Indian cities, especially of metropolitan centres.

The rapid urbanisation in conjunction with industrialisation has resulted in the growth of slums. The proliferation of slums occurs due to many factors, such as, the shortage of developed land for housing, the high prices of land beyond the reach of urban poor, a large influx of rural migrants to the cities in search of jobs etc.

In spite of several efforts by the Central and State Governments to contain the number of slum dwellers, their growth has been increasing sharply exerting tremendous pressure on the existing civic amenities and social infrastructure.

In India Slums have been defined under section 3 of Slum Areas (Improvement and Clearance) Act 1956. As areas where buildings:

(i) Area in any respect unfit for human habitation.

(ii) Area by reason of dilapidation, overcrowding, faulty arrangement and design of such buildings, narrowness or faulty arrangement of streets, lack of ventilation, light, sanitation facilities or any combination of these factors, which are detrimental to safety, health and morals.

The following criteria characterises an area as Slum:

(i) All areas notified "Slum" by state govt. under any Act.

(ii) All areas recognised as slum by state govt. which have not been formally notified as slum under any Act.

(iii) A compact area of at least 300 populations or about 60-70 households of poorly built congested tenements in unhygienic environment usually with inadequate infrastructure and lacking in proper sanitary and drinking water facilities.

Socially, slums tend to be isolated from the rest of the urban society and exhibit pathological social symptoms (drug abuse, alcoholism, crime, vandalism and other deviant behaviour). The lack of integration of slum inhabitants into urban life reflects both, the lack of ability and cultural barriers.

Thus the slums are not just huts and dilapidated buildings but are occupied by people with complexities of social networks, sharp socio-economic stratification, dualistic group and segregated spatial structures.

In India, slums are one or two-room hutments mostly occupying government and public lands. The houses in slums are built in mud or brick walls, low roofs mostly covered with corrugated sheets, tins, bamboo mats, polythenes, gunny bags and thatches, devoid of windows and ventilators and public utility services.

Slums have invariably extreme unhygienic conditions. They have impoverished lavatories made by digging shallow pit in between three or four huts and with sackcloth as a curtain, hanging in front. When the pit overflows excreta gets spread over the surrounding area and is rarely cleaned.

The children cultivate the habit of defecating anywhere in the slum area. Slums have practically no drains and are marked by cesspools and puddles. Piped water is not available to slum dwellers and they mainly depend upon shallow hand-pumps for water supply.

Such handpumps are generally dug in the middle of a stale dirty pool. People wash their clothes and utensils under the handpumps. The entire muck around the handpump percolates into the ground and contaminates the ground water. This contaminated ground water is taken out through the handpump which adversely affects the health of the slum dwellers.

Consequently people suffer from water-borne diseases like blood dysentery, diarrhoea, malaria, typhoid, jaundice, etc. These diseases stalk the people all the year round. Children with bloated bellies or famished skeletons, many suffering from polio, are a common sight. Most of the slums are located near drains (Nullahs) which contain filthy stagnant water.

Billions of flies and mosquitoes swarming over these drains cause infectious diseases. These drains are used as open lavatories by the inhabitants and are always choked. Such drains (Nullahs) pose serious threat to health of the people.

Slums are known by different names in different cities. They are called Bustees in Kolkata, Jhuggi- Jhoparies in Delhi, Jhoparpattis or Chawl in Mumbai and Cheri in Chennai.

10.7 LEGAL CHARACTERISTICS

10.7.1 Such Settlements Lack Land Ownership

From the above discussion it is clear that squatter refers to legal position of the settlement and slum refers to the condition of a settlement.

A distinction has to be drawn between squatter settlements and shanty towns. Illegality of tenure is the hallmark of the squatter settlement but shanti huts or mean dwellings are defined by their fabric. Shanty towns result mainly from massive rural-urban migration and from the inability of city authorities to provide sufficient housing facilities and employment for the vast influx of people from rural to urban areas.

Indian cities abound with slums which have been termed as 'eyesores', a 'rash' on city landscape, 'a blot on civilization' etc. But actually they are much more health hazards for its unfortunate poverty stricken inhabitants and also for the city as a whole. The most shocking aspect is that slums are growing at an accelerated rate.

Census of India, for the first time in 2001, came out with detailed data on slum population in India. According to data released by Census of India 2001, 607 towns and cities in 26 states/union territories have reported slum population.

No slum population has been reported in the remaining nine states/union territories at the time of Census 2001. Andhra Pradesh has the largest number of 76 towns reporting slum population. This is followed by Uttar Pradesh (65), Tamil Nadu (63), Maharashtra (62), West Bengal (51), Madhya Pradesh (42) and Karnataka (35).

The largest slum population of 10.6 million has been reported from Maharashtra; followed by Andhra Pradesh (5.1 million), Uttar Pradesh (4.1 million), West Bengal (3.8 million), Tamil Nadu (2.5 million), Madhya Pradesh (2.4 million) and Delhi (2.0 million).

Looking at the percentage of slum population to total population of towns reporting slum population, Meghalaya with 41.33 percent tops the list. Other states with high percentage of slum population are Haryana (33.06%), Andhra Pradesh (32.69%), Maharashtra (31.65%), Chhattisgarh (29.27%) and West Bengal (26.82%). Uttar Pradesh and Orissa are very close to the all India average of 22.58 percent.

A list of 26 million plus cities reporting slum population in 2001 (Municipal Corporation). As expected, the largest concentration of slum population is found in four major cities of Greater Mumbai, Delhi Municipal Corporation (Urban), Kolkata and Chennai. So far as percentage of slum population to total population of the cities (municipal) is concerned, Grater Mumbai with 48.88 percent of its population consisting of slum dwellers is the worst suffer.

Dharavi slum in Central Mumbai is the largest slum of Asia. Here some of the side allays and lanes are so narrow that not even a bicycle can pass. The whole neighbourhood consists of tenement buildings, two or three storey high with rusty iron stairways to the upper part, where a single room is rented by a whole family, sometimes twelve or more people. In this place of shadow-less, treeless sunlight, uncontrolled garbage, stagnant pools of foul water, the only non-human creatures are the shining black crows and long gray rats.

Dharavi was an arm of the sea that was filled by waste, largely produced by the people who have come to live there. The other cities with over 40 percent slum population to the total population (Municipal Corporation) are Faridabad and Meerut. Kolkata, Nagpur and Thane have about one-third of their population as slum population.

10.7.2 Transport

With traffic bottleneck and traffic congestion, almost all cities and towns of India are suffering from acute form of transport problem. Transport problems increase and become more complex as the town grows in size. With its growth, the town performs varied and complex functions and more people travel to work or shop.

As the town becomes larger, even people living within the built-up area have to travel by car or bus to cross the town and outsiders naturally bring their cars or travel by public transport. Wherever, trade is important, commercial vehicles such as vans and trucks will make problem of traffic more complicated.

Since most of the commercial activities of the towns are concentrated in the Central Business District (C.B.D.), the centres are areas of greatest congestion. However, other parts of the town are not free from traffic congestion.

Such areas include the roads leading to factories, offices, schools, etc., which will be thronged with people in morning and evening; minor shopping centres which grow up in the suburbs; sporting arenas, entertainment districts which will be busy at night, roads leading to residential and dormitory towns which will be busy when commuters flock to the cities in the morning to work and return home in the evenings.

Such congestion becomes greater when the centre is built up in tall skyscraper blocks whose offices sometimes employ thousands of workers, because at the end of the office hours everyone leaves the building within a short space of time to make their way home.

This puts tremendous pressure on public transport and causes journeys to take much longer period than they normally would. In most cities the rush hour or peak traffic hour lasts for about two hours and during that period buses and trains are crammed to capacity, roads are overcrowded with vehicles and the movement of traffic becomes very slow.

In other towns, the narrowness of the streets, which were built long before the motorised transport and lack of parking facilities are the main cause of congestion. Cars may be parked along the edges of the roads restricting movement to a narrow lane and the multiplicity of narrow streets, sharp comers and waits to turn into lanes of traffic may slow down the movement and thus create even greater congestion.

The traffic scenario in almost all the Indian cities presents a pathetic picture with Mumbai still having the best city transport system and Chennai, Ahmedabad and Pune being reasonably well served by local transport system. In all other cities, if one does not own a personal vehicle, great hardship is experienced in moving about in the city.

Apart from that, the level of incomes and affordability of Indian masses is very low and the citizens are not able to pay an economic fare for use of public transport system. Therefore, all city bus services sustain such heavy losses that they cannot really expand or even maintain a fleet adequately to meet the city needs.

Moreover, mixture of vehicles causes uncontrollable chaos on the roads. Free movement of stray cattle and domestic animals on the roads adds to traffic problem and often cause accidents. Heavy traffic and congestion leads to slow movement of traffic, fuel wastage environmental pollution and loss of precious time.

A study of traffic problem in Delhi will acquaint us to traffic scenario in the rest of urban India. Already there are 44 lakh vehicles on Delhi roads (in 2004) which will almost double by 2021 when the next Master Plan will be implemented. The road length, however, has not increased proportionately.

The road length per vehicle was 3 km in 1971 which reduced to 2 km in 1981, 1.3 km in 1991, 0.68 km in 1998 and 0.23 km in 2004, going in a car will take longer time than walking.

The guidelines for Delhi Master Plan 2021, allowing mixed land use, multi-storeyed structures and regularisation of 24 industrial estates will add to the city's already congested roads. Disturbing trends have also been indicated in the Status Report for Delhi, 2021 prepared by the Union Ministry of Environment and Forests.

Planning Department of Delhi Government also States that despite roads occupying 21 percent of the total area of the city, the increase of traffic on arterial roads is resulting in lower speeds, congestion, intersection delays and higher pollution level during peak hours.

Some relief is expected with the completion of metro rail. But experts fear that by the time the metro rail becomes fully operational, the demand for transport facilities will outpace the capacity of both road and rail transport.

Similar conditions prevail in most of the Indian cities. In Kolkata, metro rail and Vivekanand Setu were constructed to ease traffic flow. But traffic congestion in several old localities and near Haora bridge is almost a daily routine. In Ahmedabad, the speed of vehicles comes down to 5 km/hr on Gandhi Marg and several other roads due to congestion and overcrowding.

10.7.3 Water

What is one of the most essential elements of nature to sustain life and right from the beginning of urban civilisation, sites for settlements have always been chosen keeping in view the availability of water to the inhabitants of the settlement. However, supply of water started falling short of demand as the cities grew in size and number.

Today we have reached a stage where practically no city in India/ gets sufficient water to meet the needs of city dwellers. In many cities people get water from the municipal sources for less than half an hour every alternate day. In dry summer season, taps remain dry for days together and people are denied water supply at a time when they need it the most.

The individual towns require water in larger quantities. Many small towns have no main water supply at all and depend on such sources as individual tubewells, household open wells or even rivers. Accelerated Urban Water Supply Programme (AUWSP) was launched to provide water to towns with population of less than 20,000.

Keeping in view the increased demands for water by the urban population, Central Public Health and Environmental Engineering Organisation (CPHEEO) fixed 125-200 litres of water per head per day for cities with a population of more than 50,000, 100-125 litres for population between 10,000 and 50,000 and 70-100 litres for towns with a population below 10,000.

The Zakaria Committee recommended the water requirement per head per day 204 litres for cities with population between 5 lakh and 2 million and 272 litres for cities with population more than 2 million. This amount of water is supposed to be used for drinking, kitchen, bathing, cloth washing, floor and vehicle washing and gardening.

Sadly majority of the cities and towns do not get the recommended quantity of water. Gap in demand and supply of water in four metro cities, viz., Mumbai, Kolkata, Delhi and Chennai varies from 10 to 20 percent. The condition is still worse in small cities and towns. To meet the growing demand for water, many cities are trying to tap external sources of water supply.

Mumbai draws water from neighbouring areas and from sources located as far as 125 km in the Western Ghats. Chennai uses water express trains to meets its growing demand for water. Bangalore is located on the plateau and draws water from Cauvery river at a distance of 100 km. Water for Bangalore has to be lifted about 700 metres with help of lifting pumps.

Hyderabad depends on Nagarjuna Sagar located 137 km away. Delhi meets large part of its water requirements from Tajiwala in Haryana. Water is also drawn from Ramganga as far as 180 km. Under the proposed scheme it will meet its growing requirements of water from Tehri, Renuka, and Kishau barrages.

10.7.4 Sewerage Problems

Urban areas in India are almost invariably plagued with insufficient and inefficient sewage facilities. Not a single city in India is fully sewered. Resource crunch faced by the municipalities and unauthorised growth of the cities are two major causes of this pathetic state of affairs.

According to latest estimates, only 35-40 percent of the urban population has the privilege of sewage system. Most of the cities have old sewerage lines which are not looked after properly. Often sewerage lines break down or they are overflowing.

Most cities do not have proper arrangements for treating the sewerage waste and it is drained into a nearly river (as in Delhi) or in sea (as in Mumbai, Kolkata and Chennai), thereby polluting the water bodies.

In most Indian cities, water pipes run in close proximity to sewer lines. Any leakage leads to contamination of water which results in the spread of several water borne diseases.

10.7.5 Trash Disposal

As Indian cities grow in number and size the problem of trash disposal is assuming alarming proportions. Huge quantities of garbage produced by our cities pose a serious health problem. Most cites do not have proper arrangements for garbage disposal and the existing landfills are full to the brim. These landfills are hotbeds of disease and innumerable poisons leaking into their surroundings.

Wastes putrefy in the open inviting disease carrying flies and rats and a filthy, poisonous liquid, called leachate, which leaks out from below and contaminates ground water. People who live near the rotting garbage and raw sewage fall easy victims to several diseases like dysentery, malaria, plague, jaundice, diarrhoea, typhoid, etc.

10.7.6 Urban Crimes

Modem cities present a meeting point of people from different walks of life having no affinity with one another. Like other problems, the problem of crimes increases with the increase in urbanisation. In fact the increasing trend in urban crimes tends to disturb peace and tranquility of the cities and make them unsafe to live in particularly for the women.

Growing materialism, consumerism, competition in everyday life, selfishness, lavishness, appalling socio-economic disparities and rising unemployment and feeling of loneliness in the crowd are some of the primary causes responsible for alarming trends in urban crime.

Not only the poor, deprived and slum dwellers take to crime; youngsters from well-to-do families also resort to crime in order to make fast buck and for meeting requirements of a lavish life. Occasional failures in life also drag youngsters to crime.

The problem of urban crime is becoming more complicated in the present day world because criminals often get protection from politicians, bureaucrats and elite class of the urban society. Some of the criminals reach high political positions by using their money and muscle power.

According to study made by Dutt and Venugopal (1983), violent urban crimes like rape, murder, kidnapping, dacoity, robbery, etc. are more pronounced in the northern-central parts of the country. Even the economic crimes (like theft, cheating, breach of trust, etc.) are concentrated in the north- central region. Poverty related crimes are widespread with main concentration in the cities of Patna, Darbhanga, Gaya and Munger. This may be due to widespread poverty prevailing in this region.

However, the latest surveys show that Mumbai and Delhi figure in 35 cities that have high crime rate. As much as 31.8 percent of citizens in Mumbai and 30.5 percent in Delhi have been victims of crime. Sexual assault was higher in Mumbai (3.5 percent) as compared to Delhi (1.7 percent). Both cities score poorly in corruption, with 22.9% in Mumbai being exposed to bribery as compared to 21% in Delhi.

10.7.7 Problem of Urban Pollution

With rapid pace of urbanisation, industries and transport systems grow rather out of proportion. These developments are primarily responsible for pollution of environment, particularly the urban environment.

We cannot think of strong India, economically, socially and culturally, when our cities remain squalor, quality of urban life declines and the urban environment is damaged beyond repair. As a matter of fact, cities comprise the backbone of economic expansion and urbanization is being seen in a positive light as an engine of economic growth and agent of socio-political transformation.

The share of urban areas in the total national economic income had been estimated at 60 percent and the per capita income was about three times higher than rural per capita income. But this is not sufficient partly, due to high cost of living and partly, because of growing economic disparity in urban areas. Rich are becoming richer and poor are becoming poorer. Several steps have been initiated to meet the challenges posed by urban crisis but with little or no success.

National Commission on Urbanization (NCU) has, in its policy proposal of 1988, stressed the need for (a) the evolution of a spatial pattern of economic development and hierarchies of human settlements, (b) an optimum distribution of population between rural and urban settlements, and among towns and cities of various sizes, (c) distribution of economic activities in small and medium-sized growth centres, (d) dispersal of economic activities through the establishment of counter-magnets in the region, and (e) provision of minimum levels of services in urban and rural areas.

The other major development programmes include (i) Urban Basic Services for the Poor (UBSP) programme, (ii) the Environmental Improvement of Urban Slums (EIUS) programme, (iii) the Integrated Development of Small and Medium Towns (IDSMT), (iv) various housing and infrastructure financing schemes of Housing and Urban Development Corporation (HUDCO), (v) the Mega Cities Project, and (vi) the Integrated Urban Poverty Eradication Programme (IUPEP).

Almost all the major programmes of urban development suffer from the chronic disease of resource crunch. Right from the beginning of the planning period, urban development has been low on the development agenda with only 3-4 percent of the total plan outlay being allocated to the urban sector. The National Commission on Urbanization recommended in 1988 that at least 8 percent of the Plan outlay should be dedicated to urban sector.

10.8 SOCIAL EXCLUSION AND HEALTH INEQUITIES

Social exclusion refers to the systematic denial for certain social groups of the resources and/or recognition that could facilitate their full participation in the society. The Social Exclusion Knowledge Network (SEKN) of the World Health Organization (WHO) Commission on Social Determinants of Health advanced a comprehensive definition of social exclusion, stressing the relational and multidimensional perspective of the social systems: dynamic, multi-dimensional processes driven by unequal power relationships in the society. These processes operate along and interact across four dimensions—cultural, economic, political, and social—and at different levels including individuals, groups, households, communities, countries, and global regions. These processes result in a continuum of inclusion/exclusion and contribute to unequal access to resources, capabilities, and rights required for human development, valued recognition, involvement and engagement, social proximity, decision making, material well-being, living within environmental limits, and resisting the hazards of environmental change. This definition proposed by the SEKN provides a wider lens to social exclusion that links it to the unequal power relationships in the society and recognition that exclusion processes operate differently and have different impacts on different groups and/or societies at different times.

The proliferation of slums in LMICs highlights an additional dimension of exclusion: socio-spatial, which is defined as a process whereby different social groups residing in specific geographical areas are being excluded from access to resources and opportunities offered by the city. This concept of socio-spatial exclusion postulates that geographic areas, and in particular slums, should be investigated not as mere residences of the excluded population but as factors that contribute to further social and health risks.

The relationship between social exclusion and health inequities has not received, nor sufficient attention from researchers and policy makers in LMICs. In this context, we were able to identify examples from different parts of the world. For instance, in the east, Rashid argues that in Bangladesh, the broader political and economic conditions challenge the ability of the urban poor to improve their health and daily living context. Rashid warns that living in slum settlements is associated with shame, stigma, and spatial exclusion. In Pakistan, Hunter suggests that poor health outcomes were observed among socially excluded groups. In India, the caste system remains a major indicator for health outcomes, as reveled by the Indian National Health Survey-III (2005–2006) that documented lower levels of prenatal care, institutional deliveries, and vaccination coverage among scheduled castes; infant and maternal mortality were also higher among members of scheduled caste and scheduled tribes.

Similar examples were identified in other regions of the world. The African Population and Health Research Center reported in 2002 that in the slums of Nairobi, child mortality rates were significantly higher than those in other areas in the city; slum mortality rates exceeded rural Kenyan mortality. Additional risk included poor quality and quantity of water and sanitation, inadequate hygienic practices, poor ventilation and use of hazardous cooking fuels, transmission of infectious diseases in densely populated slums, and the elevated cost of health care, which delays, or prevents, access to health services for the poor.

In Mexico, excess mortality was evident in municipalities with very high levels of marginalization, and social exclusion contributed to "notorious" health inequalities measured in terms of mortality rates related to transmissible illnesses in childhood and produced by avoidable factors

CHAPTER

URBAN E-GOVERNANCE 11

11.1 INTRODUCTION

Electronic governance or e-governance is the application of information technology for delivering government services, exchange of information, communication transactions, integration of various stand-alone systems between Government to Citizen (G2C), Government-to-Business (G2B), Government-to-Government (G2G), Government-to-Employees (G2E) as well as back-office processes and interactions within the entire governance framework. Through e-governance, government services are made available to citizens through IT. The three main target groups that can be distinguished in governance concepts are government, citizens, and businesses/interest groups. In e-governance, there are no distinct boundaries, finance and support.

The urban area is increasingly engaged with complex digital infrastructure, which is posing a diversity of capricious and extraordinary challenges for urban governance in India. The chapter discusses the new policies and programs formulated to meet the challenges of governance in general and urban e-governance in particular. These days innovation based arrangements advance a worldview of smart urban improvement to turn out to be increasingly manageable and resilient, yet there is still a lot to do in the interest of residents themselves, to expand value being developed, equivalent access, chances to information and correspondence, and to open up the production of inventive administration.

It is based on the principles of inclusiveness, accountability, and effectiveness. The aim of this chapter is to explore the urban local bodies' role in promotion of 'e-governance' as 'smart governance'. It argues that one of the significant aspects of sustainable urbanization is e-governance. Moreover the chapter emphasizes that effectiveness of the 'smart governance' in this globalized world not only contingent on public participation but also degree information diffusion by the urban local bodies through internet.

The new millennium is getting increasingly dependent on virtual media and technology. E-media, with its speed, cost efficiency and smartness, has captivated millions of scientists, bureaucrats, planners, businesses and common citizens. Rapid urbanisation and growing spatial dependency in urban centres imply multiple challenges from urban development authorities and Urban Local Bodies (ULBs) to ensure developmental sustainability. Since the 1990s, good governance has captured the limelight in urban development dialogues and literature. With the emergence of new information and communications Technologies (ICTs), electronic interfaces are increasingly entrusted with the attainment of good governance goals. The situation demands new flexible forms of pro-citizen administration that are capable of strengthening transparency and accountability.

Academic attention to the ICT-driven urban governance and quest for an optimum model has led to the establishment of the domain of e-governance. Consequently, diverse opinions concerning ICT application in the public service delivery domain are creating a dilemma in adopting an optimum model of municipal e-governance. The purpose of this study is to conceptually appraise the various models of e-governance. This paper systematically identifies and reviews the models of e-governance as proposed by literature and suggests an alternative citizen-centric interactive model of urban e-governance, especially suitable in the context of India.

"Make all Urban Local Bodies Services accessible to the common man from anywhere, anytime through service delivery outlets and ensure efficiency, transparency & reliability of such services at affordable costs to realise the basic needs of the common man"

The Commissionerate of Municipal Administration has been introduced hybrid e-governance "Urban Tree Information system" for the benefit of citizens across all the Urban Local Bodies, which should, among other things ensure citizen-friendly, participatory, transparent, accountable Government. e-Governance would enable transformation of process and structures to facilitate political and socio-economic relationships, and particularly enhance commitment to democratic values and helping institutionalizing norms & practices, trusted services, just and equitable opportunities to all and integrity in carrying out governance functions. E-government would assist in empowering citizens and provide improved public service performance integrity.

Urban governance refers to how government (local, regional and national) and stakeholders decide how to plan, finance and manage urban areas. It involves a continuous process of negotiation and contestation over the allocation of social and material resources and political power. It is, therefore, profoundly political, influenced by the creation and operation of political institutions, government capacity to make and implement decisions and the extent to which these decisions recognise and respond to the interests of the poor. It encompasses a host of economic and social forces, institutions and relationships. These include labour markets, goods and services; household, kin and social relationships; and basic infrastructure, land, services and public safety. Large gaps often exist between poor and better-off urban residents in terms of access to social, economic and political opportunities (particularly decision-making) and the ability to participate in, and leverage, the benefits associated with urban living. According to Slack and Côté, urban governance:

- Plays a critical role in shaping the physical and social character of urban regions;
- Influences the quantity and quality of local services and efficiency of delivery;
- Determines the sharing of costs and distribution of resources among different groups; and
- Affects residents' ability to access local government and engage in decision-making, influencing local government accountability and responsiveness to citizen demands.

Urban governance involves a range of actors and institutions; the relationships among them determine what happens in the city. In managing urban transformations, government (at all levels) need to play a strategic role in forging partnerships with and among key stakeholders. While city government is the largest and most visible urban governance actor, much of what affects the life chances of the urban poor lies outside the control of city administrations. Instead, it is the market and private businesses, agencies of the central state or the collective voluntary action of civil society that determine the daily experiences of urban dwellers.

11.2 OBJECTIVES OF E-GOVERNANCE

The objectives of e-governance are as follows-

1. One of the basic objectives of e-governance is to make every information of the government available to all in the public interest.
2. One of its goals is to create a cooperative structure between the government and the people and to seek help and advice from the people, to make the government aware of the problems of the people.
3. To increase and encourage people's participation in the governance process.
4. e-Governance improves the country's information and communication technology and electronic media, with the aim of strengthening the country's economy by keeping governments, people and businesses in tune with the modern world.
5. One of its main objectives is to establish transparency and accountability in the governance process.
6. To reduce government spending on information and services.

11.3 FEATURES OF E-GOVERNANCE

It has been proven from the concept of e-governance that it is a powerful means of public service in the present era. Some of its features can be found by observing the functioning of e-governance.

1. **De bureaucratization:** Due to e-governance, the gap between the people and the government in all the services of the government is narrowing and the dependence of the people on the bureaucracy is also greatly reduced.
2. **E-Services:** Its main feature is the provision of services through the Internet. As a result, we get G2C, G2B, G2E, etc. services. This is already discussed in the section of 'types of governance'.
3. **International Services:** through e-governance, all the essential services can be delivered to the citizens who are living outside of their country for job purposes or any other reasons.
4. **It enhances the right to express to the citizens:** Using the means of e-governance anyone can share their views with the government on any bill or act or decision taken by the government.
5. **Economic Development:** With the introduction of e-governance, various information like import-export, registration of companies, investment situations, etc. are available through the internet. As a result, time is saved, procrastination decreases, and economic dynamism increases.
6. **Reduce inequality:** using e-governance tools everyone can gather information and empower themselves. In this globalized world, knowledge is power, and means of e-governance empower us by providing relevant information at minimal cost, effort, and time.

11.4 TYPES OF E-GOVERNANCE

E-Governance can be considered as the social inclusive policy for development of transparency and accountability of both people in society and administration. This policy involves providing the services to the people with collection of information through the institutional and communicational development.

It provides quality services in several ways. Those ways are also called as types of e-governance. These are mentioned below-

1. G2C (Government to Citizen)
2. G2G (Government to Government)
3. G2B (Government to Business)
4. G2E (Government to Employee)

11.4.1 G2C (Government to Citizen)

As people are the key concept of politics and government as well as governance, the government is compelled to connect with citizens through the transparent and accountable order. In this connection the government is responsible for promoting the social opportunities and public services in the field of-

- Transportation (Registration of motor vehicles, Issue of driving licenses, Issue of plying permissions, Tax and fee collection through cash and bank challans and control of pollution etc.),
- Hospitals (linking of various hospitals in different parts of the country to ensures better medical services to citizens),
- Education (availability of the e-learning modules to the citizens, right to education),
- Online job portal and various customer services.

It also ensures services such as issue of certificates, job cards, passport, ration cards, payments of bills and filing the taxes from the door step through e-governance platform. The main objectives of the G2C services are to ensure equitable distribution of information for all, acceptance of citizen's feedback, and improving welfare services.

The goal of Government-to-Citizen (G2C) e-governance is to offer a variety of ICT services to citizens in an efficient and economical manner and to strengthen the relationship between government and citizens using technology.

There are several methods of G2C e-governance. Two-way communication allows citizens to instant message directly with public administrators, and cast remote electronic votes (electronic voting) and instant opinion voting. These are examples of e-Participation. Other examples included the payment of taxes and services that can be completed online or over the phone. Mundane services such as name or address changes, applying for services or grants, or transferring existing services are more convenient and no longer have to be completed face to face.

By country

The idea behind a good sustainable Smart City is to build an ICT framework, an intelligent network of interconnected objects and machines transmitting data through wireless technology using a cloud computing architecture. Cloud-based intelligent IoT devices could receive, analyze and maintain real time data to help the city municipalities, city planners and citizens properly utilize data and make better decisions in order to improve the quality of life. The notion of Smart City e-governance started during 90's and was realized by financial support from the World Bank & United Nations. According to the UN, the world's urban population will increase approximately 1.5 times to 6 billion individuals by 2045, making up about 55% of the population (Urban Development).

Smart City governance is all about collaboration between humans and the use of ICTs for implementing city functions to get better outcomes. Cities become smart when the standard of living of its residents become better such as increasing their income level, regulating traffic, proper waste management and providing safety and security, especially to senior citizens and children. Smart cities with handling of big data and Information Technology (IT) could influence the state of how cities function. A Smart City model would enable communication between the city and its citizens/residents as the city would become an integrated system with several different connected parts. This model could be modified to achieve maximum efficiency in time or cost in different sectors such as smart transportation, healthcare and waste management, benefitting both the city and its citizens. Meanwhile, the Government of India, with a larger understanding of the use of ICT in delivering governance, acknowledges the existence of human resources and emphasizes in providing degree of excellence to its citizens' life as well as surroundings. An advancement of BIM technology could help in the improvement of the AECOO (Architecture, Engineering, Construction, Owner and Operator) industry around the world. 3D based BIM modelling technique has had a great impact for the regulatory processes in the land development industry in Singapore to embellish the range of strategies that are used to induce change in an instance of 3D digital innovation.

Banaras is leading in Smart City rankings of 100 Smart Cities mission as recently released by the Ministry of Housing and Urban Affairs (MoHUA) in India on January 15, 2021. Banaras has secured the first rank with a score of 85.25 as compared to other smart cities like Bhopal, Surat and Ahmedabad who are known to be the top scorers in the annual rankings. The city of the near future is going to be smart and more efficient for the people and entirely interconnected in order to regularly maintain traffic control, save energy, manage waste, provide uninterrupted purified water supply and all assisted by big data and artificial intelligence. Due to high volume of growth in population in the urban cities, there is also a rapidly rising number of enormous challenges i.e. more people creates more traffic which causes more pollution, higher energy consumption, more water usage, more waste and so on. Smart cities are now working to overcome these problems everywhere in places like Dubai, America, Nigeria, Malaysia, South Korea and India.

For example, cities have now started smart poles in place of regular poles which connect to their internet devices, there are driverless cars with intelligent systems and techniques which may only cause accidents very rarely and widespread use of drones and robots are helping to deliver goods. In the future, cities are going to use virtual reality for making processes more efficient. For example the loco pilots in duty can be supervised and monitored by their respective control center and Computational intelligence techniques can be used to find and correct the system errors to prevent damage before it happens. A citizen's engagement within the Smart City framework takes place in many ways using smartphones, connected vehicles, smart devices and homes by pairing devices with a city's infrastructure and services.

G2C e-governance is unbalanced across the globe as not everyone has Internet access and computing skills, but the United States, European Union, and Asia are ranked the top three in development.

The Federal Government of the United States has a broad framework of G2C technology to enhance citizen access to Government information and services. benefits. gov is an official US government website that informs citizens of benefits they are eligible for and provides information on how to apply for assistance. US State Governments also engage in G2C interaction through the Department of Transportation, Department of Public Safety, United States Department of Health and Human Services, United States Department of Education, and others. As with e-governance on the global level, G2C services vary from state to state. The Digital States Survey ranks states on social measures, digital democracy, e-commerce, taxation, and revenue. The 2012 report shows Michigan and Utah in the lead and Florida and Idaho with the lowest scores. Municipal governments in the United States also use government-to-customer technology to complete transactions and inform the public. Much like states, cities are awarded for innovative technology. Government Technology's "Best of the Web 2012" named Louisville, KY, Arvada, CO, Raleigh, NC, Riverside, CA, and Austin, TX the top five G2C city portals.

European countries were ranked second among all geographic regions. The Single Point of Access for Citizens of Europe supports travel within Europe and Europe is a 1999 initiative supporting an online government. Main focuses are to provide public information, allow customers to have access to basic public services, simplify online procedures, and promote electronic signatures. Estonia is the first and the only country in the world with e-residency which enables anyone in the world outside Estonia to access Estonian online services. One caveat of the Estonia e-residency program is that it does not give e-residents physical rights to the country. This means that unless the e-resident buys land they do not get to participate in the democratic processes. The benefit to e-residents is the opportunity to develop business in the digital European Union market. Neighboring Lithuania launched a similar e-Residency program.

Asia is ranked third in comparison, and there are diverse G2C programs between countries. Singapore's eCitizen Portal is an organized single access point to government information and services. South Korea's Home Tax Service (HTS) provides citizens with 24/7 online services such as tax declaration. Taiwan has top ranking G2C technology including an online motor vehicle services system, which provides 21 applications and payment services to citizens. India's e-governance programs have found success in regional areas. This is likely due to the ability to meet the language and literacy differences among their constituents.

Government-to-Citizen is the communication link between a government and private individuals or residents. Such G2C communication most often refers to that which takes place through Information and Communication Technologies (ICTs), but can also include direct mail and media campaigns. G2C can take place at the federal, state, and local levels. G2C stands in contrast to G2B, or Government-to-Business networks.

One such Federal G2C network is USA.gov, the United States' official web portal, though there are many other examples from governments around the world.

Concerns

A full switch to government-to-citizen e-governance will cost a large amount of money in development and implementation. In addition, government agencies do not always engage citizens in the development of their e-gov services or accept feedback. Customers identified the following barriers to government-to-customer e-governance: not everyone has Internet access, especially in rural or low-income areas, G2C technology can be problematic for citizens who lack computing skills. Some G2C sites have technology requirements (such as browser requirements and plug-ins) that won't allow access to certain services, language barriers, the necessity for an e-mail address to access certain services, and a lack of privacy.

11.4.2 G2G (Government to Government)

G2G has been referring to raising the quality of the government process by cost cutting, managing performance, and making strategic connections within government.

It enables government institutions to be more efficient and more effective by the use of IT tools such as-

- Live fingerprints scanning and verification,
- Electronic entry of reports and paperwork etc.

The major key areas in this type of e-governance are

- E-Secretariat (all the valuable information regarding the function of the government are interlinking throughout the various departments),
- E-Police (police personnel records, criminal records etc), and
- E-Court (creating a database of all the previous cases, pending and ongoing cases) and Statewide Networks.

11.4.3 G2B (Government to Business)

G2B is mainly concerned with these things-

- E-taxation,
- Getting a license from the government etc.
- Secure Electronics Transactions.

It has included the policy of government with business. According to S.P Kumar, 'the essentials for achievement of G2B services for secure and authentic transactions include: Standards for electronic transactions, a secure payment mechanism and Public key infrastructure.

11.4.1 G2E (Government to Employee)

The G2E model refers to providing information and services from government to employee and employee to government as well. It involves training through-

- E-learning methods;
- Consolidating the employee and
- Share of knowledge among the employees.

It has also facilitated the employee to access information regarding pay and benefit policies and manage their profits through online.

11.5 SMART CITIES AND E-GOVERNANCE TRENDS IN INDIA

Smart Cities Mission, an initiative launched in 2015, aims at creating the next generation cities in India. These cities would not just have an easy-to-access infrastructure but also be technologically advanced in government-citizen interaction. Technologies like Artificial Intelligence, Internet of Things, Radio-Frequency Identification (RFID), cloud computing, and many more would be used by the government to offer 'smarter' solutions. It would ease the resource-deficit burden of the country by empowering the government to do much more with less.

And when cities are becoming smarter, the traditional methods of governing would not suffice. That's why the government is taking new e-Governance initiatives that are laced with the latest technologies. Digital transformation in government is here and each government agency is taking required steps to ensure smooth e-Governance.

By e-Governance in smart cities, we mean a type of governance that aims at efficient usage of Information and Communication Technology (ICT) for improving the services offered by the government to its citizens and increasing the stakeholder participation in decision making and policy formation. This would help improve the governance of the state and move towards government digital transformation.

11.5.1 Four Stages of e-Governance

The government has crossed the most crucial stages of e-Governance, starting from having an online presence, to allowing digital interaction opportunities to citizens, and ensuring digital transactions like paying of taxes, fees, etc. Now, it aims at reaching the fourth stage of e-Governance to make the smart cities truly smart. This is the transformation stage where it seeks to improve its functioning through e-means like automation, RPA, data collection, and much more.

The inhabitants of these smart cities would get to enjoy various e-benefits like e-consultation, e-democracy, e-participation, and policymaking. As smart cities lead towards a government digital transformation, the citizens would get everything online.

11.5.2 Challenges in Smart Cities & e-Governance

The government launched the smart city initiative back in 2015 but is still fighting several odds and challenges. There have been not one, not two, but many challenges to the smart city plan and e-Governance in these cities. Whether we talk about the illiteracy of the people, their unwillingness and resistance to change, or the risk of data breach, there are several challenges that are hindering the progress of smart cities in India. Apart from that, getting the right funding for transforming your government is also a challenge.

11.5.3 Trends in Smart Cities & e-Governance

For smooth e-Governance in the country, especially the smart cities, the government needs to utilize several channels. Until now, the government has been using channels like smartphone applications, social media applications, SMS services, voice-prompted interface, and many more. In the coming times, there are several trends that are expected to change the way smart cities and e-Governance in India. Let's have a quick look:

Faster Connectivity with 5G

The buzzword in the network and communications community right now is 5G. With higher bandwidth and better performance, 5G offers much faster connectivity. Smart cities would definitely be seen utilizing this network technology to allow faster connectivity. Moreover, using 5G, e-Governance operations and processes can be completed in half the time. And it wouldn't go without saying that the latest technologies like the Internet of Things and ICT would also benefit which are acting as a foundation of transforming these Indian cities into smart cities.

Using AR & VR To Visualize Scenarios

Augmented Reality and Virtual Reality are two of the emerging technologies that will transform the way users interact with businesses. As we can see real estate using AR and VR to showcase property listings, the government would also be seen using AR & VR for visualizing different scenarios in smart cities. By visualizing scenarios like those for any emergency situation in controlled environments, the government can make better decisions for the future. It provides them with an opportunity to view structural elements that could not be performed in reality.

Moreover, they can visualize data and interact with the environment to analyze it using various perspectives. The government developers can also use AR and VR to compare different infrastructure plans and visualize their roads, highways, bridges, etc., to shuffle and see which plan would work more efficiently for development.

Quantum Computing To Detect Anomalies

Quantum computing allows scientists to calculate computational problems in a jiffy. It can be used to detect any anomaly in large bits of data to see what deviates from the 'normal'.Machine Learning can also be used for detecting anomalies. The government can use these quantum computing algorithms on the data collected from the people and offer help. In smart cities, the government can use quantum computing to see anomalies in data from domains like medicine, traffic flow, economic forecasting, tax collection, meteorology, etc. It can quickly offer a solution by detecting these anomalies and creating a solution even before the problem starts or goes out of hand.

11.5.4 Data-Driven Governance

Smart cities would have autonomous processes that would collect real-time data from everything like the traffic management to weather forecasting to make better decisions. These smart cities would follow a data-driven governance mode to ensure they are serving their citizens to the best of their capabilities. From identifying the problems to analyzing opportunities and creating solutions, data would certainly empower the government of these smart cities. Even the UN has set high-quality, timely, reliable and disaggregated data as one of its major agendas by 2030.

The government agencies would be able to utilize the data to analyze anything from specific analysis of high disease rate in certain areas to general data analysis for housing & infrastructure plan, data will help in e-Governance.

Internet of Things Making Everything Smarter

IoT aka Internet of Things is indeed connecting everything, from humans and machines to machines and machines, to make them smarter. Whether we talk about smart TVs, smart homes or smart Cars, everything is connected and can be accessed by the click on a button on your smartphone. So, how can smart cities lag behind?

These smart cities would be laced with built-in sensors in street lights, electricity grids, traffic signals, and everything else to effectively monitor and automate the data collection and distribution. By analyzing the data, these smart sensors would help the utility companies and other organizations save energy and make sustainable decisions for the cities. Things would become intelligent with edge computing and AI technology. It would be a new form of governance that would be witnessed in these smart cities because of government digital transformation that IoT and smart 'things' would bring.

Network Convergence

Network convergence means the bringing together of different networks to promise the delivery of high-speed internet. With network convergence, you can get more convenient and flexible modes for communicating and accessing information online. Smart cities will see more wire-line and wireless networking systems offering a centralized infrastructure.

Not only will this help the people of the smart cities and the government in monitoring the people here, but it would also enable the businesses to model state-of-the-art excellent business plans for the future.

11.5.5 State of The Art eDevices

If you are keeping up with the current situation, you would be aware of how the government is using drone technology to analyze the current situation in different regions. They are using drones to see if people are in their homes or still roaming the streets. Well, this is just the beginning, the government of the smart cities would be using more of these advanced and modern eDevices to monitor the people and make smart decisions.

Smart ICT and IoT devices would be on the rise in smart cities to ensure efficient monitoring of the cities and addressing real-time issues effectively.

Cloud for Centralized Monitoring

The Indian government has advanced from using client-side systems to web-based systems and is now going complete cloud to ensure stability and connectivity. Cloud-based systems will help the government in creating national-level registries that are stored centrally on cloud. Downtime and maintenance cost reduces when everything is stored on the cloud; making e-Governance easier and quicker. Cloud migration or storing data on the cloud only requires a strong internet connection and the emergence of 5G would only add icing on the cake.

The aim is to create a unified e-government infrastructure that would be based on the cloud and enable easy monitoring and also eases the concern of interoperability. Services are accessible remotely over the internet and not locally, which allows quick access to all. There are various domains in which cloud can help in centralized monitoring and easier e-Governance. These are:

- Unique Identification Authority of India
- Crime Management
- Education
- Health and land records
- Crime management
- Centralized auditing
- Legal records and case management
- Agriculture

India's National Informatics Centre has deployed an open-source eucalyptus software that acts as the foundation for its cloud approach. It allows broad-scale cloud-based e-Governance in India.

11.5.6 Better Citizen Service Through E-Governance

Smart Cities are as cities where investment in human and social capital and traditional (transport) and modern (information and communications technology - ICT) communication infrastructure fuel sustainable economic development and a high quality of life through the use of the Internet of Things (IoT) to optimize city operations' and services' efficiency.

The data gathered through the various Smart City initiatives taking place around the world provide numerous examples for ways that the increased development of Smart Cities and adoption of Smart City policy in cities around the world can improve the overall well-being of the cities' citizens and increase the efficiency of governance.

One way that some Smart Cities are trying to put technology to use to engage their citizens is through eGovernment initiatives, which serve to enhance public sector efficiency and streamline government systems to support sustainable development. eGovernment uses online, one-step gateways for citizens to access major government services quickly and easily. An example would be a local government keeping records of life events - births, marriages, deaths - in the cloud, rather than in paper files, for easy access for both government officials and employees.

Moving old paper files onto digital systems certainly improves efficiency. More documents can be processed by fewer workers, meaning citizens can get their records more quickly and undergo bureaucratic processes with more ease. The savings on labor can also lead to a reduction in costs, including processing, financial and time related costs.

An example of improved efficiency is the US state of Utah's four-day workweek initiative. The state of Utah was able to use its robust Utah.gov eGovernment services to implement a four-day workweek for its employees and cut costs by approximately $4.1 million through reduced spending on energy, janitorial services, and employee overtime payments.

According to the United Nations E-Government in Support of Sustainable Development bi-annual report (2016), there has been a sharp rise in the number of countries using e-government to provide public services online through such platforms. In 2003, just 45 countries had a one-step gateway for citizens to access information, and only 33 countries provided online transactions. In 2016, we see that at least 90 countries now provide access to some kind of e-government platform. Smart Cities can use e-government platforms to ensure that their institutions are increasingly inclusive, effective, accountable and transparent. Cities can also look to social media to engage citizens in participatory decision-making.

11.6 CHALLENGES: INTERNATIONAL POSITION

E-governance is facing numerous challenges world over. The traditional approach for introducing e-governance is just not sufficient due to the complexity from wide variety of application architecture mix from both legacy and modern worlds that need to be brought into the purview of e-governance. These challenges are arising from administrative, legal, institutional and technological factors. The challenge includes security drawbacks such as spoofing, tampering, repudiation, disclosure, elevation of privilege, denial of service and other cyber crimes. Other sets of problems include implementation parts such as funding, management of change, privacy, authentication, delivery of services, standardization, technology issues and use of local languages.

11.7 ICT BEYOND E-GOVERNANCE

Information and Communication Technology (ICT) is a broad term that covers everything that enables users to store, retrieve, manipulate, transmit or receive information electronically in a digital form, e.g. personal computers, email, smart phones. ICT is vital for the economic growth and development of our cities. Telecom sector in India started evolving with the liberalisation of the national economy in the 90s. The last two decades witnessed major transformation, with India emerging as a source of ICT services as well as a consumer. The number of internet users in India increased 9 fold between 2003-2013 and it is estimated that India's internet user base is currently overtaking the internet user base in the US. All these changes have led to a change in approach towards the use of Information and communication technology in the development sector in India.

The primary role of e-governance is to simplify the process of governance, by making it more transparent and accountable using information and communication technology.

The National Smart City Mission in India is a program that works on an incremental approach with an intense top down effort to build bottom-up stakeholder participation. It incorporates both ICT for e-Governance (from North America) and ICT for infrastructure and resource management (from Europe, Singapore etc). The mission builds a tripartite coalition between center, state and urban local body. It is competitive, convergent and driven by best practices from other previous and current missions.

11.7.1 Information and Communication Technology and e-Governance in India

Development of urban infrastructure in India was not a focus of the national policies for a long time. Jawaharlal Nehru National Urban Renewal Mission (JnNURM) of 2005 was the first programme of its kind, which addressed the issues related to building urban infrastructure. The main thrust of this programme was to ensure improvement in urban governance and financial strength of Urban Local Bodies (ULBs). JnNURM used information and communication technology as a tool for implementation of e-governance reform at the local municipal levels. Previous acts and policies before built the framework needed for this step. The New Telecommunication Policy of 1999 addressed the convergence of IT, media, telecom and consumer an identified role of IT

& Telecom in provision of good infrastructure, transparency and security of state. The Information and Technology Act of 2000 gave legal validity of electronic contracts, legal recognition of digital signatures, security procedures for electronic records and defined penalties against cyber crime. Under JnNURM, e-Governance was a of the mandatory reform for ULBs and its broad aim was to:

- Improve efficiency and effectiveness in interaction between local government and its citizens and other stakeholders,
- Improve quality of internal local government operations and management information systems to support and stimulate good governance
- Bring about transparency and accountability in urban local body operations
- Help improve reach of the delivery of services to citizens

Modules under the e-governance reforms are:

1. e-Procurement
2. Municipal accounting system
3. Property tax and user charge collection
4. Registration of birth and death
5. Public grievance redressal
6. Off Site real time monitoring system
7. Citizen service centers
8. Building permit allocation
9. Double entry accrual accounting system
10. Council management system
11. Health programme management and
12. Personnel information system

11.7.2 Evolving Role of ICT

Over the last decade, the use of ICT has remained more or less the same – focused on e-governance reform through different programmes and initiatives. But now, this focus is shifting and ICT is being looked at as an enabler and a tool for problem solving. Its purpose is widening to include engagement of social entrepreneurs, social movements, enabling crowd sourcing of data, problem solving and a sharing economy. Perhaps the greatest asset of information and communication technology as a tool for development of our cities, is its ability to have large scale impact with small incremental interventions. JnNURM was followed by the National e-Governance Plan. Approved by the Government of India in 2006, it was the first attempt at a national plan and structure for e-governance in India. It was built on a three tier structure –

- Common Service Centres (CSCs)
- Front-end delivery points for a range of citizen services
- Common and support infrastructure
- Allows information to be shared electronically between different agencies of the government and with citizens.
- Includes: the State Wide Area Networks (SWANs), which form the converged backbone network for data, voice and video throughout a state / UT and
- The State Data Centers (SDCs) which can provide common secure IT infrastructure to host state-level e-government applications and data
- 27 Mission Mode Projects (MMP)
- Transforms high priority citizen services from their current manual delivery into e-delivery
- Each MMP is owned and spearheaded by the relevant ministry/agency of the national government or by a state government
- It is called 'mission mode' because it has a definite timetable, service levels, project implementation team, and process re-engineering plans

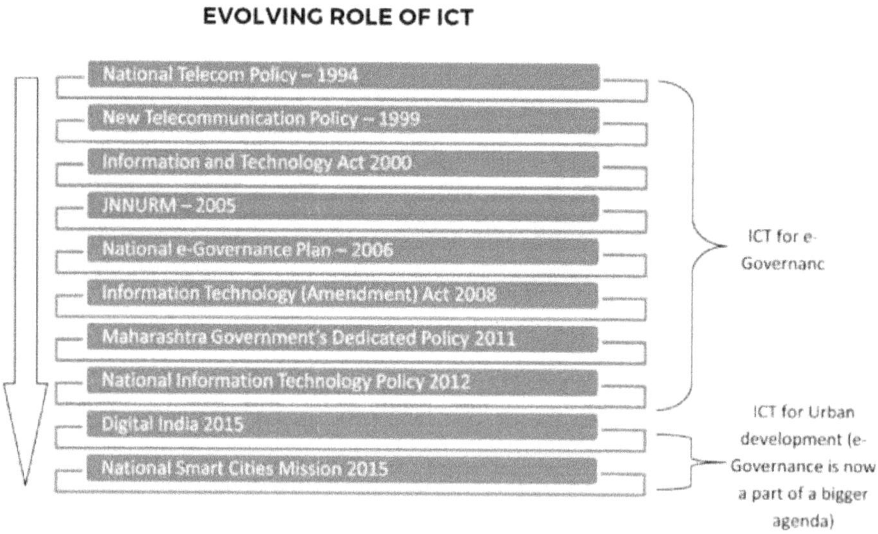

Fig. 1 Evolving Role of ICT

Rapid urbanization of our nation is putting immense pressure on our cities and its systems. Resource management and conservation will be the key to a sustainable future. Efficiency in our processes and harnessing the maximum potential of the tools at our disposal will define our growth and development. Information and communication technology is one such tool, and we must realise its true potential in the development of our urban infrastructure.

11.7.3 Role of Information and Communication Technology in Urban Development

Today, National Smart Cities Mission and Digital India programme are driving the agenda of ICT in urban development. They encourage cities to leverage the potential of ICT in improving quality of life through provision of core infrastructure to the citizens. The National Smart Cities Mission calls for a collaborative, interdisciplinary and comprehensive approach that encourages the cities to look at ICT beyond its traditional role of e-governance and citizen engagement. Elements of the national smart cities mission also act as the points of convergence with other national initiatives, providing the platform for solutions that cut across the traditional silos of the planning system. The purpose of Digital India Programme is to ensure that Government services are made available to citizens electronically by improving online infrastructure and by increasing Internet connectivity. It has three core components, the creation of digital infrastructure, delivering services digitally, digital literacy.

National Smart Cities Mission has 24 program areas. Proposals for different cities include area based or pan-city initiatives that address at 2-3 of these. The key focus here is that the implementation across the board is based on ICT as a tool. The solutions can be as simple as sensors that turn the streets lights on or off based on daylight, to complex automated tsunami warning systems. In addition to this, e-Governance and IT connectivity have dedicated program areas. Many of these solutions are already being used successfully in different parts of the world. Some of them are:

- GPS based bus tracking system that intimates the user of ETA through a digital display, a mobile app or a computer screen – New Brunswick, New Jersey
- Crowd sourced maps in disaster situations to aid in rescue operations. Nepal Earthquake, 2015
- Open Data platform that shares socio-economic and physical datasets along with their unique spatial IDs. New York City, New York
- Daylight and/or occupancy based sensor based lighting in public buildings to reduce wastage of energy
- Air quality sensors ICT has led to transformation of planning practices in the past, particularly in the field of spatial data analysis. The variety of Geographic Information Systems (GIS) platforms available today have critically reduced the time taken for scale spatial and statistical analysis. This has also led to the development of user friendly applications such as Google maps which do network analysis and computation for easy path finding and navigation.

Advances in information and communication technology have led to the propagation of this tool to widely accessible platforms over time, eventually making it an everyday tool, readily available to anybody with a smart phone, tablet or a computer. However, it is important to note that poverty and lack of education can keep some of the ICT based facilities and services from the citizens. This gap in the ability of the different segments of the society to access modern information and communication technology is commonly referred to as the Digital Divide. The Digital India programme addresses this issue through two of its pillars – universal access through mobile connectivity and public internet access programme.

Smart City Mission is about smarter ways of imagining and developing cities, using ICT is an enabler for strategic interventions. Through convergence with other missions, risk management and mitigation (citizens, finance and land), its role is to enable propagation of services and facilities across social, economic and physical barriers.

With all its given potential, use of ICT faces some significant challenges:

- Digital Divide
- Maintenance and upkeep of a complex system for all the services
- Interoperability & integration of data
- Privacy of the citizens
- Security of the system & authenticity of information
- Crossing the language barrier

However, upon addressing these issues, ICT can bring about transformation by

- Empowering citizens by bridging the knowledge gap
- Enabling data driven and evidence based planning
- By bringing about large scale impact with incremental small scale, and
- System reform, through transparency, efficiency and accountability

Information and communication technology is a powerful tool, which can empower people through propagation of information and citizen engagement. It build channels of communication and bring efficiency to systems. Indian cities have an opportunity to explore its application through the competitive process of the National Smart Cities mission. Successful implementation of ICT in urban development in India will cut across the traditional silos of the infrastructure, through sharing of data and promote a comprehensive, multidisciplinary and collaborative approach.

CHAPTER

ECO-CITIES 12

12.1 INTRODUCTION

An eco-city or ecocity is "a human settlement modeled on the self-sustaining resilient structure and function of natural ecosystems", as defined by Ecocity Builders (a non-profit organization started by Richard Register who first coined the term). Simply put, an eco-city is an ecologically healthy city. The World Bank defines eco-cities as "cities that enhance the well-being of citizens and society through integrated urban planning and management that harness the benefits of ecological systems and protect and nurture these assets for future generations". Although there is no universally accepted definition of an 'eco-city', among available definitions, there is some consensus on the basic features of an eco-city.

The world's population is continuously increasing, which puts a tremendous amount of pressure on cities due to the need for new urban development. There is an urgent need for cities around the world to adapt ecologically based urban development to work towards sustainability. The dimensions of an ecocity provide solutions to improve the living conditions in cities by solving our current unsustainable practices.

The cities around the world that face the most severe challenges associated with the world's urban population are those in developing countries. Eco-cities are commonly found to focus on new-build developments, especially in developing nations such as China, wherein foundations are being laid for new eco-cities catering to 500,000 or more inhabitants.

An ecocity is "a human settlement modeled on the self-sustaining resilient structure and function of natural ecosystems."

Here is a longer working definition by Ecocity Builders and the International Ecocity Framework & Standards (IEFS) advisory team.

An Ecocity is a human settlement modeled on the self sustaining resilient structure and function of natural ecosystems. The ecocity provides healthy abundance to its inhabitants without consuming more (renewable) resources than it produces, without producing more waste than it can assimilate, and without being toxic to itself or neighboring ecosystems. Its inhabitants' ecological impact reflect planetary supportive lifestyles; its social order reflects fundamental principles of fairness, justice and reasonable equity.

12.2 ORIGINS

The ecocity concept can be traced back to 1975 with the Berkeley-based non-profit Urban Ecology. Founded by Richard Register and a group of visionary architects and activists, Urban Ecology supported urban design concepts that integrated urban planning, ecology and citizen participation.

Furthering these experiences, Register coined the term "ecocity" in his 1987 book "Ecocity Berkeley: Building Cities for a Healthy Future."

In 1992, Register founded Ecocity Builders. Ecocity Builders remains the steward of the International Ecocity Conference Series, now known as Ecocity World Summit. This series is the longest running conference on ecocities and urban ecology.

In the first half of the 19th century, the Garden City of Ebenezer Howard, the urbanistic thought of Frank Lloyd Wright and Le Corbusier laid the foundations for a radical change of paradigm and a comprehensive transformation in the urban planning canon. Durig the post-war era, there was the spread of modernist tower blocks to satisfy the needs of urban slums.

Initial ideas behind the eco-cities can be traced back to 1975 with the formation of a non-profit organization called Urban Ecology. Founded by a group of visionary architects and activists including Richard Register in Berkeley, California, the organization worked at the intersection of urban planning, ecology, and public participation to help formulate design concepts centered around building environmentally healthier cities. Some of their efforts included initiating movements to plant trees along the main streets, promoting the construction of solar greenhouses, developing environment-friendly policies by working with the Berkeley city planning division and encouraging public transportation. Building on these strategies, Richard Register later coined the term 'ecocity' in his 1987 book titled "Ecocity Berkeley: Building Cities for a Healthy Future" describing it as a city where human beings live in harmony with nature and therefore greatly reducing their ecological footprint. Urban Ecology began publishing articles focused on similar complex urban issues that elevated the movement further with the creation of their magazine, 'Urban Ecology' in 1987. For two decades, they also publish two newsletters, 'The Sustainable Activist' and 'The Urban Ecologist' to pursue their vision.

In the 1990s, there we two important events that began the initiative for eco-city developments. The first event involved the publication of the Brundtland Report. The

Brundtland Report presented the idea of sustainable development. The second event occurred in 1992 at the United Nations Earth Summit. The members at the summit created a plan to apply sustainable development to our cities. The Eco-city's that were developed during this time period incorporated the concept of sustainable development. However, a majority of the developments were created to provide a vision of an eco-city. Eco-cities have now become a popular way to create sustainable new urban development.

12.2.1 International Ecocity Conference Series aka Ecocity World Summit

Urban Ecology further advanced the movement when they hosted the first International Ecocity Conference in Berkeley, California in 1990. The conference focused on urban sustainability problems and encouraged over 800 participants from 13 countries to submit proposals on best practices to reform cities for a better urban ecological balance.

Following this, in 1992, Richard Register founded the non-profit organization Ecocity Builders, to advance a set of goals outlined in the conference. Since its conception, the organization has been the convener of the International Ecocity Conference Series, now knowns as Ecocity World. The IECS has been the longest standing international conference series consisting of biennial Ecocity World Summits (EWS) and has been held in Adelaide, Australia (1992); Yoff, Senegal (1996); Curitiba, Brazil (2000); Shenzhen, China (2002); Bangalore, India (2006); San Francisco, United States (2008); Istanbul, Turkey (2009); Montreal, Canada (2011); Nantes, France (2013); Abu Dhabi, UAE (2015), Melbourne, Australia (2017); Vancouver, Canada (2019); and Rotterdam, the Netherlands (2022).

Other leading figures include architect Paul F Downton and authors Timothy Beatley and Steffen Lehmann, who have written extensively on the subject.

12.2.2 Ecocity Standards

Ecocity Builders facilitates a framework that enables participating cities to assess their overall ecological health and performance. Designed for a wide range of stakeholders, the Ecocity Framework & Standards charts a city's steps forward along eighteen conditions organized through four fundamental urban arenas: urban design, bio-geo-physical conditions, ecological imperatives, and socio-cultural conditions.

12.2.3 Current Trends

Criteria

An ideal eco-city has frequently been described as one that fulfills the following requirements:

- Operates on a self-contained economy that obtains resources locally
- Is entirely carbon-neutral by promoting techniques like the use and production of renewable energy

- Is established over a well-planned city layout that promotes walkability, biking and the use of public transportation systems
- Promotes conservation of resource by maximizing water efficiency and energy efficiency, while managing an ecologically beneficial waste management system that promotes recycling and reuse to create a zero-waste system
- Restores environmentally damaged urban areas
- Ensures decent and affordable housing for all socio-economic and ethnic groups and improves jobs opportunities for disadvantaged groups, such as women, minorities, and the disabled
- Supports local agriculture and produce
- Supports future progress and expansion over time.

Besides these, each individual eco-city has an additional set of requirements to ensure ecological and economic benefits that may range from large-scale targets like zero-waste and zero-carbon emissions, as seen in the Sino-Singapore Tianjin Eco-city project and the Abu Dhabi Masdar City project, to smaller-scale interventions like urban revitalization and establishment of green roofs as seen in the case of Augustenborg, Malmö, Sweden.

12.2.4 Ecocity Framework and Standards

With a growing popularity of the concept, in the last few decades, there has been an exponential growth in the number of eco-cities established around the globe.To assess the performance of these eco-cities and provide future guidance, the Ecocity Framework and Standards, established by Ecocity Builders with technical support from the British Columbia Institute of Technology (BCIT) School of Construction and the Environment, provides a practical methodology for this to ensure progress towards the intended goals of eco-cities. The four pillars in this framework include:

- Urban Design (containing 4 criteria for access by proximity)
- Bio-geo Physical Features (containing 6 criteria for the responsible management of resources and materials as well as the generation and use of clean, renewable energy)
- Socio-cultural Features (containing 5 criteria for promoting cultural activities and community participation)
- Ecological Imperatives (containing 3 criteria to sustaining and restoring biodiversity)

Using these, the International Eco-Cities Initiative recently identified and rated as many as 178 significant eco-city initiatives at different stages of planning and implementation around the world. To be included in this census, initiatives needed to be at least district-wide in their scale, covering a variety of sectors, and have official policy status. Although such schemes display great variety in their ambitions, scale, and conceptual underpinnings, since the late 2000s there has been an international proliferation of frameworks of urban sustainability indicators and processes designed to be implemented across different contexts. This may suggest that a process of eco-city 'standardization' is underway.

12.2.5 Practical Limits

Richard Register once stated that "An ecocity is an ecologically healthy city. No such city exists". Despite the conceptual ecological benefits of eco-cities, actual implementation can be difficult to attain. The conversion of existing cities to eco-cities is uncommon because the infrastructure, both in terms of the physical city layout and local bureaucracy, are often major insurmountable obstacles to large-scale sustainable development. The high cost of the technological integration necessary for eco-city development is a major challenge, as many cities either can't afford, or are not willing to take on, the extra costs.

Such issues, along with the added challenges and limits to retrofitting existing cities contribute to the establishment of newly constructed eco-cities. Along with this, the costs and infrastructure development needed to manage these large scale, two-pronged projects extend beyond the capabilities of most cities. In addition, many cities around the world are currently struggling to maintain the status quo, with budgetary issues, low growth rates, and transportation inefficiencies, that encourage reactive, coping policies. While there are many examples worldwide, the development of eco-cities is still limited due to the vast challenges and high costs associated with sustainability.

12.2.6 Related Terminologies

Eco-cities have been developed as a response to present-day unsustainable systems that exist in our cities. Simultaneously, there have been other concepts like smart cities, sustainable cities, and biophilic cities that also strive towards achieving sustainability in cities through different approaches. Owing to ambiguity in their definitions and closely related criteria defined to achieve their goals, these concepts, despite their varying approaches, are often used interchangeably.

12.3 ECO-CITIES FOR INDIA

Cities are both engines for growth and sources of concentrated environmental problems. With urbanization and economic growth, opportunities are more in the cities than in rural areas, thus encouraging immigration from rural areas to urban areas. Urbanization is increasing in many of the developing nations today and about 50 percent of the world's population lives in urban areas and about 90 percent of global urban growth now takes place in developing countries. Between the years 2000 and 2030, developing countries are projected to triple their entire built-up urban areas. This unprecedented urban expansion also means that cities, nations and the international development community face many challenges and opportunities. Improperly planned cities and lack of plans to absorb the future growth has made most cities less livable and many developing countries are now faced with the challenge of making cities both ecologically and economically sustainable.

Some critical challenges for cities in future will be: How can cities continue to harness the opportunities for economic growth and poverty reduction offered by urbanization, while also mitigating its negative impacts? How can cities cope up with the speed and the scale of urbanization, given their own capacity and constraints? How can ecological and environmental considerations be interlinked with development, so that they produce cumulative and lasting advantage for cities?

To address the above questions, Eco city concept was proposed. Richard Register first coined the term "Eco-city" in his 1987 book, Eco-city: Building cities for a healthy future. An eco-city in simple terms can be explained as an ecologically healthy city. An Eco- city builds on the synergy and interdependence of ecological and economic sustainability, and their fundamental ability to reinforce each other in the urban context. Eco-Cities are a concept to achieve this sustainability by taking the ecological principles as the central driving principles for the planning of our cities.

Eco-city aims at:

- Developing an urban ecosystem which is ecologically sound minimizing the negative impact of development on the environment.
- Reduction of ecological footprint of development thus shaping an improvement in the quality of life.
- Achieving environmental sustainability through reduced greenhouse gas emissions, utilization of renewable energy, and green transportation.
- Creation of a vibrant economy through environment-friendly production and industry that supports high capita gross domestic product (GDP) level
- Maintaining high air and water quality standards and an above-average ratio of green space per capita
- Social harmony with adequate educational and employment opportunities and a social safety net
- Protection of ecologically sensitive habitats, physical and nonphysical cultural legacies and promotion of green lifestyles and regional integration

12.3.1 Eco-City Initiatives in China and India

In 1994, China announced its 'Agenda 21' and explicitly stressed the importance of sustainable settlement. By 1996, the then State Environmental Planning Agency issued the policy document 'Guidelines for the Building of Eco-Communities (1996-2050)'. The intention was to promote the planning and construction of eco-communities across the country. Under this directive, between 2003 -2008 three Eco-cities were planned in China:Dongtan Eco-city near Shanghai, Tianjin City in northern China, and Huangbaiyu, north-east China. In Japan starting 1997 six eco-cities have been planned: Yokohama, Kitakyushu, Toyama City, Obihiro, Shimokawa and Minamata.

In India discussions on Eco cities started in 2000 and starting 2001 six medium and small Eco-cities were planned by the Ministry of Environment and Forest (MoEF) in association with Central Pollution Control Board (CPCB) and with technical assistance from German technical cooperation (GTZ). The focus of the project is pollution control, improvement of environmental quality, protection of environmental resources like rivers and lakes, improving sanitary conditions, improving the needed infrastructure and creating aesthetic environs in the chosen towns. The cities included Tirupathi, Vrindavan, Kottayam, Ujjain, Puri and Thanjavur.

The Delhi Mumbai Industrial Corridor Development Corporation (DMICDC) has also aimed to develop smart Eco-cities along the Delhi Mumbai Corridor with investment from companies in Japan. The DMICDC and the Haryana State Industrial & Infrastructure Development Corporation (HSIIDC) have planned to develop an eco city at Manesar in Gurgaon, Haryana. This is planned as a pilot model, and if it succeeds similar models will be developed in different regions of the country in the future. This has been planned based on the Japanese Eco cities of Yokohama and Kitakyushu. Sustainable eco villages and towns are also being advocated. The Prince of Wales, through his charity Foundation for the Built Environment, is planning to construct an eco-friendly community for 15,000 people outside either Bangalore or Kolkata. The design of the new shanty town is inspired by the model village of Poundbury in Dorset, which has been the Prince of Wales' pet project for thirty years.

Many of these projects are at various stages of implementation, however planning and developing an Eco-city is a tedious and uncertain process. In China failure of eco-cities (such as Dongton Eco-city) occurred mainly due to Implementation difficulties. Factors like land availability, economic growth, Infrastructure facilities, Investment, political stability and much more will play a critical role in making Eco-cities a success. Barriers and challenges have been experienced with regards to the Eco-Towns in Japan too. It would be difficult to adopt the process of the Eco-Town formation as-is to developing countries and cities because of lack of funds, differences in the social and industrial structures, and low environmental consciousness.

Some of the key aspects to be considered while designing Eco city models particularly in the developing countries are:

- Environmentally sound technology to reduce carbon emission, recycle waste and to create sustainable transport

- Land acquisition and relocation of local people
- Involvement of multi stakeholders in town planning
- Enormous financial requirement

During the last two decades most Indian cities have experienced phenomenal growth which the cities found difficult to cope with and as a consequence they are faced with problems in infrastructure, water and air pollution and other environmental problems. These problems are expected only to grow more in coming years. Social injustice and gap between urban rich and poor is also increasing leading to more urban poor. If we are to absorb and sustain the powerful wave of urbanization, while continuing to manage the existing built stock, we will need a paradigm shift on the approaches towards planning and managing cities.

Eco-cities have the potential to address many of the problems associated with urban development and failure of Eco-city models should not dampen the interest levels, however careful planning and implementation is necessary.

The evolution of cities takes many years. Each city has its own socio political, cultural and economic conditions and strategies adopted in shaping a successful eco city at one place may not necessarily work for other cities. Achieving greater sustainability in cities requires an in-depth understanding of the impacts of different urban forms on movement pattern, social conditions, environmental quality, and of their capacities to deliver future benefits. Success of Eco cities truly depends on planning taking, ecological and environmental factors into consideration.

Eco-cities cannot be formed in isolation. These projects: (a) need to ensure inter-linkages to the present city; or (b) should be aimed to develop present cities into Eco cities. Better planned eco-cities cannot be successful unless human development is taking place simultaneously. The socio cultural aspect of sustainability must also be taken into consideration while planning. With increasing economic growth in India, the growth of Indian cities can be expected to be high and it would be wise to start planning for Eco-cities today for a better sustainable future tomorrow.

12.4 BENEFITS OF DEVELOPING ECO-CITIES AND ECO-TOWNSHIPS

There are several benefits of developing Eco-cities and Eco-Townships; which among other things, are largely green and eco-friendly. These include: efficient land-use, habitat preservation and restoration, effective transport management and energy efficiency, efficient use of resources, emissions and pollution control and enhanced quality of life for the occupants as detailed below.

12.4.1 Efficient Land-Use

In today's scenario, development has become synonymous with physical expansion or growth. There is a need for significant changes in the pattern of land use and construction that will provide communities with better quality of life and at the same time conserving natural resources. Green Township rating system addresses the impacts of urban sprawl by encouraging compact, mixed-use developments and promotes higher urban densities without affecting the quality of life.

12.4.2 Preservation & Restoration

Conventional development is generally insensitive to natural environment. Such developments may scar the landscape, take prime agricultural land out of production or destroy biodiversity and natural habitats. The Green Townships rating system is designed to facilitate restoration and preservation of the natural environment by encouraging strategies that aid interface between the built environment and natural environment. This approach will not only enhance the fabric of the planned development but also provide environments conducive for living and working.

12.4.3 Efficient Transportation Management

Traffic congestion, long-distance commuting, rising levels of air and noise pollution are pressing issues in today's cities. Efforts to relieve congestion such as, constructing flyovers, road widening etc., are good initiatives but may not address issues such as fossil fuel consumption and associated emissions. 'Green Townships' rating system addresses these issues by encouraging effective and efficient transportation management strategies. Such strategies include increasing opportunities for bicycling, encouraging pedestrian-friendly networks; reduction in the number of automobile trips, promoting public transportation and use of alternative vehicles.

12.4.4 Efficient Use of Resources

Perhaps the most challenging problem facing our cities today is to meet the ever-rising demand for power, water supply and waste management. Meeting this demand requires enormous amount of investments infrastructure. Efficient and effective use of resources is thus vital in augmenting the existing infrastructure.

12.4.5 Water Efficiency

Most of the Asian countries are water-stressed, and in countries like India, the water table has reduced drastically over the last decade. Green townships encourage the use of water in a self-sustainable manner through reducing, recycling and reusing strategies and can save potable water to an extent of 30 - 50%.

12.4.6 Energy Efficiency

Green townships can reduce energy consumption of infrastructural equipment through energy-efficient street lighting, motors, pumps etc. The energy savings that can be realized by adopting this rating program in infrastructural equipment can be to the tune of 20-

30%. Further, on site power generation using various renewable energy technologies and other clean fuels can significantly reduce the load on-grid power supply.

12.4.7 Enhanced Quality Of Life

The place that we live in has profound effect on our lives. People have a natural predisposition to feel better and perform better in livable & safe environments. Green township developments are beneficial to the individual and community. Mixed land use and compact planning are the characteristic of a green development, which reduces dependency on automobiles and associated green house emissions. The outdoor air quality is enhanced by providing landscaped areas, encouraging the use of clean fuels for vehicles. Noise levels are reduced by provision of vegetative buffer. Green buildings and energy efficient infrastructure further aid in reducing the green house gas emissions. Public landscaped areas, walkable streets, bicycling lanes, community gardens and public spaces encourage physical activity and help in improving public health.

An eco-city is a city built off the principles of living within the means of the environment, an ecologically healthy city.

12.4.8 Benefits to Developers

There is a widespread perception that environmentally responsive developments are time-consuming and financially less rewarding. However, in reality well-executed green developments perform extremely well financially, as they require lower operating costs, increase health and productivity of the citizens and have higher marketability. The immediate benefits include a reduction in water and energy demand right from the initial stages of operation.

12.4.9 Eco-City and Eco-Town: Innovations and Economics

An eco-city is a city built off the principles of living within the means of the environment. The ultimate goal of many eco-cities is to eliminate all carbon waste, to produce energy entirely through renewable sources, and to incorporate the environment into the city; however, eco-cities also have the intentions of stimulating economic growth, reducing poverty, organizing cities to have higher population densities, and therefore higher efficiency, and improving health.

12.4.10 Criticism

Three Pitfalls

Looking at the patterns of progress in the last few decades of city construction towards sustainability, Valaria Saiu (University of Cagliari) poses one major criticism through the existence of a theory-practice gap caused by economic and ethical conflicts and risks that generate socio-spatial utopias. She identifies three pitfalls in the concept of sustainable cities (and therefore, eco-cities):

1. **The Idea of the City as a Business:** "Most eco-city projects are dependent on technologies available on the global market and the city is considered as a big economic affair". Often developed as techno-centric concepts, these projects seek investment opportunities by public-private partnerships leading to a top-down approach. This structure lacks democratic approaches in the decision-making process which further contributes to running high risks of failure, especially in social terms.
2. **The Oversimplification of Urban Complexity:** Due to the nature of current trends in measuring sustainability, there has been a strong focus in the quantifiable aspects of sustainability like energy-efficiency or waste-efficiency. This creates a tendency of oversimplification by neglecting the social and political aspects of the city that are unmeasurable qualitative aspects, yet significant to the fundamental concept of eco-cities.
3. **The Quest for the Ideal Community:** This section of the criticism focuses on the practical limits to merging economic goals with social goals in the urban development process. "Under the banner of green technology, inhabitants are forced to pay higher costs for their use of facilities in eco-cities."

12.4.11 Eco-Cities as Isolated Entities

Another larger conceptual criticism faced by eco-cities stems from the ambiguity in the definition of sustainability as a term. This has been further elaborated by Mike Hodson and Simon Marvin in their article titled 'Urbanism in the Anthropocene: Ecological Urbanism or Premium Ecological Enclaves' where they noted "We have tended to refer to sustainability in a generic sense, and our discussions of sustainability could be employed to anything that has sustainable as an adjective".

As a result of this, a widespread trend has been observed in the growing number of eco-cities developed over the past two decades that claim to combat our current global climate-change challenges. Many of these cities are found to be established in isolation from other existing urban centers due to the nature of their ownership. Owing to this isolation, internalization of resource-flows contribute towards a shallow sense of ecological sustainability in such cities.

12.4.12 Urban Ecological Security (UES) and the Social, Economic and Environmental Impacts of Eco-cities

Eco-cities have also been criticized to have biases towards the economic and environmental pillars of sustainability while neglecting the social pillar. The practical translations of the concept have faced criticism as eco-cities have been driven by the demand for bounded ecological security. By offering "premium ecological enclaves" factoring ecological security as an outcome of private investments driving the construction of eco-cities, the existing examples of eco-cities are criticized for not being truly sustainable solutions. On the contrary, by placing this concept in the meta-narrative of sustainable cities, these have also been further criticized for celebrating this fragmentation of society through the development of gated communities and premium ecological enclaves isolated from the real global scale of issues in today's ecological crisis. For instance, the eco-cities of Masdar and Hong Kong pose homogeneous visions, but have been criticized to be the source of fragmentation of urban society.

The term "Frankenstein Urbanism" was used by Federico Cugurullo to metaphorically symbolize this criticism of the concept that increases social stratification in exchange for ecological security, creating isolated entities that could work perfectly within themselves, but fall apart when brought in a larger view.

12.5 ECO-CITY DEVELOPMENT CRITERIA

There are currently no set criteria for what is considered an "eco-city," although several sets of criteria have been suggested, encompassing the economic, social, and environmental qualities an eco-city should satisfy. The ideal "eco-city" has been described as a city that fulfils the following requirements:

- Operates on a self-contained economy, resources needed are found locally
- Has completely carbon-neutral and renewable energy production
- Has a well-planned city layout and public transportation system that makes the priority methods of transportation as follows possible: walking first, then cycling, and then public transportation.
- Resource conservation—maximizing efficiency of water and energy resources, constructing a waste management system that can recycle waste and reuse it, creating a zero-waste system
- Restores environmentally damaged urban areas
- Ensures decent and affordable housing for all socio-economic and ethnic groups and improve jobs opportunities for disadvantaged groups, such as women, minorities, and the disabled
- Supports local agriculture and produce
- Promotes voluntary simplicity in lifestyle choices, decreasing material consumption, and increasing awareness of environmental and sustainability issues

In addition to these initial requirements, the city design must be able to grow and evolve as the population grows and the needs of the population change. This is especially important when taking into consideration infrastructure designs, such as for water systems, power lines, etc. These must be built in such a way that they are easy to modernize (as opposed to the dominant current strategy of placing them underground, and therefore making them highly inaccessible). Each individual eco-city development has also set its own requirements to ensure their city is environmentally sustainable; these criteria range from zero-waste and zero-carbon emissions, such as in the Sino-Singapore Tianjin Eco-city project and the Abu Dhabi Masdar City project, to simple urban revitalization and green roof garden projects in Augustenborg, Malmö, Sweden. Using a different set of criteria, the International Eco-Cities Initiative recently identified as many as 178 significant eco-city initiatives at different stages of planning and implementation around the world. To be included in this census, initiatives needed to be at least district-wide in their scale, to cover a variety of sectors, and to have official policy status.

12.6 PRACTICAL ACHIEVEMENTS OF ECO-CITY

12.6.1 Economic Impact

One of the major and most noticeable economic impacts of the movement towards becoming an eco-city is the notable increase in productivity across existing industries as well as the introduction of new industries, thus creating jobs. First, the movement away from carbon-producing energy sources to more renewable energy sources, such as wind, water and solar power, provides local economies with new, thriving industries. The creation of these industries, in turn, births an increase in the demand for labor; thus, not only does total employment increase, but an increase in wages also mimics increasing employment.

Moreover, one of the main priorities of a sustainable city is to reduce its ecological footprint by reducing total carbon emissions, which, economically speaking means increasing productivity. Merely increasing the rate of productivity in an industry reduces costs, both monetary and environmental; that is, as an industry becomes more productive, it can more efficiently allocate and use both its physical and human capital, reducing the time it takes to make the same amount of goods which also allows for a higher wage (because employees are doing more) and a lesser environmental impact. In all, although the initial movement towards becoming a sustainable city may be quite costly for a smaller, poorer city, the benefits of such movement are plentiful in the long-run economic model.

12.6.2 Improvement of Environmental Standards

Although, local environmental standards may differ across eco-cities, each city nonetheless has its own appropriate and practical goals and expectations that have provided the foundation for their recognition as a sustainable city. Differences in these goals and expectations are to be expected, however, due to the limitations of technology and local financing. The primary goal for all sustainable cities is to significantly decrease total carbon emissions as quickly as possible in order to work towards becoming a carbon-free city; that is, sustainable cities work to move towards an economy based solely on renewable energy.

Actions towards carbon-reductions can be seen on both the corporate and individual levels: many industries are working towards cleaner production, but individuals are also moving away from environmentally costly forms of transportation to more sustainable methods, such as public transportation or biking. On this note, another common environmental goal is to increase and make more efficient the public transportation systems. Many sustainable cities also work towards becoming more densely populated (urban density); having its citizens living closer to energy production means less environmental costs of transporting said energy to citizen households.

12.6.3 Technology and Urban Layout for Eco-City and Eco-Town

Transportation

By decreasing urban sprawl, Eco-cities and Eco-Towns decrease the residential and commercial dependence on automobiles. Concurrently, improved public transportation further decreases the demand for cars. The development of metro station and light rail transit systems provide mass transit not only within sectors of a city but between cities. Furthermore, many eco-cities are employing expanded "clean" bus routes in order to decrease the emissions from single household vehicles. Critics note, however, that the high price of "clean" diesel, CNG/LNG, hybrid electric buses, and super capacitor-powered buses may not prove "economically and operationally viable".

12.6.4 Coping with Urbanization Trend

Eco-cities as well as Eco-Towns may also seek to create sustainable urban environments with long-lasting structures, buildings and a great livability for its inhabitants. The most clearly defined form of walkable urbanism is known as the Charter of New Urbanism. It is an approach for successfully reducing environmental impacts by altering the built environment to create and preserve smart cities which support sustainable transport. Residents in compact urban neighborhoods drive fewer miles, and have significantly lower environmental impacts across a range of measures, compared with those living in sprawling suburbs ("Towards a Green Economy"; United Nations Environment Program; Retrieved 17 November, 2011). The concept of Circular flow land use management has also been introduced in Europe to promote sustainable land use patterns that strive for compact cities and a reduction of Greenfield land take by urban spraw.

In sustainable architecture the recent movement of New Classical Architecture promotes a sustainable approach towards construction, that appreciates and develops smart growth, walkability, architectural tradition and classical design. This in contrast to modernist and globally uniform architecture, as well as opposing solitary housing estates and suburban sprawl.

12.6.5 Landscape Development Focus

Eco-cities primarily employ green roofs, vertical landscaping, and bridge links as methods of decreasing the environmental impact of land use. Constructing green roofs and investing in vertical landscaping create natural insulation for residential and commercial properties as well as allows for rainfall collection. Additionally, green roofs and vertical landscaping lower urban temperatures and help prevent the heat island effect. Bridge links allow for development of a walkable city without disrupting the soil to run utility lines by connecting buildings with above-ground walkways.

12.6.6 Energy Management Priority

Eco-cities look to employ renewable energy sources, such as wind turbines, solar panels, and biogas, to reduce emissions. Wind turbines present the opportunity of being able to provide both localized districts within eco-cities and the larger region as a whole with emission-free renewable energy that can additionally supplement existing power sources. Furthermore, by designing buildings with natural ventilation systems, eco-cities reduce the need for air conditioning, thus, drastically decreasing commercial and residential energy use. The energy generated can come from large scale energy production systems such as solar farms which supply many homes and businesses or from individual buildings energizing at least in part their own energy from solar photovoltaic or small scale wind turbines or biomass. Many eco-cities additionally look to deploy solar thermal energy.

By installing solar collectors, developers will be able to provide hot water for space heating and individual and community needs while reducing dependence on gas-fueled boilers. While solar thermal energy appears to be a more efficient source of renewable energy, many urban planners also view photo-voltaic as a viable source of energy. Photovoltaic directly converts solar energy into electricity; however, the extensive costs associated with developing this technology on the city-scale may limit its use when compared to its potential payback. Biogas technology is also deployed as a source of renewable energy as the organic material from wastewater is converted into fuel.

12.6.7 Water Supply Economy

Eco-cities aim to decrease water consumption by employing technologies that reduce the amount of water that is needed for irrigation and sewage flow while also preventing black-water and grey-water runoff from entering ground water sources. Developers generally suggest installing low flow fixtures, rainwater harvesting systems, and sustainable urban drainage systems to meet eco-city standards. Additionally, advanced irrigation systems (xeriscaping) aid in maintaining green infrastructure while decreasing green space consumption of water for irrigation.

CHAPTER

URBAN MOVEMENT AND TRANSPORT 13

13.1 INTRODUCTION

Urban mobility is organized into three broad categories of collective, individual, and freight transportation. While the mobility of passengers is the outcome of individual decisions based on different rationales, freight mobility is decided in tandem between the cargo owners and transportation service providers.

Mobility is linked to specific urban activities and land use, with each type involving the generation and attraction of an array of movements. This relationship is complex but is linked to factors such as recurrence, income, urban form, density, level of development, and technology. Urban mobility is either obligatory when it is linked to scheduled activities (such as home-to-work trips), or voluntary when those generating it are free to decide on their scheduling (such as leisure). The most common types of urban mobility types are:

- **Pendulum movements.** These are obligatory movements involving commuting between locations of residence and work. They are highly cyclical since they are predictable and recurring regularly, most of the time a daily occurrence, thus the term pendulum.
- **Professional movements.** These are movements linked to professional, work-based activities such as meetings, repair, maintenance, and customer services, dominantly taking place during work hours.
- **Personal movements.** These are voluntary movements linked to the location of commercial activities, which includes shopping and recreation.
- **Touristic movements.** Important for cities having historical and recreational features. They involve interactions between landmarks and amenities such as hotels and restaurants and tend to be seasonal or occurring at specific moments.

Major sports events such as the World Cup or the Olympics are important generators of urban movements during their occurrence.

- **Distribution movements.** These are concerned with the distribution of freight to satisfy consumption and manufacturing requirements. They are mostly linked to transport terminals, distribution centers, and retail outlets. However, the growth of online transactions involves more freight movements being carried to residential areas; home deliveries.

The consideration of urban mobility, both for passengers and freight, involves their generation, the modes and routes used, and their destination:

- **Trip generation.** On average, an urban resident undertakes between 3 and 4 trips per day. Moving in an urban area is usually done to satisfy a purpose such as employment, leisure, or access to goods and services. The activity space of an individual is an important trip generation factor since it indicates the travel that needs to be undertaken. Temporal variations in the number of trips by purpose are observed, with commuting as the most prevalent pattern. Similar temporal variations are observed for freight mobility, with the majority of this mobility taking place in the morning when goods are delivered to retail outlets. This often leads to conflicts with the mobility of passengers since vehicles share the same road infrastructure, including parking space, which in urban areas is the object of capacity constraints.
- **Modal split.** This implies using a series of transportation modes for urban trips, which is the outcome of a modal choice. This choice depends on several factors, such as cost, technology, availability, preference, travel time (distance), and income. Therefore, walking, cycling, public transit, the automobile, or even telecommuting, are going to be used either as a choice or as a constraint (lack of choice). For instance, locations within five minutes of walking are considered to be readily accessible to pedestrians. Urban freight distribution can also use a variety of modes, but the van and the truck tend to dominate.
- **Trip assignment (routing).** Involves which routes will be used for trips within the city. Passenger trips usually have a stable routing. For instance, a commuter driving a car has, most of the time, a fixed route between his/her residence and place of work. This route may be modified if there is congestion or if another activity (such as shopping) is linked with that trip, a practice often known as trip chaining. The routing of freight distribution is dependent on the types of deliveries involved. For large retail outlets, direct deliveries are the norm, while for smaller stores and parcel deliveries, vehicles will accommodate flexible routing. Several factors influence trip assignment, the most important being transport costs, time, and congestion levels.

- **Trip destination.** Changes in the spatial distribution of economic activities in urban areas have caused important modifications to trip destination, notably those related to work. Activity-based considerations are essential since each economic activity tends to be associated with a level of trip attraction. Retail, public administration, entertainment, and restoration are the activities that attract the most movements per person employed. For freight movements, manufacturing, transport terminals, and retail are the activities attracting the most movements. The central city used to be a major destination for trips, particularly passengers, but its share has substantially declined in most areas, and suburbs now account for the bulk of urban trips.

Mobility is also a social issue. The share of the automobile in urban trips varies in relation to location, social status, income, quality of public transit, and parking availability. Mass transit is often affordable, but several social groups, such as students, the elderly, and the poor, are a captive market. There are important variations in mobility according to age, income, gender, and disability, with policies aiming at promoting the accessibility and mobility of groups perceived as disadvantaged. The gender gap in mobility is the outcome of socio-economic differences as access to individual transportation is dominantly a matter of income. Within households, differences in role and income are related to the respective activity range and mobility of its members. Consequently, in some instances, modal choice is more a modal constraint linked to economic opportunities.

Fig. 1 Urban Movement

Central locations generally have the most urban mobility options because private, and public transport facilities are present. However, this does not mean that mobility is easier since central areas are congested. In locations outside the central core, a share of the population not having access to the automobile is facing a level of isolation or at least a more limited access to amenities and employment opportunities. Limited public transit and high automobile ownership costs have created a group of spatially constrained (mobility deprived) people.

13.2 URBAN TRANSIT

Transit is almost exclusively an urban transportation mode, particularly in large urban agglomerations. The urban environment is particularly suitable for transit because it provides conditions fundamental to its efficiency, namely high density and significant short distance mobility demands. Since transit is a shared service, it potentially benefits from economies of agglomeration related to high densities and from economies of scale related to high mobility demands. One key advantage of public transit is the higher the demand; the more effective public transit services can be offered. Lower densities are linked with lower demand and a greater likelihood of public transit systems operating at a loss and requiring subsidies. In fact, most public transit systems are not financially sound and have to be subsidized, even if several of their core segments are profitable. Transit systems are made up of many types of services, each suitable to a specific set of market and spatial context. Different modes are used to provide complementary services within the transit system and, in some cases, between the transit system and other transport systems.

- **Bus transit.** One of the most common forms of urban transit that includes vehicles of various sizes (from small vans to articulated buses) offering seating and standing capacity along scheduled routes and services. They usually share roadways with other modes and are therefore susceptible to congestion. Bus rapid transit systems offer a permanent or temporary right of ways and have the advantage of unencumbered circulation.
- **Rail transit**. Vehicles of fixed guideways that usually have their own right of way. Light rail systems are composed of streetcars that can share the right of ways, particularly in central areas. Heavy rail systems are commonly called subways or metro since many operate underground. Another type of rail transit concerns commuter rail systems that are usually servicing central business districts and peripheral areas along specific corridors.
- **Taxi systems.** Usually private for-hire vehicles such as automobiles, jitneys, or rickshaws to offer point-to-point services. Recent technological developments have enabled car-sharing services and expanded the availability of on-demand transit.
- **Alternative transit.** Refer to transit systems that were developed to cope with specific conditions (or niche markets) using alternative modes. Ferries are the most common form of alternative transit as they are servicing cities having waterways separating different urban districts. Funiculars are also prevalent in locations that have a steep incline and enough traffic to justify their construction.

Aerial lifts are also used in some settings to connect locations difficult of access.

Contemporary transit systems tend to be publicly owned, implying that many decisions related to their development and operation are politically motivated. This is a sharp contrast to what took place in the past, as most transit systems were private and profit-driven initiatives. With the fast diffusion of the automobile in the 1950s, many transit companies faced financial difficulties, and the quality of their service declined; in a declining market, there were limited incentives to invest. Gradually, they were purchased by public interests and incorporated into large agencies, mainly to continue providing mobility. As such, public transit often serves more a social function of public service and a tool of social equity than having an economic role. Transit has become dependent on government subsidies, with little if any competition permitted as wages and fares are regulated. As a result, they tend to be disconnected from market forces, and subsidies are required to keep a level of service. With suburbanization, transit systems tend to have even fewer relationships with economic activities.

Government-owned public transit systems are facing financial difficulties for three main reasons. The first is that they are often designed to service taxpayers, not necessarily potential customers. Because of the funding base, transit systems may be spread into neighborhoods that do not provide a significant customer base. The second is that transit unions were able to extract substantial advantages in terms of wages and social benefits, increasing labor costs. The third concerns a technology fixation which incites to invest in high-cost transit (e.g. light rail transit) while low-cost solutions (buses) would have been sufficient to many transit systems, particularly in lower-density areas.

Reliance on urban transit as a mode of urban transportation tends to be high in Asia, intermediate in Europe and low in North America. Since their inception in the early 19th century, comprehensive urban transit systems had significant impacts on the urban form and spatial structure, but this influence is receding. Three major classes of cities can be found in terms of the relationships they have with their transit systems:

- **Adaptive cities.** Represent transit-oriented cities where urban form and land use developments are coordinated with transit developments. While central areas are adequately serviced by a metro system and are pedestrian-friendly, peripheral areas are oriented along transit rail lines.
- **Adaptive transit.** Represent cities where transit plays a marginal and residual role and where the automobile accounts for the dominant share of movements. The urban form is decentralized and of low density.
- **Hybrids.** Represent cities that have sought a balance between transit development and automobile dependency. While central areas have an adequate level of service, peripheral areas are automobile-oriented.

Contemporary land development tends to precede the introduction of urban transit services, as opposed to concomitant developments in earlier phases of urban growth. Thus, new services are established once the demand is deemed to be sufficient, often after being the subject of public pressure. Transit authorities operate under a service warrant and usually run a recurring deficit as services are becoming more expensive. This has led to a set of considerations aimed at a higher integration of transit in the urban planning process, particularly in cities where such a tradition is not well established.

From a transportation perspective, the potential benefits of better integration between transit and local land uses are reduced trip frequency and increased use of alternative modes of travel (i.e. walking, biking, and transit). Evidence is often lacking to support such expectations since the relative share of public transit ridership is declining across the board. There is usually a reciprocal relationship between automobile ownership and the use of public transit. Good accessibility to public transit is often associated with lower automobile use, while areas of high automobile use may impair the development of public transit systems since the automobile is already dominant.

Community and land use design can consequently have a significant influence on travel patterns. Local land use impacts can be categorized into three dimensions of relationships in terms of accessibility, the convergence of mobility and land use integration they provide. Land use initiatives are trying to be coordinated with other planning and policy initiatives to cope with automobile dependence. However, there is a strong bias against transit in the general population because of negative perceptions, especially in North America, but increasingly globally. As personal mobility is a symbol of status and economic success, public transit users can be perceived as the least successful segment of the population. This bias may undermine the image of transit use within the general population but can be subject to change with the evolution of social norms and values.

13.2.1 Possible Urban Mobility Patterns

Cities can structurally be classified as polycentric (more common) or monocentric and major flows as organized or disorganized (more common). Flows can be classified as primary, reflecting main road and transit corridors converging towards main employment and population areas, and as secondary, often linking sub-centers. Cities with a higher level of reliance on public transit tend to be monocentric with a higher level of organized flows, while cities depending more on the automobile tend to be polycentric with a more disorganized structure of flows.

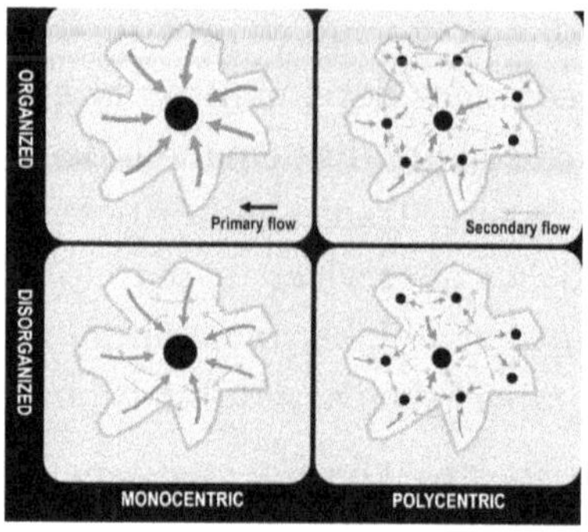

Fig. 2 Possible Urban Movement Patterns

13.3 URBAN TRANSPORT CHALLENGES

The most important transport challenges occur when urban transport systems cannot adequately satisfy the requirements of urban mobility.

13.3.1 Urban Transportation at the Crossroads

Cities are locations having a high level of accumulation and concentration of economic activities. They are complex spatial structures supported by infrastructures, including transport systems. The larger a city, the greater its complexity and the potential for disruptions, particularly when this complexity is not effectively managed. Urban productivity is highly dependent on the efficiency of its transport system to move labor, consumers, and freight between multiple origins and destinations. Additionally, transport terminals such as ports, airports, and railyards are located within urban areas, help anchor a city within a regional and global mobility system. Still, they are also contributing to a specific array of challenges. Some challenges are ancient, like congestion (which plagued cities such as Rome), while others are new, like urban freight distribution or environmental impacts.

a. Traffic Congestion and Parking Difficulties

Congestion is one of the most prevalent transport challenges in large urban agglomerations. Although congestion can occur in all cities, it is particularly prevalent above a threshold of about 1 million inhabitants. Congestion is particularly linked with motorization and the diffusion of the automobile, which has increased the demand for transport infrastructures. However, the supply of infrastructures has often not been able to keep up with mobility growth. Since vehicles spend the majority of the time parked, motorization has expanded the demand for parking space, which has created footprint problems, particularly in central areas where the footprint of parked vehicles is significant. By the 21st century, drivers are three times more likely to be affected by congestion than in the latter part of the 20th century.

Congestion and parking are also interrelated since street parking consumes transport capacity, removing one or two lanes for circulation along urban roads. Further, looking for a parking space (called "cruising") creates additional delays and impairs local circulation. In central areas of large cities, cruising may account for more than 10% of the local circulation, as drivers can spend up to 20 minutes looking for a parking spot. This practice is often judged more economically effective than using a paying off-street parking facility. The time spent looking for a free (or low cost) parking space is compensated by the monetary savings. Parking also impairs deliveries as many delivery vehicles will double-park at the closest possible spot to unload their cargo.

Identifying the true cause of congestion is a strategic issue for urban planning since congestion is commonly the outcome of specific circumstances such as the lack of parking or poorly synchronized traffic signals.

b. Longer Commuting

On par with congestion, people are spending an increasing amount of time commuting between their residence and workplace. An important factor behind this trend is related to residential affordability as housing located further away from central areas (where most of the employment remains) is more affordable. Therefore, commuters are exchanging commuting time for housing affordability. However, long commuting is linked with several social problems, such as isolation (less time spent with family or friends), as well as poorer health (obesity). Time spent during commuting is at the expense of other economic and social activities. However, information technologies have allowed commuters to perform a variety of tasks while traveling.

c. Public Transport Inadequacy

Many public transit systems, or parts of them, are either over or underused since the demand for public transit is subject to periods of peaks and troughs. During peak hours, crowdedness creates discomfort for users as the system copes with a temporary surge in demand. This creates the challenge of the provision of an adequate level of transit infrastructures and service levels. Planning for peak capacity leaves the system highly under-used during off-peak hours, while planning for an average capacity will lead to congestion during peak hours.

Low ridership makes many services financially unsustainable, particularly in suburban areas. Despite significant subsidies and cross-financing (e.g. tolls), almost every public transit system cannot generate sufficient income to cover its operating and capital costs. While in the past, deficits were deemed acceptable because of the essential service public transit was providing for urban mobility, its financial burden is increasingly controversial.

d. Difficulties for Non-motorized Transport

These difficulties are either the outcome of intense traffic, where the mobility of pedestrians, bicycles, and other non-motorized vehicles is impaired, but also because of a blatant lack of consideration for pedestrians and bicycles in the physical design of infrastructures and facilities. On the opposite side, the setting of bicycle paths takes capacity away from roadways as well as parking space. A negative outcome would be to allocate more space for non-motorized transport than the actual mobility demand, which would exacerbate congestion.

e. Loss of Public Space

Most roads are publicly owned and free of access. Increased traffic has adverse impacts on public activities, which once crowded the streets, such as markets, agoras, parades and processions, games, and community interactions. These have gradually disappeared to be replaced by automobiles. In many cases, these activities have shifted to shopping malls, while in other cases, they have been abandoned altogether. Traffic flows influence the life and interactions of residents and their usage of street space. More traffic impedes social interactions and street activities. People tend to walk and cycle less when traffic is high.

f. High Infrastructure Maintenance Costs

Cities facing the aging of their transport infrastructure have to assume growing maintenance costs as well as pressures to upgrade to more modern infrastructure. In addition to the involved costs, maintenance and repair activities create circulation disruptions. Delayed maintenance is rather common since it conveys the benefit of keeping current costs low, but at the expense of higher future costs and, on some occasions, the risk of infrastructure failure. The more extensive the road and highway network, the higher the maintenance cost and its financial burden. The same applies to public transit infrastructure that requires a system-wide maintenance strategy.

g. Environmental Impacts and Energy Consumption

Pollution, including noise generated by circulation, has become an impediment to the quality of life and even the health of urban populations. Further, energy consumption by urban transportation has dramatically increased, and so the dependency on petroleum. These considerations are increasingly linked with peak mobility expectations where high energy prices incite a shift towards more efficient and sustainable forms of urban transportation, namely public transit. There are pressures to "decarbonize" urban transport systems, particularly with the diffusion of alternative energy sources such as electric vehicles.

h. Accidents and Safety

The growth in the intensity of circulation in urban areas is linked with a growing number of accidents and fatalities, especially in developing economies. Accidents account for a significant share of recurring delays from congestion. As traffic increases, people feel less safe using the streets. The diffusion of information technologies leads to paradoxical outcomes. While users have access to reliable location and navigation information, portable devices create distractions linked with a rise in accidents for drivers and pedestrians alike.

i. Land Footprint

The footprint of transportation is significant, particularly for the automobile. Between 30 and 60% of a metropolitan area may be devoted to transportation, an outcome of the over-reliance on infrastructures supporting road transportation. Yet, this footprint also underlines the strategic importance of transportation in the economic and social welfare of cities, as mobility is a sign of efficiency and prosperity.

j. Freight Distribution

Globalization and the materialization of the economy have resulted in growing quantities of freight moving within cities. As freight traffic commonly shares infrastructures supporting the circulation of passengers, the mobility of freight in urban areas has become increasingly controversial. The growth of e-commerce and home deliveries has created additional pressures on the urban mobility of freight. City logistics strategies can be established to mitigate the variety of challenges faced by urban freight distribution.

Many dimensions to the urban transport challenge are linked with the dominance of the automobile.

13.3.2 Automobile Dependency

Automobile use is related to a variety of advantages, such as on-demand mobility, comfort, status, speed, and convenience. These advantages jointly illustrate why automobile ownership continues to grow worldwide, especially in urban areas and developing economies. When given a choice and the opportunity, most individuals will prefer using an automobile. Several factors influence the growth of the total vehicle fleet, such as sustained economic growth (increase in income and quality of life), complex individual urban movement patterns (many households have more than one automobile), more leisure time, and suburbanization (areas where mobility options are limited). Therefore, rising automobile mobility can be perceived as a positive consequence of economic development. The automotive sector, particularly car manufacturing, is a factor of economic growth and job creation, with several economies actively promoting it.

The growth in the total number of vehicles also gives rise to congestion at peak traffic hours on major thoroughfares, in business districts, and often throughout the metropolitan area. Cities are important generators and attractors of mobility, which is associated with a set of geographical paradoxes that are self-reinforcing. For instance, economic specialization leads to additional transport demands, while agglomeration leads to congestion. Over time, a state of automobile dependency has emerged, which results in a declining role of other modes, thereby limiting alternatives to urban mobility through path dependency. Future development options are locked-in because of past choices, and a city can become locked-into planning decisions that reinforce the use of the automobile. In addition to the factors contributing to the growth of driving, two major factors contributing to automobile dependency are:

- **Underpricing and consumer choices.** Most roads and highways are subsidized as they are considered a public good. Consequently, drivers do not bear the full cost of automobile use, such as parking. Like the "Tragedy of the Commons", when a resource is free of access (road), it tends to be overused and abused (congestion). This is also reflected in consumer choice, where automobile ownership is a symbol of status, freedom, and prestige, especially in developing economies. Single homeownership also reinforces automobile dependency if this ownership is favored through various policies and subsidies.
- **Planning and investment practices.** Planning and the ensuing allocation of public funds aim towards improving road and parking facilities in an ongoing attempt to avoid congestion. Other transportation alternatives tend to be disregarded. In many cases, zoning regulations impose minimum standards of road and parking services, such as the number of parking spaces per square meter of built surface, and de facto impose a regulated automobile dependency.

There are several levels of automobile dependency, ranging from low to acute, with their corresponding land use patterns and alternatives to mobility. Among the most relevant automobile dependency indicators is the level of vehicle ownership, per capita motor vehicle mileage, and the proportion of total commuting trips made using an automobile. A situation of high automobile dependency is reached when more than three-quarters of commuting trips are done using the automobile. For the United States, this proportion has remained around 88% over recent decades. Automobile dependency is also served by a cultural and commercial system promoting the automobile as a symbol of status and personal freedom, namely through intense advertising and enticements to purchase new automobiles. Not surprisingly, many developing economies perceive motorization as a condition for development. Even if the term automobile dependency is often negatively perceived and favored by market distortions such as the provision of roads, its outcome reflects the choice of individuals who see the automobile more as an advantage than an inconvenience. This can lead to a paradoxical situation where planners try to counterbalance the preference of automobile ownership supported by the bulk of the population.

The perception of automobile dependency changed over time. The second half of the 20th century saw the adaptation of many cities to support automobile circulation. Motorized transportation was seen as a symbol of modernity and development. Highways and parking lots were constructed, and streets were enlarged, often disrupting the existing urban environment by creating motorized cities. Automobile ownership levels increased rapidly. However, from the 1980s, motorization started to be seen more negatively, and cities implemented policies to limit automobile circulation, at least in specific areas, by a set of strategies including:

- **Dissuasion.** Although automobile circulation is permitted, it is impeded by regulations and physical planning. For instance, parking space can be severely limited or subject to pricing and speed bumps placed to force speed reduction.
- **Prohibition of downtown circulation.** During most of the day, the downtown area is closed to automobile circulation, but deliveries are permitted during the night. Such strategies are often undertaken to protect the character and the physical infrastructures of a historical city. They do, however, like most policies, have unintended consequences. If mobility is restrained in specific locations or during certain periods, people will go elsewhere (longer trips) or defer their mobility for another time (more trips).
- **Tolls.** Imposing tolls for parking and entry (congestion pricing) to some parts of the city has been a strategy being considered seriously in many areas as it confers the potential advantage of congestion mitigation and revenue generation. However, most evidence underlines that drivers are willing to bear additional toll costs for the convenience of using a car, especially for commuting, since it is linked with their primary source of income.

Tentative solutions have been put forth, such as transport planning measures (synchronized traffic lights, regulated parking), limited vehicle traffic in selected areas, the promotion of bicycle paths, and public transit. In Mexico City, vehicle use is allowed on specific weekdays according to license plate numbers, implying that a vehicle will be prevented from circulating at least one weekday. Affluent families have solved this issue by purchasing a second vehicle, thus worsening the existing situation. Singapore is the only country in the world that has successfully controlled the amount and growth rate of its vehicle fleet by imposing a heavy tax burden and purchasing permits on automobile owners. Since Singapore is of small size and has an extensive public transit system, this restriction did not impair mobility. However, such a command-based approach is unlikely to be possible in other contexts.

There is a growing body of evidence underlining that a peak level of car mobility is unfolding, at least in developed economies. Higher energy prices, congestion, less economic prospects, high ownership costs, and the general aging of the population are all countervailing forces to car dependency. For instance, from 2006, the number of vehicle-miles traveled in the United States peaked. There are many alternatives to automobile dependency, such as intermodality (combining the advantages of individual and collective transport), carpooling, ridesharing, or non-motorized transportation (walking and cycling). These alternatives can only be partially implemented as the automobile remains the prime choice for providing urban mobility. A significant potential change remains the development of mobile car-sharing applications enabling better utilization of vehicle assets. Although this would not reduce automobile dependency, it can offer enough flexibility for some users not to require automobile ownership.

13.4 MITIGATING CONGESTION

In some areas, the automobile is the only mode for which adequate transportation infrastructures are provided. This implies less capacity for using alternative modes such as transit, walking, and cycling. At some levels of density, no public infrastructure investment can be justified in terms of economic returns. Longer commuting trips in terms of average travel time, the result of fragmented land uses, and congestion levels are a significant trend. A convergence of traffic is taking place at major highways serving low-density areas with high levels of automobile ownership and low levels of automobile occupancy. The result is energy (fuel) wasted during congestion (additional time) and supplementary commuting distances. In automobile-dependent cities, a few measures can help alleviate congestion to some extent:

- **Ramp metering.** Controlling access to a congested highway by letting automobiles in one at a time instead of in random groups. The outcome is a lower disruption on highway traffic flows through better merging.
- **Traffic signal synchronization.** Tuning the traffic signals to the time and direction of traffic flows. This is particularly effective if the signals can be adjusted hourly to reflect changes in circulation patterns. Trucks can be allowed to pass traffic lights through delayed signals, which reduces the risk of accidents through sudden collision with a car breaking at a yellow light. Therefore, trucks are less likely to be the first vehicle at a red light, which increases capacity because trucks have lower acceleration.
- **Incident management.** Making sure that vehicles involved in accidents or mechanical failures are removed as quickly as possible from the road. Since accidents account for 20 to 30% of all the causes of congestion, this strategy is particularly important.
- **Car ownership restrictions.** Several cities and countries (e.g. Singapore) have quotas in the number of license plates that can be issued or require high licensing fees. To purchase a vehicle, an individual thus must first secure a license through an auction. Such strategies, however, go against market principles.
- **Sharing vehicles. Concerns two issues.** The first is providing ridership to people (often co-workers) having a similar origin, destination, and commuting time. Two or more vehicle trips can thus be combined into one, which is commonly referred to as carpooling. The second involves a pool of vehicles (mostly cars, but also bicycles) that can be leased or shared for a short duration when mobility is required. Adequate measures must be taken so that supply and demand are effectively matched with information technologies providing effective support.

- **HOV lanes.** High Occupancy Vehicle (HOV) lanes ensure that vehicles with two or more passengers (buses, taxis, vans, carpool, etc.) have exclusive access to a less congested lane, particularly during peak hours.
- **Congestion pricing.** A variety of measures are aimed at imposing charges on specific segments or regions of the transport system, mainly as a toll. The charges can also vary during the day to reflect congestion levels so that drivers are incited to consider other time periods or other modes.
- **Parking management.** Removing parking or free parking spaces can be an effective dissuasion tool since it reduces cruising and enables those willing to pay to access an area (e.g. for a short shopping stop). Parking spaces should be treated as a scarce asset subject to a price structure, reflecting the willingness to pay. Further, planning regulations provide an indirect subsidy to parking by enforcing minimum parking space requirements based upon the type and the density of the land use.
- **Transit.** Offering alternatives to driving can significantly improve efficiency, notably if it circulates on its own infrastructure (subway, light rail, buses on reserved lanes, etc.) and is well integrated within a city's development plans. However, public transit has its own set of issues (see the next section about urban transit challenges).
- **Non-motorized transportation.** Since most urban trips are over short distances, non-motorized modes, particularly walking and cycling, have an important role in supporting urban mobility. The provision of adequate infrastructure, such as sidewalks, is often a low priority as non-motorized transportation is often perceived as not modern despite the important role it needs to assume in urban areas.

All these measures only partially address the issue of congestion, as they alleviate but do not solve the problem. Fundamentally, congestion remains a sign of economic success, but a failure at reconciling rising mobility demands and acute supply constraints.

13.5 THE URBAN TRANSIT CHALLENGE

As cities continue to become more dispersed, the cost of building and operating public transportation systems increases. For instance, as of 2015, about 201 urban agglomerations had a subway system, the vast majority of them being in developed economies. Furthermore, dispersed residential patterns characteristic of automobile-dependent cities make public transportation systems less convenient to support urban mobility. Additional investments in public transit often do not result in significant additional ridership. Unplanned and uncoordinated land development has led to the rapid expansion of the urban periphery. By selecting housing in outlying areas, residents restrict their potential access to public transportation. Over-investment (when investments do not appear to imply significant benefits) and under-investment (when there is a substantial unmet demand) in public transit are both complex challenges.

Urban transit is often perceived as the most efficient transportation mode for urban areas, notably large cities. However, surveys reveal a stagnation of public transit systems, especially in North America, where ridership levels have barely changed in the last 30 years. The economic relevance of public transit is being questioned. Most urban transit developments had little impact on alleviating congestion despite mounting costs and heavy subsidies. This paradox is partially explained by the spatial structure of contemporary cities, which are oriented along servicing individual mobility needs. Thus, the automobile remains the preferred mode of urban transportation.

Besides, public transit is publicly owned, implying a politically motivated service that provides limited economic returns. Even in transit-oriented cities, transit systems depend massively on government subsidies. Little or no competition within the public transit system is permitted as wages and fares are regulated, undermining any price adjustments to ridership changes. Thus, public transit often serves the purpose of a social function (public service) as it provides accessibility and social equity, but with limited relationships with economic activities. Among the most difficult challenges facing urban transit are:

- **Decentralization.** Public transit systems are not designed to service low density and scattered urban areas dominating the urban landscape. The greater the decentralization of urban activities, the more difficult and expensive it becomes to serve urban areas with public transit. Additionally, decentralization promotes long-distance trips on transit systems causing higher operating costs and revenue issues for flat fare transit systems.
- **Fixity.** The infrastructures of several public transit systems, notably rail and subway systems, are fixed, while cities are dynamical entities, even if the pace of change can take decades. This implies that travel patterns tend to change with a transit system built for servicing a specific pattern that may eventually face "spatial obsolescence"; the pattern it was designed to serve no longer exists.
- **Connectivity.** Public transit systems are often independent of other modes and terminals. It is consequently difficult to transfer passengers from one system to the other. This leads to a paradox between the preference of riders to have direct connections and the need to provide a cost efficient service network that involves transfers.
- **Automobile competition.** Given cheap and ubiquitous road transport systems, public transit faced strong competition and lost ridership in relative terms and, in some cases, in absolute terms. The higher the level of automobile dependency, the more inappropriate the public transit level of service. The convenience of the automobile outpaces the public service being offered.

- **Construction and maintenance costs.** Public transit systems, particularly heavy rail, are capital intensive to build, operate, and maintain. Cost varies depending on local conditions such as density and regulations, but average construction costs are around $300 million per km. However, there are exceptions where cost overruns can be substantial because of capture by special interest groups such as labor unions, construction companies, and consulting firms. When there is inefficient regulatory oversight, these actors will converge to extract as much rent as possible from public transit capital improvements. The world's highest subway construction costs are in New York. For instance, the Second Avenue subway extension in Manhattan, completed in 2015, was done at the cost of $1.7 billion per km, five to seven times the average in comparable cities such as Paris or London. This project employed four times more labor, with construction costs 50% higher.
- **Fare structures.** Historically, most public transit systems have abandoned a distance-based fare structure for a simpler flat fare system. This had the unintended consequence of discouraging short trips, for which most transit systems are well suited for, and encouraging longer trips that tend to be costlier per user than the fares they generate. Information systems offer the possibility for transit systems to move back to a more equitable distance-based fare structure, particularly with smartcards that enable to charge according to the point of entry and exit within the public transit system.
- **Legacy costs.** Most public transit systems employ unionized labor that has consistently used strikes (or the threat of labor disruptions) and the acute disruptions they create as leverage to negotiate favorable contracts, including health and retirement benefits. Since public transit is subsidized, these costs were not well reflected in the fare systems. In many transit systems, additional subsidies went into compensation or covered past debt, not necessarily into performance improvements or additional infrastructure. As most governments face stringent budgetary constraints because of social welfare commitments, public transit agencies are being forced to reassess their budgets through an unpopular mix of higher fares, deferred maintenance, and the breaking of labor contracts.
- **Self-driving vehicles.** Developments in information technologies let anticipate in the coming years the availability of self-driving vehicles. Such a development would entail point-to-point services by on-demand vehicles and a much better utilization level of such assets. This system could compete directly with transit systems due to its convenience, comfort, and likely affordability.

Therefore, public transit systems are challenged to remain relevant to urban mobility as well as to increase their market share. The increasing volatility in petroleum prices since 2006 provides uncertainties in the costs of transit fleet ownership and operations and how effective it is to convert transit fleets to alternative energy sources. A younger generation with a preference for living in higher-density areas perceives the automobile as a less attractive proposition than the prior generations. Electronic fare systems are also making the utilization of public transit more convenient.

A recent trend concerns the usage of incentives, such as point systems (e.g. air miles with the purchase of a monthly pass), to promote public transit and influence consumer behavior. Yet, evidence underlines that the inflation-adjusted cost of using public transit is increasing, implying that the cost advantage of public transit over the automobile is not changing in a significant manner. If self-driving vehicles become a possibility, many highly subsidized transit systems may have a limited competitive advantage. Under such circumstances, the fate of many surface public transit systems will be in question.

13.6 URBAN PUBLIC TRANSPORT IN INDIA

India had the second largest population worldwide after China, more than one-third of which lived in urban areas. Public transportation had an indispensable role in daily mobility. Majority of public transport in Indian cities relied on buses, and many had developed Bus Rapid Transit System (BRTS) to improve capacities and reliability of the conventional bus system. Several metropolitan cities also developed Mass Rapid Transit (MRT) metro system for more efficient and nondisruptive urban passenger transportation.

Most of the buses and metro systems were owned and operated by local state and municipal governments. Compared to urban households, more rural households preferred using buses. In 2021, the Indian government rolled out policies to raise the share of public transport in urban areas through expansion of metro networks and city bus services. It allocated 26 billion U.S. dollars for metro expansion and pledged to spend 1.5 trillion dollars on general infrastructure, including public bus network between 2020 and 2025.

Trends influencing urban mobility in India:

- **Rapid Urbanisation:**
    - Trends and patterns present unprecedented challenges to urban mobility systems. In 2011, there were 468 cities with population above 0.1 million and 53 cities with population greater than 1 million.
- **Rapid Motorisation:**
    - Since 2000, while road network in the country has grown by 39%, the number of registered vehicles has grown by about 158%.

- o Five metro cities have vehicle registration rates in excess of 500 per 1000 people and account for 54 per cent of the total vehicles in the metropolitan cities as of 2011.

- **Dwindling share of Non-motorised Transportation:**
  - o In Indian cities, people who commute by walking, outnumber those who use private motorised transport.
  - o As cities sprawl, the share of NMT reduces drastically creating increased reliance on private modes of transport.

13.6.1 Issues Linked to Urban Transport

In most urban areas across the country, the development of public transport did not meet the rapid rise of urban population. The share of public transport in India constantly decreased since early 2000s, and personal vehicles, especially two wheelers, saw exponential growth. Growing private vehicle ownership was the main contributor of increased congestion in Indian cities, to a point where some made to the global list of the most congested cities. Furthermore, having an adequate public transportation network would possibly curb increasing road accidents and road deaths, worsening air pollution, as well as growing greenhouse gas emissions.

Problems faced by Urban Transports:

- **Road congestion:**
  - o As more people travel over longer distances on regular basis for employment and education purposes, will inevitably lead to road congestion.
- **Parking problems:**
  - o The acute shortage of parking spaces both on and off the streets in Indian cities increases the time spent searching for a parking spot and induces traffic congestion.
  - o Delhi has 14 per cent of road lengths used for on-street parking while Surat has almost 60 per cent of its road lengths blocked by on-street parking.
- **Air pollution:**
  - o According to air quality data, there is a wide variation in the pollution concentration and severity across cities.
  - o Cities are considered critically polluted if the levels of criteria pollutants (namely PM10 and NO2) are more than 1.5 times the standard.
  - o Half of the residential areas in cities monitored by Central Pollution Control Board (CPCB) are at critical levels of air pollution.

- **Deteriorating road safety:**
  - o The number of fatalities is increasing in relation to the increasing motorisation and higher slow-moving vehicles in the traffic stream.
  - o Pedestrian fatalities constitute a significant share of total fatalities and the magnitude is in fact much higher in cities that lack adequate pedestrian facilities.

Loopholes in Urban transportation management:

- **Gaps in Laws and Regulations:**
  - o There is no legislation at central, state or local level that comprehensively covers urban transport requirements of Indian cities.
  - o The Motor Vehicles (MV) Act, 1988, addresses road vehicle licensing and is inadequately equipped to deal with the problems of urban congestion and pollution leading to rampant abuse of the system.
- **Fragmented Institutional Frameworks:**
  - o Urban transport systems' functions are performed by multiple agencies under the central, state and city governments which do not necessarily work together.
  - o According to the Seventh Schedule (Article 246) of the Indian Constitution, urban development, which includes urban transport, is in the State List.
  - o While, in some states, the transport department undertakes urban transport planning, in others, it is the urban development or municipal administration at the urban level that does it.
- **Distorted land markets affecting transport infrastructure development:**
  - o About 70 per cent of delays in all infrastructure projects in 2008 were due to problems related to land acquisition.
  - o One of the factors is the heavily distorted land market, caused by zoning and development control rules in cities.
- **Human Resource challenges:**
  - o A lack of urban transport skills amongst city and state officials, along with no dedicated organisation in city or state to deal with urban transport.
- **Absence of reliable transport data:**
  - o The absence of a database with scientific management and analysis of urban transport statistics has severely constrained the ability to formulate sound urban transport plans and reliably assess the impact of the different projects carried out in the cities.

13.7 EFFORTS TOWARDS IMPROVEMENT

In 2006, the Indian government introduced the National Urban Transport Policy (NUTP), which was renewed in 2014. It was a long-term public strategy with the aim to provide sustainable transportation in the future. Two major fields of policy focus were the development of metro systems in major cities, and electrification and expansion of public bus transport fleets. With the intention to accelerate such processes, the Indian government opened metro and bus sectors to private investors.

Furthermore, the government launched the Faster Adoption and Manufacturing of Hybrid and Electric Vehicles (FAME) scheme in 2015 and a further FAME II program in 2019, which would provide incentives for electric vehicle purchases, including electric buses, and deployment of charging infrastructure.

13.7.1 Key Recommendations to Improve Urban Transportation:

- For Unified Metropolitan Transport Authority (UMTA) to be effective it should be backed by legislation and the entire funding for urban transport should be routed through UMTA.
    - o Some international case studies similar to UMTA can be considered such as:

(i) Land Transport Authority of Singapore

(ii) Unified Transport Authority of London

- For achieving effective and efficient management of traffic in the congested areas, the parking requirements have to be met appropriately.
- Intelligent Transport Systems (ITS) in Parking Facilities can help transport planners to achieve policy objectives in many different ways.
- Area Traffic Control (ATC) system for effective traffic management in city centres.
- Sustainable integration of different public transportation system: Public transportation systems should not operate in isolation, but in coordination with other modes of transport (private modes as well as other public transport modes).

In addressing complex urban mobility issues, a systems approach seems well suited for a thorough understanding of the issues and their causal linkages. It is essential to understand that mobility is a derived demand that is derived from the need for people to meet their necessary social or economic interactions. Sustainable mobility is a key enabler of economic growth and towards eliminating poverty and shared prosperity in Indian cities.

CHAPTER

URBAN SYSTEM AND ECONOMIC DEVELOPMENT 14

14.1 INTRODUCTION

The world is faced with complex interacting environmental, social, health and political challenges, which are increasingly dynamic. Traditionally, these challenges have been seen as discrete problems, to be dealt with by separate organisations with different institutional rules, competing for resources and agency. The call to researchers in this special issue was to urgently identify systemic interventions to these challenges, where the potential for generating co-benefits in health and wellbeing could be realised and replicated.

Human-induced climate change is one of the major challenges facing cities. There has been an increase in global mean temperature of 1°C above the pre-industrial average, and seven of the hottest years on record have occurred since 2014. At the current rate of greenhouse gas (GHG) emissions, children born today could live in a world 3°C warmer than the pre-industrial average. Additional environmental concerns include accelerating levels of GHGs, pollution, loss of biodiversity, as well increasing use of land and resources. Compounding these environmental challenges are interrelated social sustainability challenges, which include growing inequity in access to safe, secure, affordable housing, adequate income, and secure assets. These inequities have a marked impact on social cohesion, fairness and justice both locally and internationally.

These sustainability challenges have significant implications for the environment, as well as human health and wellbeing. The rapidly changing climate poses significant risks for food security, water availability, disease transmission and exposure to extreme events such as flooding and fires. Exposure to high temperatures and extreme heat can lead to increased morbidity and mortality from heat stress and heatstroke, as well as increases in cardiovascular and respiratory diseases. A recent study has estimated that, between 1991 and 2018, 37% of global heat-related deaths could be attributed to manmade climate change. Exposure to air pollution has been associated with cardiovascular and respiratory diseases, birth defects, poor mental health, and cognitive function. Globally, an estimated 7 million deaths from cardiorespiratory diseases are attributable to indoor and outdoor air pollution annually.

Exposure to these environmental hazards is not equally distributed across the global population. Developing countries are facing the greatest risks from the changing climate, and are also the most vulnerable to its impacts. Cumulative socio-economic disadvantages mean low-income populations in urban areas worldwide are more likely to be exposed not only to higher levels of outdoor and indoor air pollution, but also in summertime to both outdoor and indoor temperatures. Low income and marginalised communities are disproportionately affected by climate change because they lack the financial resources to choose safe locations, or to adapt to the changing climate. Higher rates of underlying health issues in low income populations can exacerbate poor health outcomes. The young, old and those suffering from chronic diseases or disabilities are at greater risk.

Transformative changes are necessary to mitigate threats to global sustainability and population and planetary health to protect 'the health of human civilisation and the state of the natural systems on which it depends'. Indeed, actions to improve sustainability can provide important co-benefits for human health, wellbeing and equity. Cities are critical arenas for making these changes because they are home to an increasing majority (54%) of the global population and economic activity (85% of gross domestic product—GDP). Cities also have a significant impact on global sustainability, with a disproportionate share of global energy (60–80%) and resource consumption (75%) and they emit around 70% of global GHG emissions. Therefore, it is essential that urban areas reduce their environmental footprints and increase their resilience to environmental hazards, while protecting and promoting planetary health.

Urban policies and sustainable development have significant—but largely unrealised—potential to improve population health and wellbeing. While it is increasingly common for urban planners to consider the environment, as well as health and wellbeing, much of the recent global increase in urbanisation in the Global South has occurred with little attention being paid to these issues. Such omissions can have serious long-term consequences, because building and urban design have environmental 'lock-in' effects for both embodied and operational carbon.

14.2 URBAN SETTLEMENTS/SYSTEMS DEFINED

A Human Settlement is a place where people live and interact through activities such as agriculture, trading and entertainment.

A rural settlement system is a community involved predominantly primary activities such as farming, lumbering and mining.

An urban settlement system engages in predominantly in secondary and tertiary activities such as food processing and banking.

There is often correlation between functions, population sizes and population density. A rural settlement tends to have small population and low population density. Urban settlement often has a large population size and high population density.

14.2.1 Site Factors for Settlement

Physical environment determine the site of a settlement depend on

1. Water supply – water is essential for human survival and agricultural activities.
2. Relief – the availability of broad flat land such as floodplains promote agricultural activities.
3. Soils- fertility of soils also another considerations for agricultural activities
4. Shelter-which are sheltered from natural elements. Windswept highland with harsh cold climate are conducive for settlement.
5. Defence– Threat from enemies and predatory animals drove people to live together for protection.

14.3 URBAN SYSTEMS IN INDIA: DEFINITIONS

Urban areas have been recognized as "engines of inclusive economic growth". Of the 121 crore Indians, 83.3 crore live in rural areas while 37.7 crore stay in urban areas, i.e approximately 32 % of the population. The census of India, 2011 defines urban settlement as :-

All the places which have municipality, corporation, cantonment board or notified town area committee

All the other places which satisfy following criteria:

1. A minimum population of 5000 persons ;
2. At least 75 % of male main working population engaged in non-agricultural pursuits ; and
3. A density of population of at least 400 persons per square kilometer.

14.3.1 Statutory Town

All Municipal areas and other urban areas declared as such by the government are statutory towns. they have their own administrative structure and in most cases can levy taxes.

14.3.2 Census Town

Census towns are areas outside the statutory towns that have distinct urban characteristics like population density (above 400 per sq. km), minimum population size of 5000 and more than 75% of male workers engaged in non agricultural activities.

These were identified on the basis of census 2001 data. Cities are urban areas with more than 100,000 population. Urban areas below 100,000 are called towns in India.

Similarly Census of India defines:-

Urban Agglomeration (UA): An urban agglomeration is a continuous urban spread constituting a town and its adjoining outgrowths (OGs), or two or more physically contiguous towns together with or without outgrowths of such towns. An Urban Agglomeration must consist of at least a statutory town and its total population (i.e. all the constituents put together) should not be less than 20,000 as per the 2001 Census. In varying local conditions, there were similar other combinations which have been treated as urban agglomerations satisfying the basic condition of contiguity. Examples: Greater Mumbai UA, Delhi UA, etc.

Urban Out Growths (UOG): An Urban Out Growth (UOG) is a viable unit such as a village or a hamlet or an enumeration block made up of such village or hamlet and clearly identifiable in terms of its boundaries and location. Some of the examples are railway colony, university campus, port area, military camps, etc., which have come up near a statutory town outside its statutory limits but within the revenue limits of a village or villages contiguous to the town.

14.3.3 Standard Urban Area(SUA)

A new concept that had been developed for the 1971 Census for the tabulation of certain urban data was the Standard Urban Area. The essential of a Standard Urban Area are :

(i) it should have a core town of a minimum population size of 50,000,

(ii) the contiguous areas made up of other urban as well as rural administrative units should have close mutual socio- economic links with the core town and

(iii) the probabilities are that this entire area will get fully urbanised in a period of two to three decades.

The idea is that it should be possible to provide comparable data for a definite area of urbanisation continuously for three decades which would give a meaningful picture. This replaced the concepts of Town Group that was in vogue at the 1961 Census. The town group was made up of independent urban units not necessarily contiguous to one another but were to some extent inter-dependent. The data for such town groups became incomparable from census to census as the boundaries of the towns themselves changed and the intermediate areas were left out of account; this concept came for criticism at one of the symposium of the International Geographic Union in Nov.-Dec.1968 and the concept of Standard Urban Area came to be developed for adoption at the 1971 Census. If data for this Standard Area were to be made available in the next two or three successive censuses it is likely to yield much more meaningful picture to study urbanisation around large urban nuclei.

While determining the outgrowth of a town, it has been ensured that it possesses the urban features in terms of infrastructure and amenities such as pucca roads, electricity, taps, drainage system for disposal of waste water etc. educational institutions, post offices, medical facilities, banks etc. and physically contiguous with the core town of the UA. Examples: Central Railway Colony (OG), Triveni Nagar (N.E.C.S.W.) (OG), etc.

Each such town together with its outgrowth(s) is treated as an integrated urban area and is designated as an 'urban agglomeration'. Number of towns/UA/OG 2011, according to Census 2011 Census are :-

1. Statutory Towns — 4,041
2. Census Towns — 3,894
3. Urban Agglomerations — 475
4. Out Growths — 981

At the central level, nodal agencies which look after program and policies for urban development are Ministry of Housing and Urban Poverty Alleviation (MoHUPA) and Ministry of Urban development. Urban development is a state subject. At state level there are respective ministries, but according to 74th Constitutional Amendment act,1992, it is mandatory for every state to form ULBs and devolve power, conduct regular election, etc. Under 12 schedule of Indian constitution, 18 such functions have been defined which are to be performed by ULBs and for that states should support the ULBs through finances and decentralization of power, for more autonomy. But this is not uniform throughout all the states and still more is need to be done to empower ULBs in India.

Urban areas are managed by Urban Local Bodies(ULBs), who look after the service delivery and grievance redressal of citizens. There are eight type of urban local government in India- municipal corporation municipality, notified area committee, town area committee, cantonment board, township, port trust and special purpose agencies.

14.3.4 Metropolis

The Constitution (seventy-fourth Amendment) Act, 1992 defines a metropolitan area in India as, an area having a population of ten lakhs or more, comprised in one or more districts and consisting of two or more Municipalities or Panchayats or other contiguous areas, specified by the Governor by public notification to be a Metropolitan area.

Here is a list of Metropolitan Cities of India.

14.3.5 Megacity

A megacity is usually defined as a metropolitan area with a total population in excess of ten million people.A megacity can be a single metropolitan area or two or more metropolitan areas that converge. The terms conurbation, metropolis and metroplex are also applied to the latter.

A megalopolis (sometimes called a megapolis; also megaregion, or supercity) is typically defined as a chain of roughly adjacent metropolitan areas, which may be somewhat separated or may merge into a continuous urban region.

The term was used by Patrick Geddes in his 1915 book Cities in Evolution,by Oswald Spengler in his 1918 book The Decline of the West, and Lewis Mumford in his 1938 book The Culture of Cities, which described it as the first stage in urban overdevelopment and social decline. Later, it was used by Jean Gottmann in his landmark 1961 study, Megalopolis: The Urbanized Northeastern Seaboard of the United States, to describe the chain of metropolitan areas along the northeastern seaboard of the US, extending from Boston, Massachusetts, through New York City, Philadelphia, and Baltimore and ending in Washington, D.C. and Northern Virginia.The latter is sometimes called the "BosWash megalopolis". The term has been interpreted as meaning "supercity".

Emerging megalopolises in the developing world include the Pearl River Delta, which when formed will constitute the core part of a Southern China Coast megaregion; the Unified Jakarta-Bandung metro area forming the core of the Northern Javamegaregion; and the Yangtze River Delta core, a part of a Central Chinese Coast megaregion.

14.3.6 Conurbation

Conurbation refers to a specific kind of geographical region. Due to rapid increase in population and industrial and technological development, the city boundary expands and one urban centre coalesces with another in a slow but continuous process of urbanisation and regional development. It is thus that conurbations are formed.

A conurbation is a region comprising a number of cities, large towns, and other urban areas that, through population growth and physical expansion, have merged to form one continuous urban or industrially developed area. In most cases, a conurbation is a polycentric urbanised area, in which transportation has developed to link areas to create a single urban labour market or travel to work area.

The term "conurbation" was coined in 1915 by Patrick Geddes in his book Cities In Evolution. He drew attention to the ability of the then new technology of electric power and motorised transport to allow cities to spread and agglomerate together, and gave as examples "Midlandton" in England, the Ruhr in Germany, Randstad in the Netherlands and North Jersey in the United States.

The word 'conurbation' has emerged from the words 'continuous' and 'urban area'. The word was used by Patrick Geddes in 1915 with reference to a continuous urban area of more than two urban centres which may have separate territorial units. C.B. Fawcett defines a conurbation as "an area occupied by a continuous series of dwellings, factories and other buildings including harbours, docks, urban parks and playing fields, etc. which are not separated from each other by the rural land..." J.C. Saoyne defines conurbation as "an area of urban development where a number of separate towns have grown into each other and become linked by such factors as common industrial or business interest or a common centre of shopping and education". R.E. Dickinson calls it an "urban tract" while Jean Gottamman refers to it as "extended city" or "Super Metropolitan Region".

14.3.7 Urban Field

The urban field is a form of urban habitat of relatively high density involving a good transportation system and a broad array of economic, social and recreational opportunities. Even though the sections of the urban field may still be in agricultural use, the area is nonetheless urbanized because anywhere within it a person is able to connect his/her home to telephones, radio and television facilities, electricity, gas, water supply systems and a network of freeway and primary roads.

14.3.8 Ethnoburb

An ethnoburb is a suburban residential as well as business area with a notable cluster of a particular ethnic minority population. Although the group may not constitute the majority within the region, it is a significant amount of the population.That can greatly influence the social geography within the area.

14.4 CAUSES OF URBANIZATION: AN ANALYSIS

Our century is the Urban Century. Soon the world will become a giant urban system. Urbanization and the fast growth of cities and towns have become one of the major characteristics of the present century. Urban life is now way of life of modern man.

- **An Efficient Agricultural System**

  The efficient agricultural system makes the surplus quantity of food possible to sustain the urban population. With surplus food production more and more people turn to activities other than primary, and the process of urbanization goes on.

- **The Invention of Power-Driven Machinery**

  The industrial revolution was a major trigger for accelerated urbanization process. Factories development accentuated the process. The concentration of production into bigger factories requiring a large number of workers.

- **The Development of Trade and Services**

  Employment in the service or tertiary sector of the labour force has been an important factor in the growth of cities. Growing trade and expansion of the service sector is a vital factor even today.

- **Development of Transportation**

  - It allows goods to be transported freely and contributed to the process of urbanization.
  - Transport makes possible for people and ideas to move freely

- o Intra-urban transportation
- o Enable mass movement of persons between home and workplaces.
- o Widespread use of automobile and truck
- o Decentralization of individuals and functions within urban areas

- ◉ **Demographic and Social Factors**

Natural Growth of Cities – About 20% of urbanization takes place through natural growth of cities. The improvement of sanitation, health care, social welfare and improvement of other infrastructural facilities added to the process.

14.4.1 Industrial Revolution

At the end of 19th century, the advent of Industrial Revolution happened and it encompassed mainly Western Europe and North America. Causes were farming mechanization, Increased agricultural efficiency, released labour forces from rural areas due to increased agricultural efficiency.The fast development of industries in cities provided many new jobs.

Now Urbanization is slowing increasing down in developed countries but increasing in developing countries.

14.5 RURAL-URBAN MIGRATION

About 70% of urbanization is attained by the rural-urban migration.

14.5.1 Origin: Rural Area

Push Factors

- ◉ Unemployment for mechanization
- ◉ A low standard of living
- ◉ Low wages
- ◉ Poor education
- ◉ Poor view of future

14.5.2 Destination: Urban Area

Pull Factors

- ◉ Many job opportunities for industrialization
- ◉ A higher standard of living
- ◉ More education
- ◉ Higher wages
- ◉ Better health care

- A better view of future

14.5.3 Origin: Urban Area

14.5.4 *Push Factors*

- Overcrowded living environment
- Pollution
- Traffic congestion
- High crime rate
- Higher land rent.

14.5.4 Destination: Rural Area/ New Town

Pull Factors

- More open space, a more pleasant living environment.
- Fresh and quiet air
- Lower land rent
- Lower crime rate
- Large modern shopping centres in the suburb.
- Large modern factories, more job opportunities

14.5.5 Subarbanization

Subarbanization is a new reality of our times. Many people migrate out of town due to the following reasons.

Causes

- Satellite and new town formation
- Improvement of living standard
- Improvement of transport between CBD and suburban
- Government policy, eg. Urban renewal programme

14.5.6 Characteristics of Urban Places

Aside from the population, urban places also have common characteristics. Think of 'urban' and many people think of roads, buildings and infrastructure like electricity cables and sewage systems. Others will think of shops, offices and busy transport hubs.

Site

Site refers to the land on which a settlement is built. For example, it might be on gently sloping land, facing south, with well-draining soil, a water supply from a local spring, and on a river bend. These features would all make it a good site for an early settlement in the pre-industrial age:

- Gently sloping land – prevents flooding
- Facing south – in the northern hemisphere, it means the settlement will be warmed by the sun during the day (also good for growing crops locally)
- With well-draining soil – so it's easy to build on
- A water supply from a local spring – for drinking etc.
- On a river bend – provides a defensive barrier from attack

As urban areas have grown, site factors have become less important. For example, most cities do not source their water locally (from springs and rivers) but receive it in pipes from elsewhere. In the modern age, most urban areas are still found in places that have good 'site' features, but this is not because of the features of the site itself.

14.5.7 Situation

Situation refers to the location of the settlement in relation to what is around it. For example, a city might be located along a coastline, at the end of a river valley, at the other end of which is a coalfield. These features would all make a good place for a modern settlement because of the things nearby:

- Located along a coastline – can build a port for international trade
- Deep sea – allows large container ships to come into the port
- At the end of a river valley – valleys are important passages through mountains as they can be used for railways and roads
- Near a coalfield – a cheap source of energy, and/or a product for export

In reality, a combination of good site and situation features are needed for a settlement to grow into a large urban area.

14.5.8 Function

Function refers to 'what the places does', or 'the reason the city is there'. Almost all settlements have more than one function, and the larger the urban area the more functions it is likely to have.

Typical functions of urban function include:

- Administrative centres – headquarters for government offices
- Industrial
- Transport – ports, railway junctions, airline hubs
- Retail

- Markets – places where agricultural products made in the surrounding area can be sold
- Residential
- Financial – headquarters of major banks, insurance companies etc.

Functions are both cause and effect of a city's growth. For example, Hong Kong grew to its current population of over 7 million because it had a historical administrative function, as well as a transport function as a sea port. Over time it has grown into a financial centre, a retail centre and an airline transport hub. This is a good example of cumulative causation.

14.5.9 Land use

Land use in urban areas is easily identifiable as not rural meaning there is little agricultural land use. (There are no farms.)

Land use is often closely linked to the function. In almost all urban areas, residential is the main land use. In industrial centres, industrial land use will be common, and so on. However there are land use types that are not necessarily a function of the urban area, but are closely linked:

- **Open space** – parks, riversides, 'empty' places awaiting development
- **Sports facilities** – informal (such as parks) and formal (such as football stadia)
- **Local transport** – roads, railways, metro systems for moving local people around
- **Government services and social amenities** – libraries, schools, hospitals
- **Infrastructure** – telephone exchanges, data centres (for internet servers), gas terminals and so on

14.5.10 Hierarchy of Settlement (Including Megacities)

A hierarchy refers to something put into a particular order, for example biggest to smallest. The settlement hierarchy is a way to classify settlements.

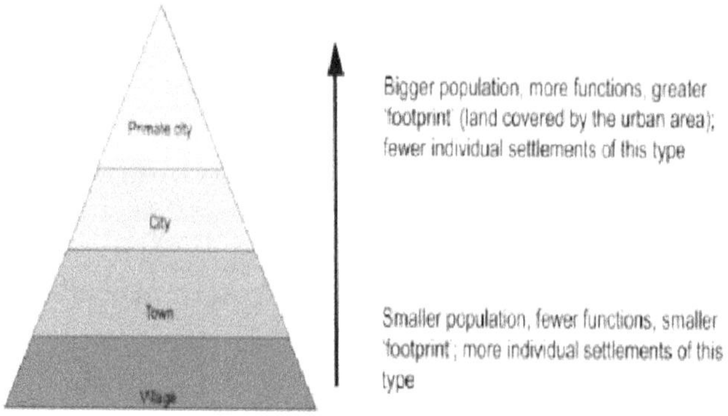

Fig. 1 An example of a settlement hierarchy.

The names and types of the settlements that can be placed in the hierarchy vary. Some common types of settlements are:

- **Isolate** – an individual dwelling e.g. a farmhouse
- **Hamlet** – a small collection of dwellings (like a very small village)
- B 14.5.1– a small settlement
- Town – varying in size from a few thousand people to several tens of thousands, depending on the characteristics of the town
- **City** – a large urban settlement
- **Capital city** – usually but not always the largest settlement in a country (e.g. London, Paris, Beijing); and/or the centre of national government (Brasilia, Canberra, Pretoria)
- **Primate city** – the name given to a country's biggest city when it is at least twice as large as the next biggest city
- **Binary city** – the name given to a country's two biggest cities when they are of roughly equal importance, e.g. Rio de Janeiro and Sao Paulo, Sydney and Melbourne, or Cape Town and Johannesburg
- **Millionaire city** – a city of at least one million people
- **Megacity** – a city of at least 10 million people
- **Metacity / hypercity / megalopolis** – a city of at least 20 million people (UN Habitat, 2006). These terms are also used to describe large conurbations such as the Pearl River Delta area in southern China.
- **Conurbation** – an urban area created when two or more large settlements grow so big that they join together

For more detail on megacities, see the 'Characteristics of megacities' page on this site.

14.5.11 A Note on Economic Settlement Hierarchies

Settlements may also be classified into a hierarchy according to the variation of economic activity in a settlement. For example, a village may have only basic services and no industry at all, while a city may have a great mixture of economic activities. This generally relates to the size of the settlement as measured by population, but not always. For this reason, the hierarchy of settlements as classified by economic activity is covered in the 'Urban Economic Activities' section of this site.

14.5.2 Growth Process (Planned or Spontaneous)

Historically cities have developed spontaneously – meaning their growth has not been planned. A combination of natural increase and in-migration from rural areas led to the increase in population in cities. This is still the case for cities in the poorest countries.

Cities in the lowest income countries frequently have rapid and spontaneous urban development. Migration and a high rate of population growth result in a shortage of adequate housing, so people are forced to build their own dwellings. These areas are sometimes known as slums, shanty towns, favelas, champas and bustees but are properly known as informal settlements. Informal settlements where the residents do not own the land and do not pay rent to a landlord are known as squatter settlements.

In general, modern cities in richest countries are carefully planned. Population growth in these High Income Countries is generally slow or in decline and is monitored by governments which then plan for the future needs of people in the cities. These cities often have strong administrations that are capable of managing changes such as infrastructure development. Over the past half century, cities have increasingly been developed through deliberate planning by governments of countries at all income levels. These new settlements are often called 'planned cities' or 'new towns'. They are defined by being largely planned in advance, rather than being additions to existing towns or cities. As existing cities became too big, governments create new cities for people to move to. They do this by denying planning permission on the edges of the bigger cities while granting it in the new town area, and by paying for the basic infrastructure in the new places (e.g. building highways, electricity and water services, schools and hospitals). However, they are often found nearby to existing cities. One of the most famous is Milton Keynes (near London, UK), but examples can be found across the world including Olgiata (near Rome, Italy), and Ma On Shan (Hong Kong). Canberra and Brasilia, the capital cities of Australia and Brazil respectively, are also both examples of planned cities.

14.6 ECONOMIC DEVELOPMENT AND THE DISTRIBUTION OF PEOPLE OVER SPACE

Urban economics is broadly the economic study of urban areas; as such, it involves using the tools of economics to analyze urban issues such as crime, education, public transit, housing, and local government finance. More specifically, it is a branch of microeconomics that studies urban spatial structure and the location of households and firms.

Much urban economic analysis relies on a particular model of urban spatial structure, the monocentric city model pioneered in the 1960s by William Alonso, Richard Muth, and Edwin Mills. While most other forms of neoclassical economics do not account for spatial relationships between individuals and organizations, urban economics focuses on these spatial relationships to understand the economic motivations underlying the formation, functioning, and development of cities.

Since its formulation in 1964, Alonso's monocentric city model of a disc-shaped Central Business District (CBD) and the surrounding residential region has served as a starting point for urban economic analysis. Monocentricity has weakened over time because of changes in technology, particularly, faster and cheaper transportation (which makes it possible for commuters to live farther from their jobs in the CBD) and communications (which allow back-office operations to move out of the CBD).

Additionally, recent research has sought to explain the polycentricity described in Joel Garreau's Edge City. Several explanations for polycentric expansion have been proposed and summarized in models that account for factors such as utility gains from lower average land rents and increasing (or constant) returns due to economies of agglomeration.

14.6.1 Urbanisation and Economic Development go Hand in Hand

Many countries have experienced sizeable shifts in the geographic distribution of their population over the past decades, in particular developing countries. Population growth paired with gradually increasing urbanisation rates have resulted in an expansion in the size of some cities, a shrinking population in others, and the emergence of entirely new cities. Population growth in Africa and Asia has been especially rapid in the previous two decades. Between 1990 and 2015, the population living in cities with more than 10 million people more than doubled, and 10 of these cities emerged, including 7 in Asia, as well as Istanbul, Lagos and Teheran. Despite the sharp increase in urbanisation in developing countries, countries with higher per capita GDP (purchasing power parity [PPP], 2011 USD) still tend to be more urbanised. This finding is visible at different points in time between 1975 and 2015. More specifically, building on data covering 168 countries in 2015, a positive and statistically significant correlation is consistently observed between national GDP per capita and measures of urbanisation. The latter include the share of people living in metropolitan areas, in cities only or cities, towns and semi-dense areas. The term "metropolitan areas" in this publication is used as a shorthand for Functional Urban Areas (FUAs), i.e. cities plus their commuting zones. While national GDP per capita is positively correlated with all three measures of urbanisation, the correlation is strongest when the share of the population in metropolitan areas is considered. Also, from a theoretical point of view, it would seem most appropriate to look at the connection of urban development with the share of people that are part of cities' labour markets, i.e. live in metropolitan areas.

Consequently, this chapter aims to describe the relationship between economic development and the spatial distribution of the population across metropolitan areas (with more than 50 000 inhabitants). This relationship is examined at different spatial scales and through different lenses, depending on the question of interest. The fact that urbanisation and economic development often tend to go hand in hand raises the question of which one drives which. There are good reasons to believe that causality runs both ways. It is generally argued that while the fact that countries urbanise does not necessarily imply that they will develop, sustainable economic growth does not occur without urbanisation.

14.6.2 Higher Levels of Development Correspond to Larger Metropolitan Areas

As countries develop, the share of people living in metropolitan areas increases and does so relatively more in the largest ones. The average share of a country's total population living in metropolitan areas is only 34% in low-income countries, but 71% in high-income countries. Between low- and high-income countries, the relative difference in the population share in metropolitan areas increases with the size of the metropolitan area. While the share of the population living in metropolitan areas with up to 1 million inhabitants is roughly the same across the different income classes, the population share living in metropolitan areas with more than 1 million is roughly 4 times higher in high-income countries than in low-income countries. Even more striking, the share of inhabitants of metropolitan areas with more than 5 million inhabitants is roughly 13 times as high in high-income than in low-income countries. More precisely, when moving from low- to high-income countries, the average share of people living in metropolitan areas with more than 5 million people increases from 2% to 22%. By contrast, the average share of population living in metropolitan areas with less than 250 000 people shrinks with economic development: it halves from 14% in low-income countries to 7% in high-income economies. This pattern is robust to controlling for potentially different average country sizes in the four income groups.

Overall, larger metropolitan areas tend to be relatively more related to national economic development. To some extent, this might be the result of a structural transformation process happening at a higher pace in these areas.

The positive relationship between economic development and the share of the metropolitan population holds also over time. A panel analysis based on GDP and population data in 1990, 2000 and 2015 confirms that, over time, higher economic growth and a larger share of the metropolitan population are correlated. This correlation between economic development and share of the metropolitan population is particularly strong in countries at low and middle stages of development relation.

14.6.3 Economic Theories on why Urbanisation is Connected to Economic Development

Traditional economics has long viewed the process of urbanisation through the lens of structural transformation. When countries develop, they experience structural changes in their economic sectors and therefore urbanise. In particular, the spatial distribution of the population, employment, and production changes with economic development. Theories on structural transformation state that the transition to modern economic growth is intrinsically linked to rural-urban migration through a declining share of employment in rural agriculture and a shift toward manufacturing and service industries in cities.

Thus, urbanisation occurs for two concurrent reasons when economies develop. First, productivity gains in agriculture release rural labour and push employment toward cities. Second, the rise of industrial sectors (for example those driven by enhanced international trade), pull resources towards cities.

Other views stress the importance of technological progress in allowing for urbanisation, citing, for example, the role of sewage, surges in agricultural productivity, high-rise buildings or urban transport as factors allowing for increasing levels of urbanisation. Agglomeration benefits and increased innovation arising in decently-functioning cities would then drive economic development. This would be accompanied by a sectoral transformation as when people move out of rural areas into urban centres, they typically shift from agricultural to manufacturing and service activities. These non-agricultural sectors have high rates of productivity growth and agglomeration effects in urban areas also promote further economic growth.

Causality likely runs both from economic development to urbanisation, and from urbanisation to economic development. That is, as countries develop, they become more urbanised, which in turn drives economic development. Examples of such a virtuous circle encompass countries in Europe and the United States since the 18th-19th centuries, Japan and South Korea since the mid-20th century, as well as China and India more recently.

14.6.4 Evidence on Structural Transformation at Play

The structure of countries' economies differs with their economic development, as previously mentioned. Across the world, higher GDP per capita is associated with a lower share of agriculture, and higher shares in manufacturing and services in the economy. In fact, the macroeconomic literature documents that modern economic growth is intrinsically linked to the transition of the economy from agriculture to manufacturing and services. While the decrease of agriculture and the increase in the service sector is steady with economic development, manufacturing behaves differently. Its share in the economy follows a hump shape. It increases for lower levels of development, peaks at some moment in time, and then starts to decrease for higher levels of development. This transition of the economy between different sectors implies that the spatial distribution of the population, employment and production processes change, and some of these shifts will favour urbanisation. Most of today's developed economies have undergone this path during the first wave of urbanisation that accompanied the Industrial Revolution. Improvements in agricultural productivity driven by innovation occurred in Great Britain before the 18th century and spread to other parts of Europe, which coincided with a move of workers out of agriculture and into manufacturing. As a larger population could be sustained from farming the same amount of land, the number of people that could live in cities increased leading to fast urbanisation. In this context, however, it is also important to note that urbanisation itself typically leads to shifts in sectoral activity, for example by resulting in an increasing weight of the service sector in the economy.

14.6.5 Market Forces in the Development of Cities

Market forces in the development of cities relate to how the location decision of firms and households causes the development of cities. The nature and behavior of markets depend somewhat on their locations therefore market performance partly depends on geography. If a firm locates in a geographically isolated region, its market performance will be different than a firm located in a concentrated region. The location decisions of both firms and households create cities that differ in size and economic structure. When industries cluster, like in Silicon Valley in California, they create urban areas with dominant firms and distinct economies. By looking at location decisions of firms and households, the urban economist is able to address why cities develop where they do, why some cities are large and others small, what causes economic growth and decline, and how local governments affect urban growth. Because urban economics is concerned with asking questions about the nature and workings of the economy of a city, models and techniques developed within the field are primarily designed to analyze phenomena that are confined within the limits of a single city.

14.6.6 Land use

Looking at land use within metropolitan areas, the urban economist seeks to analyze the spatial organization of activities within cities. In attempts to explain observed patterns of land use, the urban economist examines the intra-city location choices of firms and households. Considering the spatial organization of activities within cities, urban economics addresses questions in terms of what determines the price of land and why those prices vary across space, the economic forces that caused the spread of employment from the central core of cities outward, identifying land-use controls, such as zoning, and interpreting how such controls affect the urban economy.

14.6.7 Economic Policy

Economic policy is often implemented at the urban level thus economic policy is often tied to urban policy. Urban problems and public policy tie into urban economics as the theme relates urban problems, such as poverty or crime, to economics by seeking to answer questions with economic guidance. For example, does the tendency for the poor to live close to one another make them even poorer?.

14.6.8 Transportation and Economics

Urban transportation is a theme of urban economics because it affects land-use patterns as transportation affects the relative accessibility of different sites. Issues that tie urban transportation to urban economics include the deficit that most transit authorities have and efficiency questions about proposed transportation developments such as light-rail. Megaprojects such as this have been shown to be synonymous with unexpected costs and questionable benefits.

14.6.9 Housing and Public Policy

Housing and public policy relate to urban economics as housing is a unique type of commodity. Because housing is immobile, when a household chooses a dwelling, it is also choosing a location. Urban economists analyze the location choices of households in conjunction with the market effects of housing policies. In analyzing housing policies, we make use of market structures e.g., perfect market structure.,

BIBLIOGRAPHY

- Abbott, Carl. "Urban History for Planners," Journal of Planning History, Nov 2006, Vol. 5 Issue 4, pp 301–313
- Armus, Diego and John Lear. "The trajectory of Latin American urban history," Journal of Urban History (1998) 24#3 pp 291–301
- Beachy, Robert and Ralf Roth, eds. Who Ran the Cities?: City Elites and Urban Power Structures in Europe and North America, 1750-1940 (2007)
- Bennett, Larry. The Third City: Chicago and American Urbanism. (U of Chicago Press, 2015), 241 pp
- Borsay, Peter. The eighteenth-century town: a reader in English urban history 1688-1820 (Routledge, 2014)
- Clark, Peter, and Paul Slack. English Towns in Transition 1500-1700 (1976)
- David N. Livingston (2000). "History of Geography." In R. J. Johnston, Derek Gregory, Geraldine Pratt, and Michael Watts, eds., The Dictionary of Human Geography, Fourth Edition. Oxford: Blackwell, 304-308
- Davies, Gary W. "The rise of urban history in Britain c. 1960-1978" (PhD dissertation, University of Leicester, 2014) online, With detailed bibliography pp 205-40
- Denecke, Dietrich, and Gareth Shaw, eds. Urban historical geography: recent progress in Britain and Germany (Cambridge UP, 1988).
- Derek Gregory (2009). "Geography." Derek Gregory, R.J. Johnston, Geraldine Pratt, Michael Watts, and Sarah Whatmore, eds., The Dictionary of Human Geography, Fifth Edition. Chichester, U.K.: Wiley-Blackwell, 287-295.
- Emerson, Charles. 1913: In Search of the World Before the Great War (2013) 526pp short essays on 21 major world cities in 1913, including London, Washington, Winnipeg, and Constantinople etc.
- Emmen, Edith. The Medieval Town (1979)

- Engeli, Christian, and Horst Matzerath. Modern urban history research in Europe, USA, and Japan: a handbook (1989) in GoogleBooks
- Epstein, S. E. ed. Town and Country in Europe, 1300-1800 (2001), a major anthology of scholarly articles
- Frost, Lionel, and Seamus O'Hanlon. "Urban history and the future of Australian cities." Australian Economic History Review (2009) 49#1 pp: 1-18.
- Gillette Jr., Howard, and Zane L. Miller, eds. American Urbanism: A Historiographical Review (1987) online
- Goldfield, David. ed. Encyclopedia of American Urban History (2 vol 2006); 1056pp; excerpt and text search
- Handlin, Oscar, and John Burchard, eds. The Historian and the City (Harvard U.P., 1963)
- Harvey, David, Consciousness and the Urban Experience: Studies in the History and Theory of Capitalist Urbanization (1985), a Marxist approach
- Haynes, Barry. Register of European Urban History (Leicester University, 1991)
- Haynes, Douglas E., and Nikhil Rao. "Beyond the Colonial City: Re-Evaluating the Urban History of India, ca. 1920–1970." South Asia: Journal of South Asian Studies (2013) 36#3 pp: 317–335. Online
- Hays, Samuel P. "From the History of the City to the History of the Urbanized Society," Journal of Urban History, (1993) 19#1 pp 3–25.
- Isin, Engin F. "Historical sociology of the city' in Gerard Delanty & Engin F. Isin, eds. Handbook of historical sociology (2003). pp. 312–325. online
- Lees, Andrew. "Cities, Society, and Culture in Modern Germany: Recent Writings by Americans on the Großstadt." Journal of Urban History 25.5 (1999): 734–744.
- Lees, Andrew. "Historical perspectives on cities in modern Germany: recent literature." Journal of Urban History 5.4 (1979): 411–446.
- Lees, Lynn Hollen. "The Challenge of Political Change: Urban History in the 1990s," Urban History, (1994), 21#1 pp. 7–19.
- McKay, John P. Tramways and Trolleys: The Rise of Urban Mass Transport in Europe (1976)
- McManus, Ruth, and Philip J. Ethington, "Suburbs in transition: new approaches to suburban history," Urban History, Aug 2007, Vol. 34 Issue 2, pp 317–337

INDEX